W9-DEC-593

THE BIOLOGY AND EVOLUTION
OF AUSTRALIAN LIZARDS

THE BIOLOGY AND EVOLUTION
OF AUSTRALIAN LIZARDS

by

Allen E. Greer

Illustrations by

Phyllis A. Koshland
and
Deborah S. Kent

Published by

Surrey Beatty & Sons Pty Limited

This book is copyright. Apart from any fair dealing for purposes of private study, research, criticism or review, as permitted under the Copyright Act, no part may be reproduced by any process without written permission. Enquiries should be made to the publisher.

The National Library of Australia
Cataloguing-in-Publication entry:
The Biology and Evolution of Australian Lizards.
Bibliography.
Includes index.
ISBN 0 949324 21 3.

1. Lizards — Australia. I. Title.

597.950994

Published December, 1989

PUBLISHED IN AUSTRALIA BY

SURREY BEATTY & SONS PTY LIMITED
43 Rickard Road, Chipping Norton, NSW 2170
Printed in Hong Kong and Australia

Dedicated to Phyllis A. Koshland,
for all those lizards caught,
all those lizards drawn,
and all those kilometres
travelled together.

FRONTISPIECE: Thermoregulating dragon, *Ctenophorus clayi* from 64 km south of Exmouth, Western Australia.

CONTENTS

LIST OF ILLUSTRATIONS

Introduction

THIS book is an attempt to summarize the biology of Australian lizards against the background of their evolutionary relationships. There are several reasons why such an attempt might be useful. First, Australian lizards have evolved in relative isolation in an ecologically diverse continent. Hence there are numerous species showing great diversity in morphology, physiology, behaviour and ecology. These provide abundant opportunity to study the relationship between biology and environment. Second, the evolutionary history of Australian lizards has recently become much better known as a result of greater interest in the subject and also because of recent conceptual and technical advances in deducing a group's phylogeny. This provides an opportunity of placing the present-day relationship between biology and environment into an historical context. And third, despite the fact that Australian herpetology has enjoyed the recent exponential growth in knowledge that has characterized most fields of science, there has been no review of Australian lizard biology since Robert Bustard's (1970d) magnificent little book "Australian Lizards", published nineteen years ago and long out of print. I found Bustard's book very stimulating, and I would count this book a success if it inspires only a few people as much as his book inspired me.

In gathering the material for this book, I have been impressed by two aspects of the current state of knowledge of Australian lizard biology, and indeed Australian herpetology in general. The first is the role of amateurs, and the second is the amount of information still locked away in peoples' heads or in unpublished theses and reports. I would like to elaborate briefly on both points.

Amateurs have contributed substantially to the development of Australian herpetology, as will be evident from the references to their work in this book. Their importance derives primarily from their observation of live animals: captive specimens, a species seen during a field trip, or a local fauna. They publish without the inhibitions of having to preserve a professional image, and if they fail to understand the full ramifications of what they have seen, at least they have put into the literature an observation that can be taken-up and developed by someone else.

The journals that publish most of the amateurs' work are as important as the amateurs themselves for they not only provide an outlet for the observations, but also a safe haven for beginning authors to develop their expository skills. The major journals here are *Herpetofauna*, and to a lesser extent, the *South Australian Herpetology Group Newsletter*, the *Victorian Naturalist*, and the *Western Australian Naturalist*. The dual role of these journals — as an outlet for incidental or only partially developed observations and as a testing ground for new authors — make them extremely important national assets and hence worthy of support.

Unfortunately the role of amateurs in Australian herpetology is now under threat from certain state authorities. Rules and regulations make it increasingly difficult to collect, keep, or even disturb in order to observe or photograph native fauna. For many amateurs who are often either too young to fight or have no affiliation with another, protective bureaucracy like a museum or university, this means giving up or going "underground". As a result, much opportunity for knowledge and personal development through investigation, discovery and discussion is lost.

The amount of information known about Australian herpetology — often widely known, but not written down is immense. And if one adds to this, unpublished work such as theses and reports, i.e., knowledge that is virtually inaccessible to the world at large, the amount of "discovered" but unavailable information is staggering. This produced but undisseminated knowledge represents both a waste of resources and of opportunity; a waste of resources because it cost money and time to produce and of opportunity because accomplishments in science are most readily judged and appreciated on the basis of published results.

Finally, a problem peculiar to one section of the Australian herpetological literature deserves special mention. This is the unsupported assertion about an animal's biology, usually its food habits or brood size, in the literature designed for a popular audience. In reading these observations one has the haunting feeling that they may not be first hand with the author and worse, that they get taken up, supplemented by "observations" of a similar nature and republished. In some cases the "observations" seem to be several generations old. To deal with the basic uncertainty in such observations, I have decided to accept only those that have the ring of first-hand authenticity to them. Better to leave out a few good but poorly stated or documented observations rather than perpetuate the rot.

Australian lizards, like all other forms of life, have had an evolutionary history. Knowledge of this history as an actual event in nature is not only an appropriate end in itself but is also a powerful tool with which to organize new information and to ask new questions. When we discover or suspect something new about an animal or a group of animals we can immediately ask how widespread is it in the evolutionary tree and when, at least in a relative sense, did it appear in the history of the group. This helps us perhaps to ask further questions about how and why the novelty arose and why it was subsequently modified even further. At the same time, the distribution of the new feature in the tree helps us to test the accuracy of the tree and modify it if necessary.

This raises the question of how we infer the evolutionary history of a group of animals. We know that there was one, and only one, history, but how do we study a unique event in the past?

There are at present two ways. The first is based on the recognition of new features that have arisen in the past in a population and have been passed on to its descendants, therein marking them indelibly as having all shared, uniquely, this common ancestor. This ancestor plus all its descendants are a branch, or lineage, of the evolutionary tree, and it can be a big trunk or a small twig. Such a lineage may also be called a monophyletic (from the Greek words "monos" — one and "phyle" — tribe) or natural group. In contrast, a group consisting of two or more lineages that themselves do not share a common ancestor is said to be a polyphyletic or unnatural group, or simply an assemblage.

The strength of this method is that the novel, or "derived", features in lineages can often be related to some environmental factor and this relationship may allow further insight into how the lineage became established and persisted. The evolution of a spectacle and its significance for small-sized animals living in arid habitats (p. 120) and the evolution of live-bearing habits

and its significance for life in cool climates (p. 126) are but two examples. However, there are several drawbacks. First, except for lineages in which there is a fossil of known age, only relative time, not absolute time, can be associated with the tree. Second, there is always the possibilty that novel features used to recognize lineages have evolved more than once, thereby causing us to confuse two or more lineages as one. Third, there is the problem of recognizing what is novel and what is old. This usually requires knowledge of the closest relatives — or "outgroup" — of the actual group within which one wishes to work out relationships. This group is necessary because novel traits within the group of interest — the "ingroup" — are by definition traits that arose in that group and hence are recognized by being absent in the nearest relatives. Recognition of the "outgroup", or even the suite of possible outgroups is not always easy. Recognition of lineages on the basis of novel features is the heart of a methodology called phylogenetic systematics — perhaps the most important conceptual advance in modern, morphologically-based systematics (Wiley 1981).

The second way to study a group's evolutionary history is based on certain molecules, possessed by all organisms, that are thought to change at more or less constant rates, based on randomly occurring mutations that are unrelated to the animal's environment and hence unexposed to selection. Because these changes are thought to occur at a statistically constant rate they allow us to draw an evolutionary tree based on the principle that the more closely related two species are, the more similar they should be in the molecules of the particular system being studied. Furthermore, if we can equate the rate of molecular change to real time, using fossils of known age, then actual dates can be associated with each branch. This is a very useful method when we are dealing with groups in which novel features are difficult to recognize or when they are suspected of having arisen more than once (for a recent review of the "molecular clock" — and its own special problems, see Ayala 1986).

Unfortunately to date there is only one study of the broad phylogenetic relationships of any Australian lizards based on this principle of ever-diverging molecular structure. The study involved a very rough immunological comparison within skinks (Hutchinson 1981, 1983), and its primary conclusion confirmed an earlier result based on morphology (Greer 1979b).

As readers of this book may require background or supplementary information in related areas, it will be worthwhile to make a few recommendations in these directions. To begin with, there is a good deal of anatomy in this book — a reflection of my own interest as well as the state of knowledge, and I have assumed a basic understanding of vertebrate anatomy throughout. Any textbook on this subject should serve to remind a reader on any forgotten detail; however, for osteological structure, which figures heavily here, I recommend going straight to Romer's (1956) classic text.

Also on a few occasions, but especially in the last chapter, I have found it useful to make a comparison with lizard groups not found in Australia. For an introduction to these groups, any of the following three introductory texts in herpetology would suffice: Porter (1972), Goin et al. (1978) and Halliday and Adler (1986).

Within the Australian fauna itself some readers may want to know something about the distinguishing features or the distribution of a species, or they may just want a picture of an animal to obtain a more vivid impression of it while reading this book. For this they should consult either the single comprehensive guide for all Australian reptiles and amphibians (Cogger 1986) or one of the two currently available, partial regional faunas, i.e., for South Australia (Houston 1978) and Western Australia (Storr et al. 1981, 1983). All give descriptions, distinguishing features, identification keys, distribution maps and figures.

Other readers may desire some background in the principles of lizard behaviour, ecology and physiology. For these, I recommend Heatwole (1976) and Bradshaw (1986) which are good introductions with many Australian examples. Finally, still other readers may be interested in how to care for live animals in captivity. For this, they should consult the only keepers guide written expressly for Australian animals and conditions (Banks 1980) and any general work on the diseases of reptiles which, of course, tells one how to prevent, as well as treat, disease (Marcus 1980; Cooper and Jackson 1981; Frye 1980).

Before turning to the discussion of the lizards themselves, it is worth making two general observations about Australian environments as they relate to lizards. First, in general there are two seasons for lizards in Australia, an active season centred over summer and a quiescent season centred over winter (Fig. 1). The active season coincides with the warm part of the year in the temperate south and with the wet part of the year in the tropical north. Reciprocally, the quiescent season is the southern winter and the northern "dry".

Second, there is often a great deal of environmental variability at any one locality, and animals have to cope with this variability in both the short-term behavioural/physiological sense and the long-term evolutionary sense. For example, flooding may follow a prolonged drought; searing hot may follow freezing cold, and fire may reduce a habitat to a charcoal field in a few minutes. An observer may only see the animals under one set of circumstances, generally favourable, but he should remember that the animals have to cope with a range of environmental conditions that he may only vaguely appreciate.

MATERIALS AND METHODS

Where species names have changed since the initial observations, I have used the current name and supplied in brackets the name used by the original author. Where the identity of the species is ambiguous, I have generally ignored the observation or used it at the lowest reliable taxonomic level possible, usually the genus. In considering new species "legally" described but based on inadequate diagnoses, I have accepted the names of only those species which my own work or the work of colleagues indicates to be sound.

In discussing an observation on a particular taxon, I have introduced it at the taxonomic level where I believe it is likely to have greatest applicability. Occasionally this is at a higher taxonomic level than that of the original report. If subsequent information shows that I "missed" the appropriate level, the responsibility, of course, is mine and not the original author's.

In general I have rounded off decimals to the nearest whole or tenths decimal place, rounding down after odd numbers and up after even. However, for snout-vent lengths I have always rounded up to the nearest whole number.

Statistical levels of significance are indicated by asterisks: .05 — *; .01 — ** and .001 — ***.

Abbreviations used for original observations and calculations throughout the book are as follows:

AM — Australian Museum.

C — observation based on a captive individual.

CTMax — critical thermal maximum temperature. The body temperature at which a heat-stressed animal loses its righting reflex but from which it fully recovers upon being removed to a lower temperature.

CTMin — critical thermal minimum temperature. The body temperature at which a cold-stressed animal loses its organized locomotory abilities but from which it recovers upon being removed to a higher temperature.

BT — field body temperature: core body temperature of an animal active in the field.

HL — head length; measured from the tip of the snout to the centre of the external ear opening or in the case of earless forms, the deepest part of the auricular crease or depression.

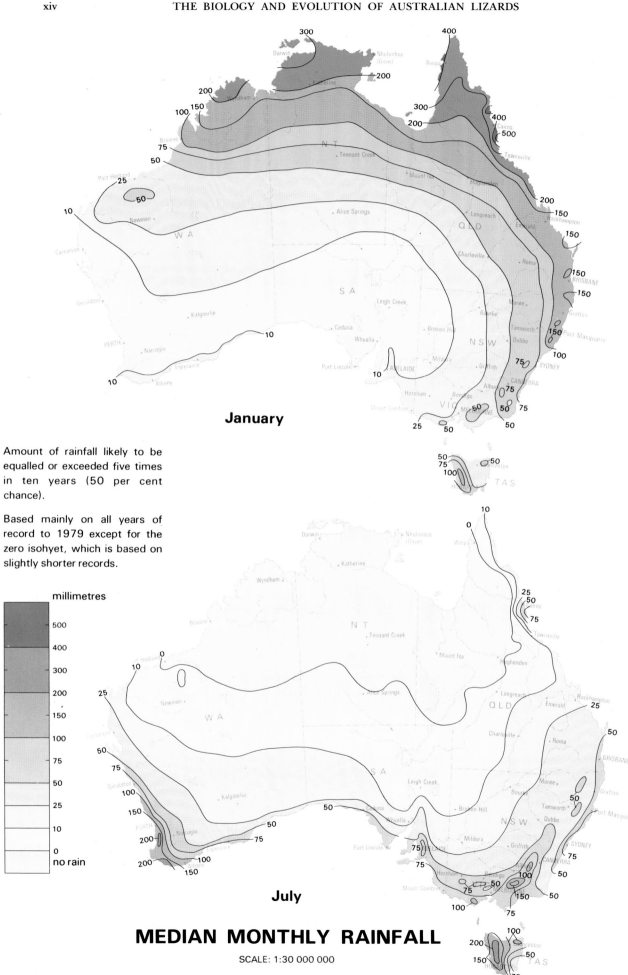

January

Amount of rainfall likely to be equalled or exceeded five times in ten years (50 per cent chance).

Based mainly on all years of record to 1979 except for the zero isohyet, which is based on slightly shorter records.

millimetres

	500
	400
	300
	200
	150
	100
	75
	50
	25
	10
	0
	no rain

July

MEDIAN MONTHLY RAINFALL

SCALE: 1:30 000 000

SOURCE : Maps specially prepared by Bureau of Meteorology, 1981.

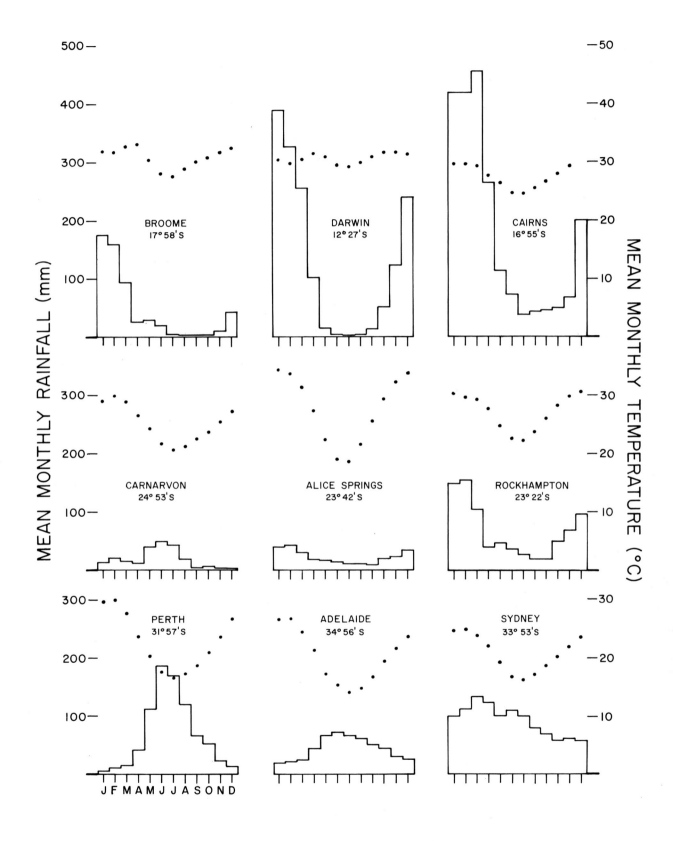

Fig. 1. Climate in Australia. Top: mean monthly rainfall for nine localities. Left: mean monthly rainfall throughout the continent for January and July. Note especially the strong wet/dry seasonality of the north. Maps Crown copyright; reproduced by permission of the General Manager, Australian Surveying and Land Information Group, Department of Administrative Services, Canberra.

HLL — hind limb length; in dragons, measured along the posterior edge of the outstretched leg from the midline of the precloacal area to the tip of the claw; in all other families measured along the anterior edge of the leg from the base of the tail to the tip of the claw.

N — number of specimens.

n — number of observations, if different than N.

PBT — preferred body temperature: body temperature of an animal in a laboratory thermal gradient.

r — correlation coefficient.

RCM — relative clutch mass. This is one of several direct or indirect measures of clutch (or litter) mass relative to female mass. The ratio is thought to estimate the relative reproductive effort, or energy investment, of the female in her clutch (Vitt and Congdon 1978). The indirect measures are usually either volume or calories. The ratio is usually expressed in one of two ways: clutch/female, or clutch/female and clutch. The former is more rigorous statistically but the latter is widely reported.

Comparisons can be made using any definition of RCM as long as only one definition is used in any particular comparison.

The definitions of RCM used in this book are as follows:

RCM (1): fresh wet clutch mass/fresh wet female mass.

RCM (2): fresh wet clutch mass/fresh wet female mass and clutch mass.

RCM (3): preserved wet cluch volume/fresh wet female and clutch mass. This definition pertains only to the work of Pianka and Pianka 1976 and Pianka 1986.

RCM (4): preserved wet cluch mass/preserved wet female and clutch mass. This definition pertains largely to James 1983.

SD — standard deviation.

SE — standard error.

SVL — snout-vent length; measured from the tip of the snout to the posterior edge of the preanal scales.

t — value for Student's t test.

TL — tail length; measured from the posterior edge of the preanal scales to the tip of the tail.

W — observation based on a wild individual.

Finally, I have had to choose a cut-off point for incorporating recently published literature and I have chosen 31 March, 1987.

ACKNOWLEDGEMENTS

A number of people have read either all or parts of the manuscript at various stages of production. They have read primarily for accuracy and sense but along the way have consistently picked up little errors of spelling and grammar that have eluded me for reasons of ignorance and fatigue (I wish more of the latter!). These readers were: H. Ehmann, G. Hunt, D. Kent, P. Koshland, R. Sadlier, L. Schwarzkopf, G. Shea and R. Shine.

Many people have contributed original unpublished observations and interpretations which I have woven into the text with, I hope, proper acknowledgement in all cases. Very often this information was completely new and filled large holes in the overall picture. These contributors were: M. Bamford, J. Coventry, H. Ehmann, P. Harlow, D. Kent, R. Sadlier, G. Shea, L. Smith, S. Wilson and B. Miller.

Other people have contributed photographs from their personal files. These photographers were: R. Braithwaite and S. Wilson.

In addition to these specific acknowledgements, I would also like to express general thanks to the Trustees and management of the Australian Museum for providing me with the opportunity to work in Australia; to R. Sadlier and G. Shea for giving me the benefit of their extensive field experience and in many cases saving me from embarrassing errors; to D. Kent and J. Nancarrow for bearing the brunt of the technical work required to produce the manuscript, and to P. Koshland for years of companionship, encouragement and good advice.

The drawings are the work of Phyllis A. Koshland (Figs 1, 4, 8, 10, 24-26, 29-32, 34-35, 37-40, 43-44, 46-48, 56-58, 63-64, 67-70, 72-74, 76-77) and Deborah S. Kent (Figs 2, 5-6, 9, 19-20, 22, 28, 36, 42, 51, 62, 82-83, 87-88).

The Phylogenetic Relationships and Geographical Origins of Australian Lizards

LIZARDS and snakes together form an evolutionary lineage called Squamata or, in the vernacular, squamates, a reference to their squamous or scaly skin. However, despite the name, a scaly skin does not serve to identify them as a lineage because this feature occurs widely in other reptiles, e.g., crocodilians, the tuatara and at least some dinosaurs. What does unite them as a lineage are, among other things (Table 1), the absence of a bony strut bordering the lower temporal opening of the skull (Fig. 2), and two penises (Fig. 3). The presence of two penises is particularly convincing as an indication of the monophyly of the group because it is a complex structure otherwise unique in vertebrates and hence unlikely to have evolved more than once.

The closest living relative of squamates is the tuatara, *Sphenodon punctatus*, of New Zealand. This species is the sole survivor of a diverse group, the sphenodontians, known as far back as the early Triassic approximately 225 million years ago (Benton 1986 and references therein). Sphenodontids and squamates comprise the group known as the Lepidosauria.

Major evolutionary relationships within squamates are only poorly known. Some hypotheses propose a basic dichotomy between one of the limb-reduced or limbless groups, such as snakes, and all other squamates (Underwood 1970; Rieppel 1978; Rage 1982), while others propose a dichotomy that puts both limbed and limbless groups into each of two basic lineages — usually called the Ascalabota and the Autarchoglossa (Camp 1923; McDowell and Bogert 1954; Northcutt 1978). I believe the latter hypothesis is more likely to be true.

There are five families of lizards in Australia comprising approximately 476 species: Agamidae, Gekkonidae, Pygopodidae, Scincidae and Varanidae (Table 2). The agamids belong to the Ascalabotan branch of the basic squamate dichotomy and the others to the Autarchoglossan branch (Fig. 4). Within the Ascalabotans, the agamids are thought to be related to the chamaeleons which today occur only in Africa, Madagascar, southern Europe, southwest Asia and India. The agamids and chamaeleonids, in turn, are thought to be related to the iguanids, a basically New World, Madagascan and Fijian family similar in general appearance to agamids. These three families are the only living ascalabotans.

Within the autarchoglossans, geckos and pygopodids have long been known to be each other's closest relatives (McDowell and Bogert 1954), and recent work even suggests that pygopodids may be most closely related to the gecko subfamily endemic to the Australian Region, the Diplodactylinae (A. Bauer, pers. comm.). The relationships of skinks are obscure, but the best guess is with a group of African-Madagascan lizards called gerrhosaurids (pers. obs.; p. 117). The closest living relatives of varanids are two small groups: the rare, monotypic Lanthanotidae of Borneo and the venomous Helodermatidae (two species) of North America. Some people have suggested that this group of three families is the closest living relative of snakes (McDowell and Bogert 1954).

The Australian agamids comprise one large endemic lineage, informally called the amphiboluroids, and three species, currently placed in two widespread genera, *Gonocephalus* and *Physignathus*. The relationships of the amphiboluroids are obscure, as are also those of the *Gonocephalus* and *Physignathus* species — despite their generic allocations.

All available evidence suggests that agamids evolved initially on the northern landmass and entered the southern continents, including Australia, only relatively late in their history.

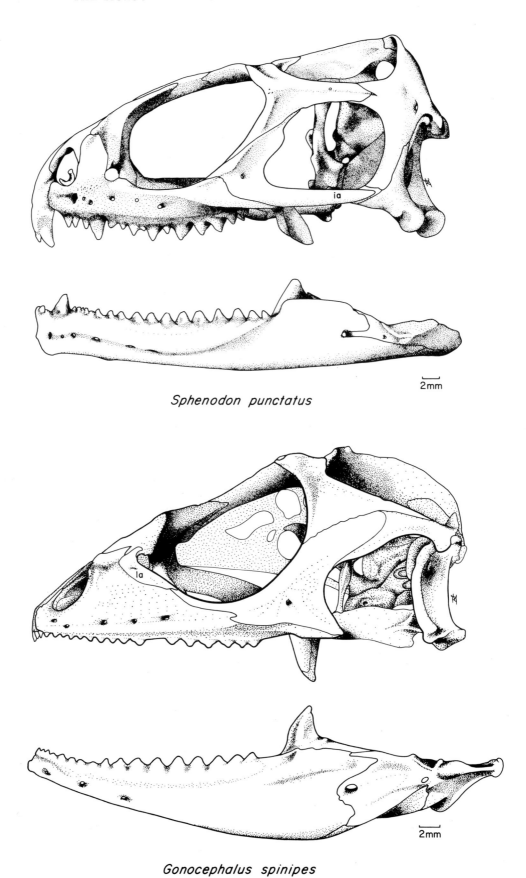

Sphenodon punctatus

Gonocephalus spinipes

Fig. 2. Skulls of *Sphenodon punctatus* (AM R 123339) and *Gonocephalus spinipes* (AM R 93471) to show the loss of the infraorbital arch (ia), a derived character for squamates, in the latter.

Fig. 3. The hemipenes of *Varanus gouldii.* The hemipenis on the right side is fully everted but flacid while the hemipenis on the left side is only slightly everted. Hemipenes are a derived character for squamates. Photo: P. R. Rankin.

The earliest fossils (Cretaceous) are northern; the two most generally primitive living taxa (*Leiolepis* and *Uromastyx*) are northern, and there are no records of agamids on any of the southern continents prior to the Miocene, a time by which the northward drifting southern fragments were within easy reach of northern groups.

The fossil record of dragons in Australia goes back to the Middle Miocene on the basis of a specimen with acrodont dentition (Estes 1984). More recent records include another acrodont specimen from the Pliocene (Archer and Wade 1976) and several from the Pleistocene (Bennett 1876; Smith 1976, 1982).

Three species of amphiboluroids occur in southern New Guinea in savanna woodland habitats similar to those occupied in Australia: *Chlamydosaurus kingii, Diporiphora "bilineata"* and *Amphibolurus temporalis.* Presumably the Australian and New Guinea populations of these species have probably been separated from each other since at least the latest rise of sea-level, but whether any differentiation has occurred between the populations in the two areas has yet to be determined.

Heterochrony (the different rate of development of a body part in a descendant compared to an immediate ancestor) has been advanced as a possible explanation for the "explosive adaptive radiation" of certain amphiboluroids (Cogger 1961 as cited by Cogger and Heatwole 1981). However, aside from the fact that heterochrony is involved in many morphological differences between species, there is as yet no published evidence indicative of the relative importance of heterochrony in amphiboluroid evolution. Nor for that matter is there any reason to think their rate of speciation has been "explosive".

Although dragons are widespread in Australia, the best places to see the largest number of dragon species in a relatively small area are the region just south of Exmouth Gulf and a region in the Great Victorian desert (Cogger and Heatwole 1981: fig. 12; Cogger 1984: fig. 3).

The Australian geckos comprise one large lineage, the subfamily Diplodactylinae, that is endemic to Australia, New Caledonia and New Zealand, and seven genera of the widespread subfamily Gekkoninae: *Christinus, Cyrtodactylus, Gehyra, Hemidactylus, Heteronotia, Lepidodactylus* and *Nactus.* Two of these genera, *Christinus* and *Heteronotia,* are endemic to Australia, but the remaining five are widespread on the islands of the southwest Pacific and, in some cases, beyond.

The relationships of the diplodactylines are obscure, other than for the possibility of a relationship with the indigenous pygopodids (p. 1). There are also no fossils by which to set even an upper limit on the time of origin for the group, and hence to relate it to the then current configuration of Australia, New Caledonia and New Zealand.

The same problems apply to the seven gekkonine genera. Relationships are not well enough understood to infer the geographic history of any of the genera in any meaningful way and fossils are non-existent. To say that the two endemic genera, *Christinus* and *Heteronotia,* probably evolved in Australia and that the Australian representatives of the three widespread southwest Pacific island genera, *Cyrtodactylus, Hemidactylus* and *Lepidodactylus,* probably entered Australia after having evolved elsewhere is only to state the obvious under the probably not too realistic proposition of "all else being equal". However, stating this will no doubt provide comfort to some historical zoogeographers. *Gehyra* and *Nactus,* which are about equally well represented in Australia as elsewhere, are not even amenable to this simplistic anaylsis.

Currently, geckos reach their highest species densities in the arid centre and in certain coastal and near-coastal areas of northeastern Australia (Cogger and Heatwole 1981: fig. 10; Cogger 1984: fig. 2).

All pygopodids, with the exception of the New Guinea *Lialis jicarii,* occur in Australia, and it is likely the family evolved here. However, when they arose is unknown as there are no fossils. The relationships of the pygopodids are also uncertain. According to a long established view, their closest living relatives are geckos as a group, but a more recent, but as yet unpublished, view realted them to the large endemic gecko lineage, the diplodactylines (p. 1).

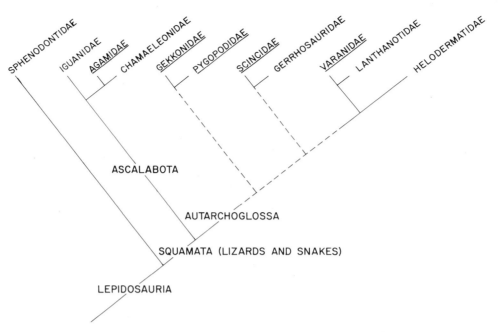

Fig. 4. Highly abbreviated diagram showing the probable relationships of the five families of Australian lizards. Not all autarchoglossan families are shown nor are problematic groups such as snakes and other limb-reduced lizards.

Pygopodids reach their highest species and generic diversities in the coastal regions north of Perth in Western Australia (Cogger and Heatwole 1981: fig. 11).

All the Australian skinks are members of the largest lineage in the family, the Lygosominae, a group that counts for nearly two-thirds of all the species and much of the family's distribution. Within this lineage there are two large lineages, the *Eugongylus* and *Sphenomorphus* groups, and one small assemblage, or possible lineage, the *Mabuya* group. The *Eugongylus* and *Mabuya* groups are thought to be closely related (Greer 1979b; Hutchinson 1981, 1983). All three groups are represented in Australia by largely endemic genera whose relationships with each other as well as with other members of their group are generally obscure.

As a family, skinks have a distinctly "northern look" about them. The most generally primitive skinks, in the genus *Eumeces*, are northern; the earliest (Oligocene) totally believeable skink fossil (*Eumeces*) is northern (Estes 1983), and the earliest fossil in any southern continent — Australia — is only Middle Miocene. Furthermore, the most generally primitive living lygosomines, the only subfamily represented in Australia, are certain species of *Mabuya* living today in south-east Asia.

The Middle Miocene record for skinks in Australia is based on material fairly certainly identified as *Egernia* and *Tiliqua* (Estes 1984) of the *Mabuya* group. These same two taxa, plus *Eulamprus*, are also reported from the Pleistocene (Smith 1976, 1982).

Many taxa in the *Eugongylus* group occur on islands not inhabited by either the *Mabuya* or *Sphenomorphus* group. If this difference is due to some biological feature that allows them to cross water more readily than other lygosomines, then it would suggest that representatives of this group may have been the first skinks to reach Australia.

At present, skink species diversity is concentrated in the arid centre, the southwest coast and enclaves along the north and central east coast (Cogger and Heatwole 1981: fig. 14; Cogger 1984: fig. 4). The Australian varanids comprise 24 of the approximately 30 living species and of these 20 are endemic to Australia and four shared with islands to the north. Despite attempts to identify groups of varanids using various approaches, morphological, electrophoretic and chromosomal, I believe there is as yet no clear indication of the lineages within varanids (p. 195), and hence no basis for discussion of their origins in Australia other than at the family level.

All evidence to date indicates that varanids evolved in the north and later spread south into some but not all of the southern continents. The closest living relatives of varanids are northern, the oldest varanid fossils (Upper Cretaceous) are northern, the most generally primitive extinct (*Saniwa* — Estes 1983) and living taxa (*Varanus salvator* — Mertens 1942a) are northern, and the earliest varanid fossil on any southern continent is only Miocene (Estes 1983). In Australia itself, the fossil record of varanids goes back to the

Middle Miocene (Hecht 1975; Estes 1984), with more recent records in the Pliocene (Archer and Wade 1976) and Pleistocene (Hecht 1975; Smith 1976, 1982; Estes 1983).

Varanids probably would have had little trouble in reaching Australia over water, as there is a strong aquatic component to both their own biology and that of many of their close relatives. For example, some of their certain relatives, e.g., lanthanotids, as well as their probable close but extinct relatives, e.g., aigialosaurids, dolichosaurids and mosasaurids, are or were aquatic as are many of the living species, including the most primitive, *Varanus salvator*.

Varanid species numbers show a marked north-south decline, with a high of ten species in broad sympatry in the Kimberleys and the Top End of the Northern Territory and a low of two along the south coast (Cogger and Heatwole 1981: fig. 13; Pianka 1981: fig. 10).

To summarize, it would appear that at the family level, at least, Australian lizard families evolved in the northern hemisphere and then moved into the southern hemisphere, including Australia. The biggest unresolved question in each case is when this occurred: while Australia was still attached on the large far southern megacontinent 53 million years ago; during its northward migration between 53 and 14 million years ago, or shortly after its arrival in approximately its present position 14 million years ago. The earliest lizard fossils, all Middle Miocene, are just a fraction too recent to be informative because they occur in the record just after Australia reached its present position. Hence resolution of this question awaits either the discovery of earlier fossils, or estimates of times of origin of the relevent taxa based on the molecular clock.

Although nine of the world's 16 lizard families occur in the probable source area for Australian lizards, i.e., southeast Asia, the four lineages (counting geckos and pygopodids together) that have actually made the crossing, are the ones that could have been expected. For example, all are species rich and, hence, would have had a numerically superior chance to disperse. Furthermore, all four possess at least some ecological attributes that would facilitate crossing open seas (arboreality, occupation of littoral or near-littoral habitats, heliothermy and association with large bodies of fresh water) and, perhaps most telling of all, all are proven water crossers in other contexts, i.e., to other islands. In contrast, the four remaining, south-east Asian families are represented by only a few species, some of which are rare (anguids and lanthanotids) and others of which are dependent on moist,

shaded conditions (dibamids and lanthanotids). Perhaps the only slightly surprising omission is the lacertids, the one southeast Asian species of which (*Takydromus sexlineatus*) is a basking, climber.

Most of what we have been discussing above falls under the discipline of historical zoo-geography, and as it is my view that this is the most speculatively practised aspect of natural history in Australia (and elsewhere), I would like to make a few critical comments, using examples from Australian lizards. These criticisms are mainly for the benefit of students who are often fed zoogeographic conclusions in a very uncritical fashion.

Put simply, historical zoogeography is nothing more than phylogenetics placed in a geographical context, and from this it follows that it can be no better than the weakest aspect of our understanding of each of those disciplines. Zoologists have little critical knowledge of geography (in its widest sense to include geology), but they do have, or at least should have, knowledge of phylogenetics. Unfortunately, however, they often neglect the hard core phylogenetics and jump directly into the zoogeography. They too often rely on weak or very poor phylogenetics in whole or in part — rather than simply foregoing zoogeographic speculation or spending their time developing better phylogenetics, or they use their own (Horton 1972; Schuster 1981a-b) or others' unpublished, and therefore basically inaccessible, phylogenetics as the basis of their zoogeographic speculations. In some cases they cite published works which themselves only cite the unpublished work (e.g., Heatwole 1976, citing Horton 1972), thus creating the impression that the phylogenetics have been published. Surprisingly, editors turn a blind eye to this practice.

Historical zoogeographers not only play free and easy with their hard data, they are also given to generalizations that are breath-taking in their distance from what other biologists would call reality. Perhaps their favourite generalization is that species numbers and morphological diversity are directly proportional to time (Keast 1959; Cogger 1967; Cogger and Heatwole 1981; Pianka 1981; Schuster 1981; Witten 1982). The underlying assumption is that the rates of morphological evolution and speciation are equal in similar organisms, in this case all lizards or at least families of lizards. This concept has led some zoogeographers to identify waves of immigration in different groups (e.g., dragons and geckos — Cogger and Heatwole 1981) or to date, in rough terms of geological time, the origin of a family in Australia (e.g., dragons —

Witten 1982).[1] The validity of the underlying assumption, i.e., equal rates of morphological evolution, has always been suspect and is now almost totally discredited (think of little-changing groups like *Sphenodon*'s lineage and crocodilians and the relatively recent appearance of that most bizarre of animals — man). It is not clear what factors determine rates of gross morphological differentiation and speciation, but it is clear that rates differ enough to make them unreliable indicators of either relative or absolute time. No doubt zoogeographers were anxious to grasp the concept of equal evolutionary rates because in the absence of critical fossils (which is the norm) and before the development of an understanding of the molecular clock, they had no other way to estimate evolutionary time in either a relative or absolute sense. But in their zeal for some sort of dating method, they embraced a basically implausible assumption (equal rates of change in a highly variable natural world) and did so in the face of some obvious counter-examples.

Perhaps the second favourite generalization of historical zoogeographers is to evoke competition between major faunal groups to explain broadly complementary distributions between these groups. This, when ecologists — the people closest to the heart of the matter — are still debating whether competition even exists between closely related species or if it does exist how it can be measured. Here are some of the major interactions proposed: hylid frogs vs diplodactyline geckos in mesic eastern Australia (Storr 1964c); colubrid snakes vs pygopodids (Storr 1964c), *Ctenotus* skinks vs "*Amphibolurus*" dragons (Storr 1964c), and birds vs lizards in the arid zone (Pianka 1986). In this same vein historical zoogeographers and those dabbling in the subject are given to making some very off-hand ecological, or niche, analogies, e.g., the skinks *Menetia* and some *Lerista* as insects (Pianka 1969a, 1981, 1986) other *Lerista* and the pygopodid *Aprasia* as worms (Pianka 1969a, 1986), *Tiliqua* spp. as terrestrial tortoises (Milewski 1981; Haacke 1982); agamids as lacertids (Haacke 1982) or iguanids (Pianka 1969); pygopodids as snakes (Underwood 1957; Pianka 1969a, 1981, 1986), *Ctenotus* skinks as cnemidophorine teiid lizards (Vitt and Congdon 1978), and varanids as carnivorous mammals (Storr 1964c; Pianka 1973, 1981, 1986; Haacke 1982).

Third on the list of cherished concepts might be the proposition that present day centres of species abundance are past centres of evolution (Cogger and Heatwole 1981; Cogger 1984). That these centres of abundance could also simply represent the subsequent overlapping of the species' habitats seems not to have been considered. However, imagine that during mesic periods, arid-adapted populations are isolated in arid coastal enclaves where they speciate but then spread toward the centre and overlap in range when more arid conditions return. The obverse pattern would have the same effect.

A fourth favourite pastime of historical zoogeographers is to hypothesize dispersal pathways or corridors; some of which actually end up as lines on maps in a kind of historical road map of "which way they went" (Keast 1959; Storr 1964; Horton 1972; Cogger and Heatwole 1981). There are several remarkable aspects to this. First, such speculation ultimately has a phylogenetic basis, and zoogeographers are as willing here as they ever are to accept uncritically whatever taxonomy is hanging on the rack at the time of writing. Second, this speculation is often heavily based on contemporary geography and habitat distributions and largely ignores the historical changes that geographies and habitats have undergone. Third, the arguments are usually couched in terms of the animals dispersing, and yet this is only one way animals move about. Perhaps equally, if not more important, may be the geographical shift of habitats along with their resident species. Could not a contemporary "corridor" be nothing more than a much reduced remnant of a previously widespread habitat?

Why biologists let themselves go so when it comes to zoogeography is difficult to understand. But let go they do — and therein perhaps lies the greatest usefulness of historical zoogeography: as a collecting ground for unexamined current assumptions about the way evolution works. Students take note.

To conclude this chapter on the phylogenetic relationships and geographical origins of Australian lizards, we might briefly consider two other, related topics often discussed in the literature: desert lizard species diversity and geographical barriers in speciation. It is often stated that Australian deserts have more lizard species than other deserts and that some Australian localities have the highest desert lizard species densities in the world. This situation, often called "extraordinary" or "remarkable", has been analyzed on several occasions, and a list of "explanatory" factors produced which boil down to: more empty niches; milder, more

---

[1]There seems to be some difference of opinion amongst historical zoogeographers as to the polarity of the relationship between morphological diversity (as measured by the number of genera in a lineage) and time, that is whether it is direct or inverse. Compare these two quotes: "the number of distinct endemic gecko genera (8-9) confirmed other evidence that the family has had a long history in Australia" (Keast 1959) and "at the generic level, Australian deserts are somewhat richer (about 23 genera) than either the Kalahari (13 genera) or North American deserts (12 genera). A relatively recent burst of speciation is thus suggested in the Australian desert saurofauna" (Pianka 1981).

equable climate; greater habitat heterogeneity; recognition of finer habitat subdivisions, and impoverished nutrient base. However, there are several problems with all this. First, measurements and comparisons have rarely gone beyond the anecdotal stage, other than for a basic tally of taxonomic units. Second, different definitions of species diversity seem to be involved: species number at one locality, species number in the arid regions in general[1], and perhaps, by implication, species numbers on the continent as a whole. Furthermore, morphological diversity as measured by generic richness seems to be an additional measure of diversity that is often ambiguously linked to the issue of species diversity (Pianka 1981; see quote in footnote on p. 6). And third, there would also seem to be a sampling problem at least amongst deserts — the level at which the discussion has been limited to date — because only three are involved (Australia, southwest Africa and southwest North America), and any ecological variable displaying three states has a one in three random chance of having these states paralleling (directly or inversely) the number of lizard species (Australia > Africa > North America). There is no doubt that there is widespread feeling that there are real differences involved amongst the world's deserts with regard to the number of lizard species, but more rigorous comparisons both amongst and within deserts are required to identify the environmental factors that determine species richness in these areas.

There has been much speculation as to the geographical barriers that may have contributed to speciation in Australian lizards, and the current list of possible barriers includes most of the usual ones: rivers (Cogger 1961, for dragons, as quoted in Heatwole 1976; Cogger and Heatwole 1981); lakes and lake systems, in conjunction with low ranges (Kluge 1967b, for geckos); sea ways (Rawlinson 1974a, for Tasmanian reptiles); "mountain" ranges (Horton 1972 and in Heatwole 1976, for *Egernia*), and specific vegetation/soil habitat types (Pianka 1969, 1972, for desert lizards in general; Chapman and Dell 1985, for certain agamids). These too

are satisfying ideas, but they have the usual shortcomings of their kind: they fail to pinpoint, let alone quantify, the operative factors and how they work on local populations, i.e., those at the edge of the range. Perhaps future work in this area, if there is to be any, will concentrate more on specific environmental parameters and will include some experimental studies as well.

One important potential use of these hypothetical speciation barriers would be to help date the time of separation between populations and thereby to estimate the rate of divergence between them. So far, only sea barriers have all the factors necessary for providing routinely reliable chronologies (Rawlinson 1974a), but unique opportunities may occasionally arise in other areas as well.

Table 1. Derived characters of squamates (lizards and snakes) in comparison to all other reptiles.

Soft anatomy
   Intromittent organs present and paired (hemipenes)
Skull and hyoid
   Postparietal absent
   Tabular absent
   Quadratojugal absent (hence lower temporal arch absent)
   Pterygoids separated from vomers
   Pterygoids separated from one another (secondarily in contact in some skinks)
   Quadrate has external conch
   Teeth pleurodont or subpleurodont
   Prearticular and articular fused
   Retroarticular process elongate, well-developed
Postcranial skeleton
   Sacral ribs fuse to centra early in development
   Rib head single

Table 2. Composition of the Australian lizard fauna (after Cogger 1986).

Family	Number of Species	Percentage of Species
Agamidae	64	13.5
Gekkonidae	87	18.3
Pygopodidae	29	6.1
Scincidae	271	57.0
Varanidae	24	5.1
	475	100.0

[1]"In terms of the number of lizard species they support, the Australian deserts are the richest in the world. Here one can find as many as 40 different species of lizards occurring together. Such extraordinarily high species densities raise important ecological and evolutionary problems." (Pianka 1969).

CHAPTER 2

Agamidae — Dragon Lizards

DRAGON lizards or just "dragons" is the common Australian name for the family of lizards technically known as agamids or Agamidae. They are a conspicuous and hence well known group throughout their range which extends from Africa and south-west Asia east through Asia and the Indonesian Archipelago to the Solomon Islands and Australia. Worldwide there are approximately 320 living species in as many as 53 or as few as 34 genera, depending on author (Moody 1980 and Wermuth 1967, respectively); in Australia there are 64 species (Cogger 1986) in 13 genera. While few species probably remain to be discovered either in Australia or elsewhere, much remains to be discovered about the group's relationships and higher taxonomy.

Although dragons are easy to distinguish from other Australian lizards, they are, as a family, often difficult to distinguish, on superficial glance, from certain other lizards from other parts of the world. In Australia they are recognizable by their diurnal habits, rough scales, well-developed limbs and slightly rounded heads. But in all these regards they are similar to iguanids, a family of American, Madagascan and Fijian lizards. The only ready way to distinguish the two groups is to examine the teeth. Iguanids, like most other lizards, have their teeth set individually in small indentations in the inner side of the jaw bones whereas agamids have all but the front teeth confluently fused to the sides of the jaw; the front teeth are like other lizards. Furthermore, iguanids, and most other lizards, replace their teeth at regular intervals whereas agamids do not, except for the front teeth. The replaceable teeth in dragons and other lizards are called pleurodont and the irreplaceable, permanently fixed teeth acrodont (Fig. 5). The technical features that diagnose agamids as a lineage are given in Table 1 while some standard measurements and proportions, and vertebral counts of Australian species are given in Tables 2 and 3, respectively.

It is currently thought that dragons are related to the chamaeleons of southern Asia and Africa, a group with which they share acrodont teeth, and that these two families taken together as a group are related to iguanids (Moody 1980; Estes 1983; Fig. 4). However, many of the features used to support these groupings, especially the ones used to ally all three families, seem to be based largely on retained primitive characters, and future work based on yet-to-be-discovered derived characters may lead to alternative associations, some of which might even involve other lizard groups.

Dragons in Australia occupy a wide variety of habitats. Forest dragons (*Gonocephalus*) occur in rainforests; water dragons (*Physignathus*) live along rivers, creeks and pools; certain species of *Tympanocryptis* inhabit stony deserts; some species of *Ctenophorus* occur on loosely consolidated sand dunes, often just a few metres from the sea, while others live on bare rock outcrops, and one inhabits the bed of dry salt lakes.

All dragons are terrestrial or arboreal; none are soil, sand or litter swimmers. All dragons are also primarily diurnal. They are also all heliotherms, i.e., they seek to raise their body temperature above that of the air by basking in the sun (p. 12).

Most dragons move at ordinary speeds using ordinary four-legged locomotion, but a few species show variations. Two quite different species, the almost exclusively arboreal *Chelosania brunnea* and the exclusively terrestrial *Moloch horridus*, seem capable of moving at only a very slow and deliberate pace, even when in danger. This slow motion may serve to avoid attracting a predator's attention. *Chlamydosaurus kingii* may occasionally hop like a kangaroo (G. Krefft, in Bennett 1876). Such a hop is also often employed by cornered dragons turning to the offensive, perhaps in one last desperate effort to startle the intruder.

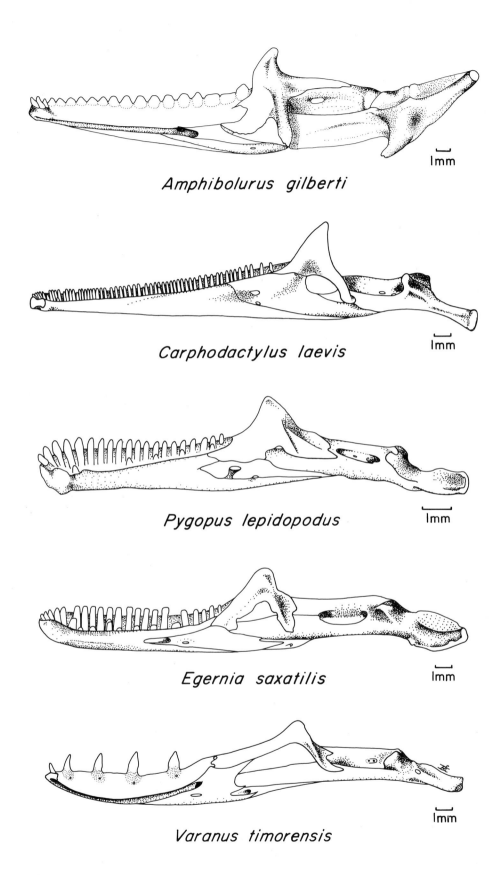

Amphibolurus gilberti

Carphodactylus laevis

Pygopus lepidopodus

Egernia saxatilis

Varanus timorensis

Fig. 5. The right lower jaw in lingual view of representatives of the five families of Australian lizards to show differences in shape, proportions and type of tooth implacement.

Several dragons are capable of running very rapidly on their hind legs only — a habit that has earned some of them the name of bicycle dragon. This bipedal running has been observed in *Amphibolurus longirostris* (Baehr 1976); *A. muricatus* (Waite 1929); *A. temporalis* (Cogger 1973); *Chlamydosaurus kingii* (Woodward 1874; Saville-Kent 1897; Stirling 1912; Harvey and Mover 1978; Wendt 1987: photo pp. 52-53); *Ctenophorus caudicinctus* (pers. obs.); *C. cristatus* (Rau, in Stirling 1912; Houston 1978; pers. obs.); *C. isolepis* (Slater and Lindgren 1955, as *Amphibolurus maculatus* but probably this species); *C. ornatus* (pers. obs.); *C. rufescens* (Stirling and Zietz 1893); *C. scutulatus* (pers. obs.); *Gonocephalus spinipes* (F. Hersey, in Webb 1984); *Physignathus lesueurii* (Worrell 1958; Wilson 1974; Jenkins and Bartell 1980); *Pogona barbata* (Carpenter *et al.* 1970) and seems to be used only when the lizard flees over a great distance very quickly. *Amphibolurus longirostris* fleeing bipedally has been estimated to average 20-24 km/hr (Baehr 1976; N = 3)[1]. Bipedal walking has been reported for a few dragons, e.g., *Chlamydosaurus kingii* (C. Coxen, in Bennett 1876; DeVis 1884) and *Amphibolurus nobbi* (Fyfe 1981b) but these observations require confirmation. Similarly, there are a few observations of bipedal standing, e.g., *Chlamydosaurus* (Bustard 1970d), *Ctenophorus nuchalis* (Bradshaw 1986) and *Tympanocryptis* (Swanson 1976: fig. 5).

Most dragons eat a wide variety of arthropods but some extend their diet beyond this or show restricted specializations (Table 4). One striking contrast to other lizards is the amount of ants consumed; most lizards either avoid them or eat only a few, but many dragons incorporate a high percentage of them in their diet (e.g., Pianka 1986, appendix E3), and one, *Moloch*, feeds almost exclusively on them. Why other lizards avoid ants while dragons eat them in abundance is unclear. If other lizards avoid them because of the large amounts of formic acid they contain, there may be some special mechanism in dragons to overcome this problem. In Australia, many dragons seem to prefer ants of the genus *Iridomyrmex* (e.g., *Ctenophorus ornatus* — Bradshaw and Shoemaker 1967; *Moloch* — Davey 1923; Paton 1965; Pianka and Pianka 1970), although the reasons for this are unknown. Most dragons, perhaps all but *Chlamydosaurus* and *Moloch*, eat some vegetable matter and the larger and more sedentary species (e.g., *Ctenophorus nuchalis* and *Pogona*) eat relatively large amounts of this food (more than 20% of the diet by volume). The very large dragons will also occasionally eat small vertebrates[2].

Dragons are primarily sit and wait predators, that is, they sit quietly in one spot watching for prey which they then pursue (Huey and Pianka 1981). Dragons seem to almost never actively search for prey as many other lizards do.

The two kinds of teeth seen in dragons (p. 9) may be related to the diversity of the food they eat. The replaceable teeth in the anterior part of the mouth are conically pointed whereas the more posterior permanent teeth are laterally compressed (Figs 2, 5-6, 9). The anterior teeth appear to comprise a grasping and piercing dentition perhaps best suited to obtaining and killing animal prey, whereas the posterior teeth are a cutting and slicing dentition perhaps best suited to chopping both animal and plant matter.

Like most other basking reptiles, dragons prefer to have body temperatures in the mid-30's (°C) for most of their activities (Table 9-10). This body temperature is close to that of mammals and birds, but unlike these groups, reptiles can not achieve it from internal sources by physiological means; instead they achieve it

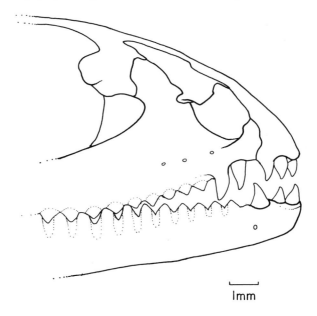

Ctenophorus decresii

Fig. 6. Dentition in the anterior part of the mouth of the dragon *Ctenophorus decresii* (AM R 81641) to show the anteriorly placed pleurodont and the posteriorly placed acrodont teeth.

---

[1]The absence of any small dragons from the list of bipedal runners is slightly odd (the smallest is *Ctenophorus isolepis* with a maximum SVL of 70 mm). One wonders whether this absence relates to the biophysics of bipedal running or to the inability of human observers to see what smaller animals are doing.

[2]Although there are many studies of the food items found in the stomachs of wild-caught lizards in Australia, inferences as to the selectivity of prey by populations or species based on such studies is rarely warranted. This is because the studies rarely include observations of lizards faced with choices in nature or of the spectrum of prey seen by the lizard in its habitat. A sample of lizards with mostly termites in its guts (e.g., *Caimanops amphiboluroides* — Pianka 1986, appendix E3) might mean that the species eats nothing but termites, but it might also mean that termites were swarming in abundance just prior to the sample being collected, i.e., that termites were the only item on the menu! For example, in a study of the food of *Ctenophorus isolepis* "nearly half the termites [recorded] were found in the stomachs of only 26 lizards [of 511 examined], collected shortly after cyclone Elsie (Jan. 1967) when isoptera were swarming (Pianka 1971c).

from external sources either passively or by behavioural means. Because dragons are as conspicuous in their behavioural thermoregulation as they are in their other activities, this aspect of their biology has been well-studied (Bradshaw and Main 1968; Bradshaw 1977a; Cogger 1978) and can serve as a model for other less well known lizard groups.

When dragons first emerge from their nocturnal retreats they are usually dark in colour, slow moving and with a body temperature close to ambient. They usually move directly to their basking sites and turn their backs to the sun, often with their sides expanded. Hence, the colour, shape and orientation of the lizards all serve to maximize absorption of sun's radiation and hence heating of the lizards. Often the lizards' basking site will be a substrate that is a poor thermal conductor, i.e., wood or foliage, so that they will not lose the absorbed heat to the substrate.

As the lizards heat up, they lighten in colour (Rice and Bradshaw 1980) and begin changing position to fine tune their temperature and prevent overshooting. As a suitably high temperature is approached they may begin other activities such as defecation, feeding, or a variety of social interactions. As the day progresses and the general environmental temperatures rise, the lizards must increasingly avoid becoming too hot, and often the difference between suitable and too hot is just a few degrees (Table 11). To keep from overheating, they may become very light in colour and turn to face the sun, thereby minimizing the amount of body surface area exposed to the sun and limiting that exposure to the more reflective light underbody. They may also move to the semi-shade, up a shrub or tree or onto a substance that has a slower rate of heat gain itself, like wood. If it is still too hot, they may go higher on their perches, or if on the ground, they may raise their toes and tails off the substrate (Francis 1981), which is usually quite warm by now; they may also simply enter deep shade or retreat to shelter. If caught out in severe heat they may, as a last resort, open their mouths and begin panting (Heatwole et al. 1973; Chong et al. 1973; Cogger 1974; Heatwole et al. 1975; Parmenter and Heatwole 1975; Firth and Heatwole 1976) — but a panting dragon is very close to being a cooked dragon.

As the sun begins to decline and the air temperature cools, dragons try to retard their rate of cooling. Because various components of the environment now retain heat, they use these as well as the rays of the declining sun as heat sources. For example, they may press themselves against the warm soil and rocks. Finally, near sundown and with the imminent loss of light and the declining air temperature the dragons retire, and if the retreat also retains heat, their rate of cooling can be retarded even further, perhaps gaining them extra time to digest food or grow.

In addition to these obvious behavioural and physiological changes which occur in thermoregulation, dragons also change the pattern of blood flow within the head and body to help control their temperature. For example, when dragons first emerge in the morning, they may shunt blood to large sinuses in the head. This is evident by the slightly bulging eyes of early basking dragons (Mitchell 1973; pers. obs.). The purpose of this is probably to heat the brain and sensory organs in the head to maximize function. Also, while heating with the back to the sun, the dorsal blood vessels dilate and the heart rate rises, presumably to facilitate absorption and circulation of the heat. In addition, although it hasn't been proven, during basking, dragons may shunt the cool blood of the core to the superficial layers of the back in order to heat it quickly, and, conversely, in order to retard heat loss, they may impede the flow of warm blood from the core to the periphery (Bradshaw 1981).

Dragons are the most visually oriented of all Australian lizards, a fact evident in a number of aspects of their biology. First, they all are primarily diurnal and carry out most of their activities in the open; second, they have been observed to sight prey, conspecifics and predators over long distances, e.g., up to 50 m in Ctenophorus maculosus (Mitchell 1973); third, most species have some degree of sexual difference in adult colouration at least during part of the year and these differences almost certainly exist to be perceived visually by conspecifics; fourth, many of their social interactions consist of body movements that can only be visual signals, and fifth, they rely very little on tongue flicking — the behaviour mediating the smell/taste sensory perception of Jacobson's organ located in the roof of the mouth of lizards and snakes — to investigate their surroundings.

Although dragons are primarily diurnal, there are occasional reports of nocturnal behaviour that seem to be indicative of more than just the odd animal chased out of its nocturnal retreat, e.g., Amphibolurus nobbi, Ctenophorus fordi (Morley and Morley 1985), Diporiphora sp. (Fyfe 1981b), Pogona minor (Bush 1983), and Tympanocryptis (Fyfe 1981b). The report for Amphibolurus nobbi is absolutely fascinating; in one instance they were observed "walking on their hind legs only". This kind of observation not only makes one wonder what they were doing, but also to realize that more attention should be paid to "odd time" occurrences such as these; they might reveal greater behavioural capacities than now realized.

Most dragons have the ability to change colour over short periods of time (i.e., a few minutes) by expanding or contracting the dark pigment, melanin, in special skin cells called melanophores. These changes are associated with thermoregulatory, social, and predator avoidance behaviour, and perhaps also with reproductive state (Kent 1987). Two types of colour change occur in dragons: a general lightening and darkening of the overall colour, and a specific lightening and darkening of local pattern elements, such as spots and bands. General colour change is most often associated with temperature regulation (p. 12) whereas pattern change is generally associated with social interactions, its variable control perhaps allowing the lizard to vary the intensity of the "message". Most dragons can change general colour and many can change pattern. Only *Ctenophorus ornatus* appears to have lost the ability to change any aspect of its colour (Bradshaw and Main 1968) and only *Moloch* and *Tympanocryptis* have lost the ability to control local colour patterns (pers. obs.).

As mentioned above, most dragons are sexually dichromatic as adults, and the colour differences are usually permanent. However, the fixed patterns may be enhanced, and other colours and patterns added, on a seasonal basis. One of the most common sexual colour differences in dragons is a permanent black mark or patch on the venter, usually on the throat or chest. There is also usually a positive correlation between the relative size of this black area and the size of the species within a group, e.g., *Ctenophorus*. The function of this or any other sexual colour difference is unknown in dragons. Only a few species appear to lack sexual dichromatism, e.g., *Ctenophorus nuchalis* (Storr et al. 1983), *Diporiphora*, *Moloch*, and perhaps some *Tympanocryptis* (pers. obs.).

Most dragons have the tongue and lining of the mouth a pale pink colour due to the colour of red blood cells streaming through the capillaries (Fig. 7). A few species, however, have other colours. For example, in *Amphibolurus muricatus* (Peters 1974, 1975), *A. norrisi*, *Chlamydosaurus* (the western population), *Pogona barbata* (Peters 1974, 1975), *P. minor* (the southwestern population, *fide* Bradshaw and Main 1968; Storr 1982c), *P. nullabor* and *P. vitticeps* the tongue and mouth lining is a pale to dark yellow colour; in *P. henrylawsoni* it is said to be bright orange (Wells and Wellington 1985), and in *Rankinia diemensis* (at least the northern population) the tongue is mango yellow and the mouth lining grayish blue (pers. obs.). What these colours derive from and how they function are unknown.

Data on age at sexual maturity in dragons is skimpy but reveals some interesting differences.

Fig. 7. Mouth colour in dragons: the usual and primitive pink colour (*Gonocephalus spinipes*) and the derived yellow colour (*Pogona minor*).

There is direct (mark and recapture) and in-direct (age classes in preserved material) evidence that some dragons, such as *Amphibolurus norrisi*, *Ctenophorus caudicinctus*, *C. femoralis*, *C. fordi*, *C. isolepis*, *C. maculatus*, *C. nuchalis*, *C. ornatus* (some) and *Rankinia adelaidensis*, may reach sexual maturity in one year or less (Tyler 1960; Storr 1965, 1967b; Pianka 1971c; Cogger 1974, 1978; Baverstock 1979; Bradshaw 1975, 1981; Witten and Coventry 1984). Other dragons, either as individuals, populations or species, take longer, e.g., individuals of *Ctenophorus ornatus* may take two to three years, *Pogona minor* two years (Davidge 1979) and the large *Physignathus lesueurii* five years, at least in captivity (Hay 1972). The rate of growth necessary to achieve sexual maturity in one year in some of the larger species must be quite high. This has been measured in *Ctenophorus nuchalis* at a maximum of 25 mm/month (Bradshaw 1981). There may be a positive relationship between age at maturity and average life-span (Bradshaw 1981).

Courtship and/or mating has been observed in only a few Australian dragons (e.g., *Ctenophorus fordi* — Cogger 1978; *C. maculosus* — Mitchell 1973; *Pogona barbata* — C. S. Sullivan, in Kinghorn 1931; Carpenter *et al.* 1970; Brattstrom 1971a) which is surprising considering the conspicuousness of the animals. During mating, the male bites and holds the female by the nape; this is the primitive mating grip for lizards (Böhme and Bischoff 1976).

As far as is known, and the record is fairly complete (Table 5), all Australian agamids are oviparous (some overseas species are ovoviviparous) and clutch size in all species, except perhaps *Ctenophorus femoralis* which lays only two eggs in a clutch, is variable. Within species with variable clutch size and for which there is a large enough sample for a fair test (N≥12), there is a trend for clutch size to increase with female size. The only exceptions to this generalization seem to be *Diporiphora bilineata*, *Moloch horridus* and *Tympanocryptis intima* (Table 7). Amongst species, there is also a positive correlation between mean female size and mean clutch size (r = 79.2***, N = 28). Most species, for which there is information, lay more than one clutch per season. In the few species that have been examined, there is no correlation between female size and egg size within species, but there is, of course, between species (James 1983). The variable clutch size within dragon species is in contrast to the constant clutch size (of two or one) in geckos, pygopodids and some skinks, and the multiple clutches per season are in contrast to most skinks and varanids.

As far as is known, females of all species dig a nest burrow in earth in which they deposit the eggs, and very often several "test burrows" may be dug by the female before she decides to lay her eggs (e.g., *Ctenophorus fordi* — Cogger 1978; *C. maculosus* — Mitchell 1973). In *C. fordi* rejected test burrows are left open and the chosen burrow is filled in and covered-over in a camouflaged fashion (Cogger 1978). It has been claimed that a day or two before hatching a *Pogona vitticeps* female dug away part of the covered hole "to make it easier for the hatchlings to dig their way up to the surface" (Smith 1974; species identification *fide* H. Ehmann); however this observation is unique for dragons and requires confirmation.

Reproduction is centered over the spring-summer period (September-March) for dragons inhabiting the temperate parts of Australia (e.g., Pianka 1971a-c; Mitchell 1973; Cogger 1978; Bradshaw 1981; James and Shine 1985). This timing is typical of temperate lizards in Australia in general. Reproduction in most tropical dragon species (e.g., *Amphibolurus gilberti*, *A. temporalis*, *Ctenophorus caudicinctus* in Arnhem Land, *Diporiphora* spp.) is centered over the wet (September-May) (James and Shine, 1985; pers. obs.), and hence is more or less synchronous with reproduction in the south. However, at least one tropical species, i.e., *Chelosania brunnea*, is known to reproduce during the dry (July-September), although its activities during the

rest of the year are unknown (Anonymous 1973b; Pengilley 1982; James and Shine 1985).

Unfortunately, the timing of reproduction in dragons inhabiting the arid interior parts of Australia is little studied and then largely in peripheral species/populations. *Amphibolurus minor* in the northern part of its range, i.e., central western Australia is said to breed between July and September, i.e., before late spring and summer when conditions might be too hot for critical stages such as eggs in nests. However, in central western Australia (vicinity of Giles meterological station), *Tympanocryptis lineata centralis* females are gravid with oviducal eggs in late September (pers. obs.) suggesting partial synchrony with most other dragons. *Ctenophorus caudicinctus* in the Pilbara breeds in February-March, a period of cyclonic rains, but its activities at other important times of the year, i.e., November-January, are unknown (Bradshaw 1981, 1986). *C. nuchalis* in the same area breeds during the summer as it does further south at Shark Bay, but in both areas the "stress" of breeding under these extreme conditions appears to lead to very high mortality (Bradshaw 1986). *C. femoralis* females in the Exmouth Gulf region can carry enlarged yolking ovarian eggs and oviducal eggs in mid-October, again suggesting synchrony with most other dragons (pers. obs.). Despite these partial and peripheral insights the potentially most interesting work on reproduction in arid zone dragons has yet to be done, i.e., year-round monitoring of events in a population living under the full force of the climatic extremes typical of this area.

Although all dragons, or at least the females, dig burrows for their eggs (Fig. 13), and hence are clearly capable of digging, only a few species actually dig burrows as a permanent shelter and a retreat from predators. The known burrowers are *Ctenophorus clayi* (Storr 1966a); *C. cristatus* (pers. obs.); *C. gibba* (Houston 1974); *C. nuchalis* (Storr 1966a); *C. pictus* (Mayhew 1963; Mitchell 1973; pers. obs.); *C. salinarum* (Storr 1966a); *C. reticulatus* (Bush 1981), and *Rankinia adelaidensis* (H. Ehmann, pers. comm.). There is also some evidence that a few non-burrowers are capable of burying themselves in loose soil to avoid harsh surface conditions (e.g., *Ctenophorus fordi* — Cogger 1974, *C. maculosus* — Mitchell 1973 and *Diporiphora linga* — Houston 1977). In burrowing, some species (e.g., *Ctenophorus clayi* and *C. fordi*) have been observed to use both legs on one side simultaneously. Some burrowers back plug their burrows when inside (p. 28).

Dragons provide some of the best cases of substrate colour matching amongst Australian lizards, a fact perhaps not too surprising given the degree to which they are active out in the open during the day. Dragons inhabiting areas of red sand are mostly rusty red (e.g., *Ctenophorus femoralis, C. maculatus badius, C. rubens*) while those in yellow sand are often beige or tan (e.g., *C. isolepis citrinus* and *C. m. maculatus*). Dragons arboreal on trunks and branches are often grey (e.g., *Amphibolurus, Caimanops,* and *Pogona*) while those in foliage are often green (e.g., *Diporiphora superba*). Rock dragons also often match their varied substrates (e.g., *Ctenophorus caudicinctus* and *C. rufescens*).

Almost all dragons ultimately respond to predators by fleeing, usually to some form of retreat such as a burrow, rock crevice, hollow log, dense vegetation, or in the case of *Physignathus*, the water itself. However, before fleeing a number of species show other behaviours. Resting absolutely motionless, "freezing", is adopted by some species such as *Chlamydosaurus, Moloch, Pogona* and *Tympanocryptis*. Turning around to the far side of whatever projection the lizard might be sitting on or next to is a common response in *Gonocephalus* (usually a tree trunk; Bevan 1983), the *Ctenophorus maculatus* group (usually a clump of hummock grass), the *C. nuchalis* group (usually an upright twig, rock or large earth clod). Upon closer approach a few species will flash a sudden display that has a certain shock value, at least to the human observer. The bearded dragons of the genus *Pogona*, or at least the larger species, will open the mouth and extend the "beard", while the frilled-lizard, *Chlamydosaurus*, will open the mouth and erect the frill. When grasped, almost all dragons except *Moloch* will bite, and most of the larger species will lash with the tail, e.g., *Chlamydosaurus* and certain *Pogona* (Saville-Kent 1897).

During climatically harsh seasons, dragons retreat to burrows, holes, hollow logs, leaf litter, or rock crevices to await the return of better conditions. All southern dragons retreat from winter cold, and many northern ones, although especially the larger species for unexplained reasons, retreat during the winter dry. For example, *Chlamydosaurus* is about as common during the summer in the north as are *Pogona barbata* in the south, and it is about as rare during the winter as well, even though other lizards are active. What little information there is suggests that the first animals to emerge from dormancy are adult males (e.g., *Ctenophorus fordi* — Cogger 1978; *C. maculosus* — Mitchell 1973). Early emergence permits these males to establish their territories before the emergence of potential mates.

Social behaviour in dragons involves a variety of changes in body shape, appendage position, colour and pattern. The head is frequently bobbed in rapid succession in varying patterns which are often relatively invariant and species-specific; the throat may be distended, the body raised, the sides compressed or depressed (*Pogona*), the mid-dorsal crest raised, the tail raised (these last four often reveal "hidden" colours and/or patterns), and a foreleg rotated rapidly. This latter behaviour, sometimes called circumduction, gives the impression that the lizard is waving and has led to various species being called ta-ta lizards, after the colloquialism for "good-bye" (Thurston 1973). Some species that lack an actual physical mid-dorsal crest can raise the skin of the mid-dorsal line into a distinct ridge or crest, e.g., *Ctenophorus pictus* (Swanson 1976: fig. 83; Hudson 1979), and *Tympanocryptis lineata* (pers. obs.).

Aggression displays between conspecific males and general "assertion displays" are the best known form of social behaviour in dragons; these have been studied or remarked on in *Amphibolurus muricatus* (Carpenter *et al.* 1970); *Ctenophorus decresii* group (Gibbons 1977, 1979); *C. nuchalis* (Carpenter *et al.* 1970, as *Amphibolurus inermis*); *C. maculosus* (Mitchell 1973); *C. pictus* (Mayhew 1963); *C. reticulatus* (White 1949), and *Pogona barbata* (Carpenter *et al.* 1970; Brattstrom 1971a). Displays to predators involve most of the same components seen in displays to members of the same species. A voice has been recorded in dragons but only infrequently; *Chlamydosaurus kingii* has been observed to hiss when confronted by a predator (Saville-Kent 1897; Swanson 1976) and *Tympanocryptis lineata* was once heard to give a high pitched squeak in confrontation with another species of dragon (D. Kent, pers. comm.).

Two dragon species have unusual female displays that appear to act as inhibitory signals to reproductively active males. Female *Ctenophorus maculosus* roll over on the back and lie immobile when approached (Mitchell 1973), and gravid *C. fordi* present the posteroventral part of the body, rear legs and tail by turning the back toward the male and raising the back legs and tail (Cogger 1978). The latter behaviour occurs in some other lizards, including another, overseas agamid, but the former is unique amongst lizards.

Territoriality, i.e., defense of a discrete area against individuals of the same species, is claimed to exist in several dragon species (e.g., *Amphibolurus longirostris* — Baehr 1976; *Ctenophorus inermis* — Heatwole 1970; Pianka 1971a; *C. ornatus* — Bradshaw 1971; *C. pictus*; *C. reticulatus*, and *Pogona barbata* — Brattstrom 1971a), but this is largely on the basis of casual observation of spacing patterns in nature or of intraspecific aggression seen either in nature or in confinement. The only well-studied case of territoriality in dragons to date is in *Ctenophorous maculosus* (Mitchell 1973). Some well-known dragon species appear to lack territorial behaviour, e.g., *Amphibolurus nobbi* — Witten 1974; Witten and Heatwole 1978; *Ctenophorus fordi* — Cogger 1978, but clearly this is an area where more work is needed.

One of the distinctive morphological features of agamids is the loss of the line of weakness in the tail vertebrae across which the tail is easily broken when grasped. This autotomy line, or fracture plane, is a primitive lizard feature retained by most but not all groups of living lizards. When the tail is broken along one of these planes the free piece is left actively wriggling and presumably distracts whatever animal may have broken the tail in the first place, presumably a predator, allowing the rest of the lizard to escape. In time a new tail grows to replace the old, but it differs from the old in now being supported by a central cartilaginous rod instead of a column of vertebrae, and usually in being covered with scales of a different size and colour.

Although agamids lack caudal autotomy planes, the tail can be broken if enough force is applied, but it is clear from the difficulty of the procedure this is not as common a mode of defense and escape in agamids as it is in most other lizards. Agamids can also grow a new tail supported by a cartilaginous rod. Amongst Australian agamids regenerated tails have been observed in *Amphibolurus temporalis* (Arnold 1984), *Ctenophorus caudicinctus* (Arnold 1984), *C. femoralis* (pers. obs.), *C. ornatus* (pers. obs.), *C. rubens* (pers. obs.), *C. rufescens* (pers. obs.), *Diporiphora bilineata* (Bustard 1970d), *Physignathus lesueurii* (Loveridge 1934; P. Rankin in Anonymous 1976; Hardy and Hardy 1977) and *Rankinia adelaidensis* (Tyler 1960). *Physignathus lesueurii* can regenerate a second tail if the first is only partially broken — leading to a bifid tail, and it can regenerate from a break in a regenerated tail (P. Rankin, in Anonymons 1976).

Differences in the shape of dragons, both within species during growth and among species by age group, have recently been studied with a view to determining general patterns (Witten 1982a, 1985). Intraspecifically, dragons show the classic vertebrate trends with regard to change of shape during growth, i.e., in comparison to snout-vent length, the head and limbs become relatively shorter and the tail relatively longer[1]. Interspecifically differences in certain aspects of the shape of adult dragons show a clear relationship to their behaviour and ecology. There is a wide range of shapes, but the two ends of the spectrum are formed by the active runners and climbers which have relatively long legs and tails and the more sedentary species

which have relatively short legs and tails. For example, single large males of nine species of largely terrestrial active runners had hind limbs ranging 95–110 percent of SVL whereas single large males of seven species of largely terrestrial sedentary species had hind limbs ranging 60-79 percent of SVL (pers. obs.)[2]. Furthermore, maximum tail length, a general indicator of adult tail length ranged 245–317 percent of SVL in the active species but only 86–196 percent in sedentary species (Table 2). Perhaps the only non-obvious aspect of these relationships is the relatively long tail of active animals; in this case the tail probably acts as a counterbalance to the body, which may be carried slightly up while running[3]. Interestingly, in contrast to adult shape which appears to "track" behaviour and ecology, juvenile shape appears to concord more closely with taxonomic group (Witten 1982a, 1985).

The shape of the pelvis is also said to correlate with behaviour in agamids. Active climbers and long distance runners have the pubic bones projecting anteriorly to form an angular pelvis in ventral view, whereas more sedentary species have the pubic bones projecting directly medially to form a more box-like pelvis (Mitchell 1965a; Moody 1980; Fig. 8). The functional significance of this osteological difference is not clear.

The postsacral diapophyses, i.e., the lateral projections on the postsacral vertebrae, vary in number (Table 3), thickness and angle of projection amongst Australian dragons. The taxonomic and ecological significance of this variation would well be worth exploring.

The size range within species of Australian dragons is fairly well known (Table 2). In general, hatchlings range 23-50 mm snout-vent length (SVL) and adults 46-258 mm. The ratio of maximum to minimum SVL ranges from about 2 to 6.5, i.e., some species are able to grow maximally to only about twice their hatchling size whereas others grow up to about six and a half times the size they start life. The relationship between minimum and maximum SVL amongst species appears to be linear with 1 mm of hatchling SVL corresponding to approximately 5.2 mm

---

[1]Specifically, for head length *vs* snout-vent length, 32 of 34 Australian agamid species and subspecies had negative coefficients of allometry (log-log) whereas two had positive coefficients ($X^2 = 26.4***$); for hind limb length 27 of 37 had negative coefficients and 10 positive ($X^2 = 7.8**$), and for tail length 28 of 37 taxa had positive coefficients and nine negative ($X^2 = 9.7**$)(Witten 1982a). Not surprisingly, a similar set of relationships was demonstrated when body mass was used instead of SVL in a single agamid species — *Ctenophorus nuchalis* (Garland 1985).

[2]The active runners were: *Ctenophorus caudicinctus* — 110; *C. cristatus* — 105; *C. fermoralis* — 112; *C. fordi* — 95; *C. isolepis* — 100; *C. maculatus* — 100; *C. ornatus* — 95; *C. rubens* — 98, and *C. scutulatus* — 106. The sedentary species were: *Ctenophorus clayi* — 73; *C. nuchalis* — 69; *C. reticulatus* — 74; *Moloch horridus* — 60; *Tympanocryptis cephala* — 71; *T. intima* — 79, and *T. lineata* — 74.

[3]It should be noted that in contrast to this interspecific association between shape and running behaviour in dragons, there was no intraspecific relationship in the one species studied to date — *Ctenophorus nuchalis* (Garland 1985).

of maximum adult SVL. What controls growth in reptiles is little understood but if hatchling size (or, more likely, whatever controls hatchling size, perhaps yolk volume) has an effect, only a small increase at this near-starting point of growth can have a large influence on the ultimate end point, especially at the lower end of the size range.

The scales on the undersurface of the digits, i.e., the subdigital lamellae, show relatively little variation amongst species of dragons, which is surprising given the different surfaces with which different species come in contact. Basically there is a single row of scales along the mid-ventral area of each digit and each scale in the row bears a low spiny keel on each side, thereby forming a paired row of such keels along the digit. The only noticeable variation in this basic pattern is the relative length of the preaxial keel in some species. The one exception to this basic arrangement is *Physignathus lesueurii* which has a series of small unkeeled scales along the mid-ventral area of each digit instead of the single, bicarinate row. The functional significance of these subdigital scale morphologies is unknown, but presumably *Physignathus'* aquatic habits are the basis of its distinctive morphology.

Hybridization between species of Australian dragons has been recorded twice, both times as experimental crosses. The first cross involved a *Pogona barbata* (female) and a *Pogona vitticeps* (male) from 240 km apart across their contact zone (the species have abutting or only slightly overlapping ranges) and the second cross involved two individuals of the same species/sex combination as above but from only 100 km apart. Compared to the controls (clutches from single species matings and incubated along with the hybrid clutches), the near-term embryos in the first cross showed major deformities (but all clutches failed to hatch due to low temperature) and the hatchlings in the second cross showed minor, non-lethal deformities (Badham 1976).

Most Australian dragons have a well developed external ear opening which consists of a relatively large opening and a short auditory canal at the bottom of which is a thin tympanum or ear drum to which the stapes or ear bone attaches internally. However, in two lineages of Australian dragons — *Ctenophorus maculosus* and *Tympanocryptis* — the external ear has been lost in that it now consists of nothing more than scaly epidermis. Internally, two further modifications are evident. In the more primitive condition the tympanum is well developed and lies in loose but intimate association with the overlying skin (its position is generally indicated by a dimple in the skin); furthermore, the stapes maintains a fairly normal relationship with the tympanum. This condition occurs in *Ctenophorus maculosus* and the *Tympanocryptis parviceps* species group. In the more derived condition, the tympanum, indeed if it can even be recognized as such, lies very deep to the skin, the intervening space having been filled in by the muscle which opens the lower jaw (and no dimple is evident in the skin); furthermore, the stapes has now assumed a firm connection with the quadrate, the bone which supports the anterior part of the tympanum in species retaining this structure. This condition occurs in the *T. lineata* species group. To judge from the morphology, this more advanced degree of ear loss probably has had a more profound effect on hearing than has the former. Indeed one even wonders if the members of the *T. lineata* species group can hear much of anything.

THE LINEAGES OF AUSTRALIAN DRAGONS

Discussion of the lineages and ecological assemblages of Australian dragons can begin with the two most generally primitive groups, *Gonocephalus* and *Physignathus*. Morphologically, they retain from their early agamid ancestry a well developed crest along the midline of the back and neck, a large lacrimal bone in the skull (Fig. 9), and what is thought to be the primitive chromosome number and morphology, 12 large bi-armed pairs and 24 small pairs. Ecologically, they are also quite distinct from other Australian dragons in that they occur only in relatively moist forests and woodlands along the east coast.

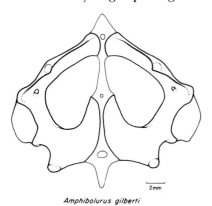

Amphibolurus gilberti

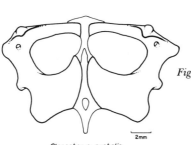

Ctenophorus nuchalis

Fig. 8. The pelvic girdle in ventral view in two dragons to show the anteriorly projecting pubic bones in runners (*Amphibolurus gilberti* — AM R 118506) and the medially projecting pubic bones in more sedentary species (*Ctenophorus nuchalis* — AM R 118507).

The genus *Gonocephalus* is very poorly defined but includes forest dragons from the Indonesian Archipelago, New Guinea and the Solomon Islands as well as the two species in Australia. It is uncertain whether the two Australian species are each other's closest relatives as this problem is part of the larger unsolved problem of the relationships of *Gonocephalus* species in general. The two Australian species, *G. boydii* and *G. spinipes*, are quite different in appearance from one another and are widely separated geographically. *Gonocephalus boydii* has strong dorsal and gular crests, large flat temporal scales in males and occurs in the rainforests of northern Queensland, while *G. spinipes* has smaller crests, no large flat temporal scales and inhabits the forests and closed woodlands of southeastern Queensland and northeastern New South Wales south to Gosford.

Very little is known of the life history or habits of either species. Both species are unusual in collections and are only sporadically encountered in the field. Two aspects of their behaviour may make them especially difficult to detect. First, they both often "freeze" in the face of danger, perhaps relying on their cryptic colouration to conceal them. *G. boydii* will sometimes even "freeze" to the point of letting itself be picked up by an observer, and I have driven a vehicle over a *G. spinipes* "frozen" on a gravel forestry road and looked back, fearing the worst as the dust cleared, only to see it still sitting in the same pose in the same place. Indeed, *G. spinipes*, at least, seems to have a vast capacity for sitting nearly perfectly still for long periods of time (Longley 1943; Webb 1984). Second, when climbing, *G. spinipes* is very adept at keeping the trunk of the tree between itself and the observer, thus concealing itself from view.

Nothing is known of the population structure of either species. However, *Gonocephalus spinipes* is said to form aggregates, that is, several individuals will be found together in a relatively small area of a seemingly widespread, uniform habitat, but whether the aggregations are real or just sampling artifact is uncertain.

Both species of *Gonocephalus* have been reported to take a wide variety of invertebrate and vertebrate prey (Table 4), but there is apparently no observation as yet of herbivory. Whether this reflects a lack of opportunity for the captive animals on which most observations have been made or the real absence of an otherwise widespread dragon feeding habit is unclear.

There is an intriguing report of *Gonocephalus spinipes* "drawing water into the mouth in a similar manner to that of snakes when drinking, and not lapping with the tongue, as is the case of the skinks and many other lizards" (Longley 1943), including other dragons (pers. obs.). This observation requires confirmation.

Like all other Australian dragons, the two Australian species of *Gonocephalus* are oviparous. However, clutch size is known for only one specimen of one species — *G. boydii* with three eggs (Table 5). *G. spinipes* is unique amongst other dragons in apparently laying communally, a nest of 42 eggs having been discovered (P. Webber in Webb 1984).

Both species of *Gonocephalus* entirely lack dark pigment in the parietal peritoneum, the thin membrane lining the body cavity. This is in strong contrast to most other Australian dragons which have a dark peritoneum. A dark peritoneum is thought to protect the many rapidly dividing cell types in this region of the body from the damaging effects of short wave length radiation (Porter 1967). Perhaps the heavily shaded habitats (and retiring habits?) of the forest dragons render such a screen unimportant, in contrast to the open and sun-loving habits of other Australian dragons.

The genus *Physignathus* consists of two rather different and geographically widely separated species: *P. cocincinus* on the mainland of southeast Asia and *P. lesueurii* in eastern Australia.

Physignathus lesueurii, the water dragon, is the largest dragon in Australia, with adults attaining snout-vent lengths of 245 mm and total lengths of nearly a metre. As their common name implies they are almost always found in the immediate vicinity of water, usually flowing water such as creeks, streams or small rivers and usually in well vegetated areas. They enter the water to escape predators and perhaps to forage, and when startled or persued in trees will often jump into water from several metres or more (Chisholm 1924; Worrell 1958; Day 1982). They also occur in the intertidal area, at least in the Sydney region, where they forage on algae and crabs (Mackay 1959).

In eastern Victoria, the large size and aquatic habits of water dragons has led to their being called "crocodiles" or "alligators" (Kershaw 1927; Barrett 1931; Brazenor 1932). It is also known in some areas as the salamander, or "sally" in the nationally-beloved diminutive.

Water dragons have a number of morphological and physiological features associated with their aquatic habits. Morphologically, they have deep, laterally compressed tails which they use to help propel them under water, crocodile fashion, that is, with the arms and legs tucked back alongside the body and tail while lateral waves are passed posteriorly (Worrell 1958). They also have dorsally placed nostrils, possibly to facilitate breathing when in the water.

Physiologically water dragons show a feature when they dive that is typical of most, if not all, diving vertebrates — a pronounced slowing of the heart rate (Courtice 1985). As in other diving vertebrates, this is presumably an oxygen-conserving response. They also appear to have a limited capacity to give off through the skin CO_2 accumulated in the blood (Courtice 1981b-c). Voluntary submersions can last as long as two hours (Krefft 1866; Courtice 1981a).

Studies done on both free-living and captive specimens (Table 4) indicate that juveniles are exclusively predaceous but adults virtually omnivorous. The species apparently even maintains the ant-eating habits of its smaller relatives (Rose 1974). Whether the species forages and feeds underwater is unknown; the only evidence that it may do so is the observation of animals moving their jaws as they emerged from the water (Rankin, in Anonymous 1976).

Water dragons are said to reach sexual maturity in captivity at an age of 5 years (Hay 1972). Like other aquatic and semi-aquatic oviparous reptiles, they often lay their eggs away from their usual haunts along the water's edge (Barrett 1931; Longley 1947b). Presumably this lessens the chances of a nest being flooded. Clutch sizes range 6-18 (Table 5) and 1-2 clutches may be laid per season (Hay 1972; P. Harlow in Anonymous 1976). Nesting females will place "mislaid" eggs into the egg chamber by rolling them with the snout or picking them up and carrying them in the mouth. They may also be unusually aggressive when disturbed during nesting (Giddings 1983).

The preferred body temperature (PBT) of *Physignathus lesueurii* is much lower than that of any other Australian agamid measured to date; its mean PBT is 30.1°C versus a range of means of 34.6-39.0°C for 12 other dragons (Table 10). This is an intriguing dichotomy: one of the two generally primitive, closed habitat dragon versus what appears to be a distinct radiation of more open habitat dragons (see below). In this context it would be interesting to have comparable data for the only other generally primitive, closed habitat dragon, *Gonocephalus*, and for some of the cool climate species of the open habitat group, e.g., *Rankinia diemensis*. It would also be worthwhile to try to establish the physiological basis of these temperature differences.

The aquatic habits and low temperature preferences of water dragons seem to carry over into another, curious aspect of their natural history: their willingness to remain "out" during warm rain (M. Maddocks, in Anonymous 1976; Smith 1979). Under similar circumstances most other dragons would seek shelter.

Water dragons apparently use burrows as retreats (various observers, in Anonymous 1976), but it is unclear whether they construct these burrows themselves or simply appropriate those of other animals. During summer, water dragons often sleep in vegetation above water (M. Anstis in Anonymous 1976).

Like other large reptiles, water dragons are relatively long-lived, spans of 14 (Hay 1972), 11 and 10 (Giddings 1983) years having been recorded for captive specimens.

The remaining Australian dragons, with the possible exception of *Chelosania* (page 38), appear to form a natural group, that is, they share a common ancestor apart from all other dragons. This, plus their restricted Australian distribution, make it likely that their evolution has taken place entirely within Australia. The morphological features distinguishing this group are a lacrimal bone which is much reduced in size or, most often, absent (Fig. 9), and a karyotype consisting of 12 macrochromosomes and only 20 microchromosomes instead of the primitive 12 "macros" and 24 "micros". The members of this group are sometimes rather informally referred to as the amphibo-luroids, a name used here for convenience.

Unfortunately, relationships within the amphiboluroids are not well understood. However, there is a basic dichotomy on the basis of the morphology of the preanal and femoral pores (Humphries 1972; Houston 1978; Witten 1982a-b) that may have some phylogenetic significance. These are the pores on the underside of the thighs and in the precloacal area through which flows a waxy secretion, which as it sits in the pore looks not too unlike a small bit of toothpaste squeezed from the tube. The pores are usually better developed in males than in females and are probably under androgen (male hormone) control (Fergusson *et al.* 1985) The pores occur in a number of other lizard families. The functional significance of the pores' secretion is unknown (Cole 1966).

Those dragons that retain pores may show one of two conditions: the pore either arises from within a single scale or from between two or more scales (Witten 1982a-b; Storr *et al.* 1983: fig. 2). The problem is to decide which condition is primitive and therefore without taxonomic significance and which is derived and therefore significant. The two most generally primitive agamids are each characterized by one of the conditions: *Uromastix* from north Africa and south-west Asia has the pores between the scales and *Leiolepis* from south-east Asia has them within. Hence the question of the polarity of the two conditions cannot be resolved by

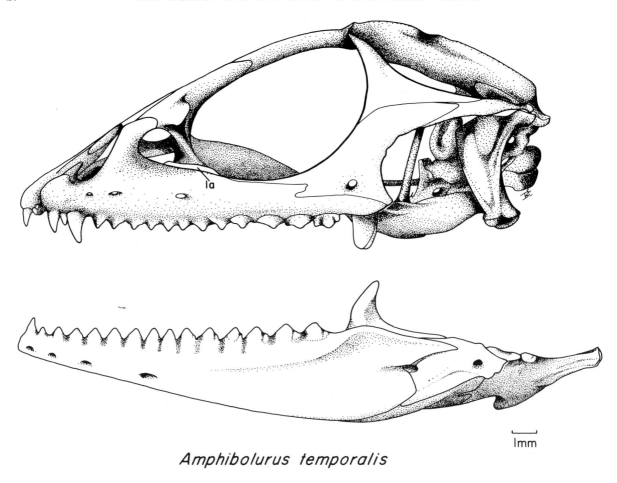

1mm

Amphibolurus temporalis

Fig. 9. Skull of *Amphibolurus temporalis* (AM R 75457) to show the reduced lacrimal bone (1a) which is characteristic of the Australian amphiboluroid lineage (compare with *Gonocephalus spinipes* of Fig. 2).

reference to these genera alone. The only other dragons with pores "outside" the amphiboluroids are *Hydrosaurus* and *Physignathus*, and both have the pores entirely within the scales. On this basis we can tentatively accept pores within the scales as primitive and pores surrounded by scales as derived.

Interestingly, the assemblage which has each pore contained entirely within a single scale also has a slightly more primitive look to it in that certain species retain, uniquely, the primitive traits of a lacrimal bone and 24 presacral vertebrae (instead of the more usual 23 or fewer). The genera included in this assemblage are *Amphibolurus, Caimanops, Chlamydosaurus, Cryptagama* and *Diporiphora*.

Within this assemblage there is a further assemblage of species which has no special feature that would allow us to recognize it as a lineage, but which nonetheless brings together species which may be morphologically very similar to the ancestors of the group with this pore type and perhaps even to the ancestors of the entire amphiboluroid radiation. The generic name available for this assemblage is *Amphibolurus*. It

includes eight species, all of moderate size and slightly to strongly arboreal habits: *A. burnsi, A. centralis, A. gilberti, A. longirostris, A. nobbi, A. norrisi, A. muricatus,* and *A. temporalis* (Houston 1978). At least three species in this group retain the lacrimal — *A. longirostris, A. nobbi, A. temporalis* and two of these retain 24 presacral vertebrae — *A. longirostris* and *A. temporalis*.

The possible relationships of the *Amphibolurus* species remain obscure but it may noted that *A. muricatus* and *A. norrisi* share yellow mouth colour (instead of the more usual pink) and *A. nobbi* has a colour pattern very similar to many species of *Diporiphora*, i.e., a pink or rose flush to the base of the tail and yellowish sides.

Although all the species of *Amphibolurus* would lend themselves to long-term population studies, most of our information to date is limited to incidental natural history observations and selected laboratory studies. *Amphibolurus nobbi* is one of the more thoroughly researched species (Witten 1972) and to judge from the literature, has a number of remarkable aspects to its biology (Witten 1974). For example, it is said to have distinct areas where individuals converge during

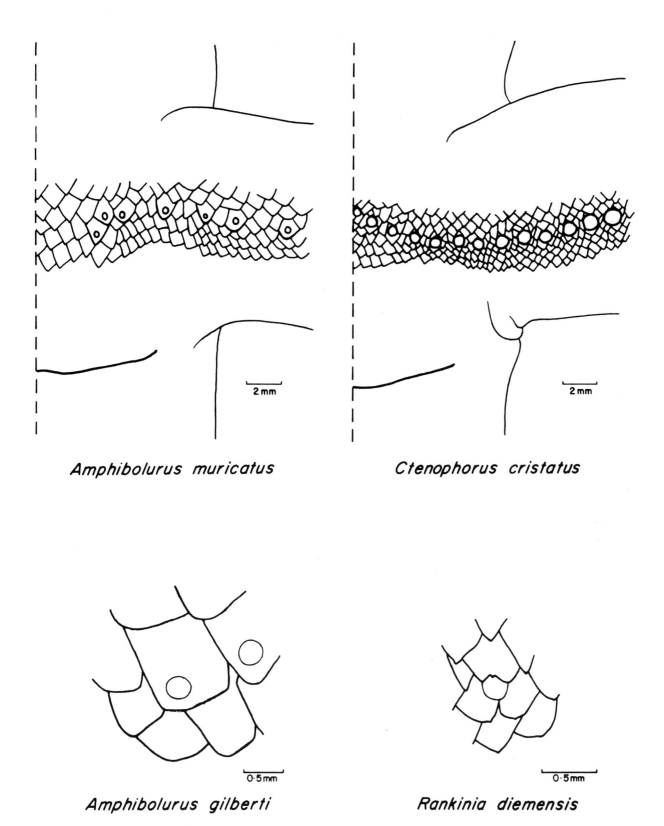

Amphibolurus muricatus   Ctenophorus cristatus

Amphibolurus gilberti   Rankinia diemensis

Fig. 10. The two different pore types in dragons: pore entirely within one scale (*Amphibolurus muricatus:* AM R 107344 and *A. gilberti:* AM R 110526) and pore surrounded by two or more scales (*Ctenophorus cristatus:* AM R 67956 and *Rankinia diemensis:* AM R 67956). The former condition is inferred to be primitive and the latter derived. Lower two drawings from left preanal area.

certain times of year. These areas have been called breeding areas but could equally well be called hibernation areas as the animals come to them in the late summer and early autumn, presumably hibernate there, and then reappear and stay around in the spring before dispersing in the summer. It has been inferred that individuals may move up to 15 kms from these sites. No other dragon, except *Gonocephalus spinipes,* is known to form seasonal aggregations.

It is uncertain when mating actually takes place or when egg-laying occurs, but the appearance of what appear to be hatchlings in the spring in the aggregation areas suggest that at least some eggs are laid in autumn, overwinter and then hatch in spring. No other dragon is known to have overwintering eggs.

Amphibolurus nobbi has also been interpreted as lacking territorial behaviour (Witten 1974; Witten and Heatwole 1978). This would be in strong contrast to at least certain other *Amphibolurus* species where enough has been seen of spacing patterns and behaviour to suggest that males are strongly territorial, e.g., *A. longirostris* (pers. obs.).

A single triploid male *Amphibolurus nobbi* has been reported (Witten 1978), but its significance, if any beyond that of a single aberrant specimen, is unknown.

Mention has already been made of *Amphibolurus nobbi* having been collected on a summer's night "walking on their hind legs only" (Morley and Morley 1985). To date, this behaviour defies interpretation.

Three species of *Amphibolurus* which are often associated with water courses have in the past been considered congeneric with the water dragon *Physignathus.* Correctly separated from this latter genus, they were assigned for a while to their own genus *Lophognathus.* However, there are no derived characters to support this concept and most recently they have been relegated to the generally primitive assemblage *Amphibolurus* (Cogger and Lindner 1974). The three species in this assemblage (within an assemblage within an assemblage!) are *A. gilberti, A. longirostris* and *A. temporalis.*

Amphibolurus gilberti is widespread throughout the northern part of the continent. It is largely arboreal, but when inactive during the dry season may bury itself in the thick leaf litter below trees (pers. obs.). On Barrow Island, Western Australia, it occurs in coastal mangroves (Smith 1976). In the Northern Territory, the species apppears to be reproductively active during the wet season but quiescent during the dry (based on gonadal state in the both sexes; James and Shine 1985).

Amphibolurus longirostris is widespread throughout central and northwestern Australia and is the only species in the *Amphibolurus* assemblage to penetrate well into the arid zone. In the Centre it is usually found near water courses, wells, and in gardens and can be quite common locally (Brunn 1980; Martin 1975). Females gravid with oviducal eggs have been collected in northwestern Australia in mid-spring (11 October) and late summer (9 February); the one female in the former case also carried yolking ovarian eggs, indicating a second clutch (pers. obs.).

Amphibolurus temporalis occurs disjunctly in the Top End of the Northern Territory and on Cape York Peninsula in Queensland; it also occurs in southern New Guinea. It is largely arboreal. The species has been reported as being attracted to a stream of liquid hitting the ground (a man urinating), as feeding on green ants, and as undergoing large fluctuations in population size over the space of a few years (Cogger and Lindner 1974). Limited information on the species in the Top End indicates reproductive quiesence in males, at least, in the mid-dry season but activity in both sexes in the late dry (James and Shine 1985); a female collected in the mid-wet (December) carried oviducal eggs (pers. obs.).

Amphibolurus muricatus and *A. norrisi* occur in south-east Australia. The former is a familiar lizard because of its large size and proximity to major population centres, but like its distant relative with the same attributes, *Pogona barbata,* while it has been extensively studied in the laboratory, it has been almost totally neglected in the field except for some field body temperatures (Heatwole *et al.* 1973). My experience is that it inhabits woodland and is often found on down dead timber. The physiological parameters of panting and high temperature stress have been extensively studied (Heatwole *et al.* 1973; Chong *et al.* 1973; Heatwole *et al.* 1975; Parmenter and Heatwole 1975; Firth and Heatwole 1976; Heatwole and Firth 1982), but otherwise they are only some incidental observations on captive specimens (Groom 1973a-b). The only information for the recently described *A. norrisi* is some data on reproduction and diet (Witten and Coventry 1984).

The remaining species in the *Amphibolurus* assemblage is the long recognized but only recently described *A. burnsi* from central and western Queensland and northwestern New South Wales. The species inhabits woodlands on blacksoil and sand plains and is said to be fast moving, difficult to catch and a good climber. Gravid females have been found in November (Bustard 1968e, as *Amphibolurus* n. sp.; Wells and Wellington 1985). As far as is known the karyotype of the species is unique among amphiboluroids, consisting of 40 chromosomes instead of 32 (Witten 1983; King 1985).

The genus *Diporiphora* is characterized by a number of subtle but collectively convincing features. All the species are relatively small in size with gracile limbs and long tails (as maximum % of SVL for species: 232-413); they have the vertebral crest reduced to only a small nuchal component, and they have a reduced number of preanal and femoral pores or none at all. Most species, perhaps all but *D. winneckei*, also have a distinct axillary/shoulder blotch, the expression of which is under local control (Fig. 11). This blotch is grey/black in *D. albilabris*, *D. australis*, *D. bennettii*, *D. bilineata* and *D. magna* and light green in *D. superba*. There are 14 species, many very similar to one another (Storr 1974c, 1979a; Houston 1977). The genus is widespread throughout central and northern Australia and southern New Guinea, but most species occur in the Top End and in the Kimberleys.

Very little is known about the biology of the genus. Most species are semi-arboreal, but their climbing appears to be restricted to tussock grasses, shrubs and small trees (Houston 1977, 1978; pers. obs.). At least one species, *D. bennettii*, goes against a general trend in lizards, in having positive allometry for the limbs (Witten 1985); the functional significance of this atypical growth pattern is unclear. Another species, *D. superba*, is unique amongst Australian dragons in its brilliant lime green colour (Smith and Johnstone 1981). *D. bilineata* is known to flee into burrows (Cogger and Lindner 1974), but this seems to be an unusual trait in the genus, most species fleeing into vegetation or under surface cover. It is not known whether *D. bilineata*

Fig. 11. *Diporiphora australis* to show aspects of the colour pattern typical of the genus: yellow wash on sides of body, red wash over tail and black shoulder spot — all of which can be expressed or concealed by expansion or contraction of melanophores.

makes its own burrow. Observations on captive specimens have led to the suggestion that *D. linga* may bury itself in the sand and remain dormant at certain times, e.g., mid-summer (Houston 1978).

Food habits have been studied only in *Diporiphora winneckei*, and in some individuals a surprisingly large proportion of plant material ("up to 75% in some cases") was found (Houston 1978).

Diporiphora is oviparous with a clutch size of 1-8, the generic range being set by one species, *D. albilabris* (Table 5). The two tropical species for which there are data from appropriate times of the year, *D. albilabris* and *D. bilineata*, appear to breed only during the wet season (pers. obs. and James and Shine 1985, respectively). In *D. albilabris*, McKinlay River specimens collected in July-September had reproductively quiescent gonads but two females from other localities in the Top End collected in January were gravid. Furthermore both were carrying shelled oviducal eggs and enlarged follicles, indicating that at least two clutches would have been laid. *D. bennettii* also appears to breed during the wet as four females from the Mitchell Plateau were gravid with oviducal eggs in the period 6-8 January (G. Shea, pers. obs.).

Diporiphora winneckei (as *D. bilineata*) appears to be one of the most heat resistant lizards known. For example, specimens caught on the tops of various forms of vegetation at midday when the air temperature was 45.5°C had body temperatures ranging from 42.3-46.0°C (mean = 44.3, N = 12); these temperatures would have been lethal for most other lizards. All the lizards were facing the sun in a manner minimizing exposure to the sun's rays, and all were very light in colour. Perhaps they were also wishing they were like their burrow-inhabiting neighbours, *Ctenophorus nuchalis*, one of whom had a body temperature of 39.2°C in its burrow when the air temperature was 45.0°C and the ground temperature 60°C (Bradshaw and Main 1968). In the laboratory, specimens of *Diporiphora winneckei* tolerated exposure to an ambient temperature of 49°C for nearly 30 minutes (mean time for six individuals acclimated to 40°C was 29.5 minutes, SE = ± 2.31 minutes), perhaps a record for lizards (Bradshaw and Main 1968; Table 11). The species also appears to be fairly resistant to evaporative water loss at high temperatures (Warburg 1966 as *D. australis*).

Probably the best known Australian dragon is the frilled lizard, *Chlamydosaurus kingii*, the common name of which derives from its most distinguishing feature: the large Elizabethan-like ruff or frill which it raises abruptly around the neck when alarmed (Fig. 12). Frilled lizards occur across the far northern parts of Australia and southern New Guinea. They inhabit woodlands and are most often seen sitting up alertly on the ground or head up on some low projection such as a fence post, fallen tree or termite mound. They are quite conspicuous during the summer wet but virtually disappear during the winter dry. Where they pass the dry season has yet to be discovered. At present only one species is recognized, but colour differences between the Queensland populations and the Western Australia-Northern Territory populations suggest there could be two (Swanson 1976).

A number of observations have been made on the food of *Chlamydosaurus*, although primarily in captivity (Table 4). The list of items eaten includes a variety of invertebrates, small mammals and pieces of meat but no mention of vegetable matter. This latter is unusual because almost all other dragons, and especially the larger species, eat some plant matter. One wonders if the ommission of plant material from the known diet of *Chlamydosaurus* reflects the real situation or just bad luck in observing.

Fig. 12. Chlamydo-saurus kingii, the frilled lizard, with the frill folded back along the side of the neck and body (Kununurra, W.A.; 25 April 1981) and extended in display (U.D.P. Falls area, N.T.; 8 October 1983). Photos: S. Wilson.

Surprisingly little is known about reproduction in *Chlamydosaurus*. Clutch size ranges 7-14 (Table 5; Frauca 1966), although eight has been also given as an average (McPhee 1979). Egg-laying has been recorded during the early wet (mid-November to early December, N = 3; Harcourt 1986).

When fleeing, frilled dragons usually run up a tree, or more rarely to a hollow log. If pursued up one tree they show no hesitation about leaping from it to the ground and running up another tree (Broom 1897).

The mouth colour of frilled dragons has been given or shown as either pink (Longley 1946b; Worrell 1963) or yellow (Cunningham, in Bennett 1876; Saville-Kent 1897; Bustard 1970d; Schmida 1973a). As there is only one other well-known case of variation in mouth colour within a dragon species (southwestern *Pogona minor* have yellow mouths), the situation in *Chlamydosaurus* would bear closer examination. Could it vary with the geographic colour differences mentioned above?

The frill in *Chlamydosaurus* is composed of a U-shaped collar which is "open" on the nape. Two elongate bony rods, the first cerato-branchials from the hyoid arch, extend back into the lower lateral edge of the collar on each side. Presumably when these rods are brought down and out by muscles attaching at their base, the lower half of the frill is also brought down and forward extending the loose and folded flaps of the frill into an erect ruff or frill. The first ceratobranchials are the bony supports for throat extension in all lizards, but in *Chlamydosaurus* they are proportionally longer than in other lizards (Beddard 1905; for a comparison between the gular pouch of other dragons and the frill of *Chlamydosaurus*, see p. 31). Many authors state that the frill cannot be erected without the mouth opening as well (Saville-Kent 1897), but no one seems to have tried the obvious experiment with a rubber band.

During the years 1982-1984 the frilled lizard enjoyed widespread fame in Japan and for a while became as much of a symbol of Australia for the Japanese as the koala and kangaroo. The immediate reason for this fame was said to have been a popular car commercial on television that featured the frilled lizard in display. The more fundamental reason for the popularity of the commercial and the dragon was apparently its appeal to the Japanese sense of bluff and display; the Japanese apparently respect an elaborate display of power and position in personal encounters. As an example of how popular the frilled lizard was at this time, rumour has it that the Australian two-cent coin which features an artist's interpretation of the species were being sold for a dollar a piece in Japan.

Caimanops is a monotypic genus of uncertain affinities restricted to Western Australia (Storr 1974c). Its biology is poorly known because its distribution is centered over the less frequented interior of southwestern Australia and it is also either genuinely rare or very furtive, perhaps both. What is known is that it is associated with shrublands (often acacia on red soils), and it is both terrestrial and arboreal. One specimen was found on the end of a twig on a dead shrub at 2000 hrs in mid-October (field notes in Western Australian Museum) which either indicates nocturnal activity or a sleeping position, probably the latter. Another specimen which was dissected for an anatomical study had the stomach crammed full of termites (D. Kent, pers. obs.).

The rarest genus of dragons in Australia is *Cryptagama*, a monotypic taxon known from only three specimens discovered in a small area on the south-east edge of the Kimberleys (Storr 1981a). The specimens were originally placed in the genus *Tympanocryptis* due to a general similarity to two of the species of this genus. This generic allocation was made despite the fact the species retains an external ear opening (heretofore the loss of an ear opening was a diagnostic character for *Tympanocryptis*), a feature alluded to in the species name *aurita*. However, recent work has shown that the species also retains four phalanges in the fifth toe of the rear foot (the primitive dragon number), unlike *Tympanocryptis* which has lost one phalange. Consequently it has been removed from this genus and placed in a genus of its own, this latter decision due to its being "not clearly related to any other agamid genus" but having to have some generic name under the rules of nomenclature (Witten 1984).

Morphologically the species is distinctive for its small size, dumpy body and short tail, features seen elsewhere in Australian dragons only in *Tympanocryptis cephala*, one of the species to which it was originally compared. Ecologically, virtually nothing is known other than that the original specimens were found associated with *Triodia* on stony soils (Storr 1981a).

The amphiboluroids which have each preanal/femoral pore situated between two or more scales (Fig. 10) include the genera *Ctenophorus*, *Moloch*, *Pogona*, *Rankinia* and *Tympanocryptis*, and if the polarity of the pore types inferred above is right, this is a true phylogenetic lineage within the amphiboluroids.

Unfortunately, *Ctenophorus*, the largest genus in the group, is probably only an assemblage and not a lineage. However, the lizards in this genus fall into three clear ecological categories which are a useful basis for discussion and may even have some phylogenetic relevance. The categories relate primarily to where the species shelter: rocks, burrows or vegetation.

Seven species of *Ctenophorus* are associated with rocks: *C. caudicinctus*, *C. decresii*, *C. fionni*, *C. ornatus*, *C. rufescens*, *C. vadnappa* and *C. yinnietharra*. The most extensively studied members of this group are *C. ornatus* and the three species of a complex consisting of *C. decresii*, *C. fionni* and *C. vadnappa*.

Ctenophorus ornatus occurs in south-west Australia where it is strongly associated with the low, open dome-like granite outcrops which are characteristic of that area. Because of its accessibility, both in terms of proximity to Perth and its open habitat, the species is one of the better studied dragons, and perhaps the overall best-known rock dragon (Bradshaw and Shoemaker 1967; Bradshaw and Main 1968; Bradshaw 1970, 1971; Baverstock 1975, 1978; Baverstock and Bradshaw 1975). Now one might think that a habitat as extreme as an open rock slab in the interior of Australia would have elicited a variety of special adaptations. However, there are relatively few features in its biology that are not fairly typical of dragon biology in general, showing that species may be able to enter new and seemingly demanding habitats with actually very little modification. Indeed, perhaps the most pronounced difference between this species and other dragons is its shape; it has a very depressed head and body, a feature that may be related to the thin crevices beneath the exfoliating slabs on the granite domes that are critical refuges in an otherwise exposed habitat. The crevices are used as daily and seasonal shelters from extremes of weather and also for protection from predators, especially birds of prey. All other studied aspects of its biology are typical of numerous other dragon species in different habitats and hence can not be said to be related to the special features of its own habitat.

However, there are two aspects of the biology of *C. ornatus* that, although they are probably fairly typical of dragons in general, are especially well known in this species. First, it is apparent that in some populations there are large differences in the growth rates between different individuals and that these probably have a large genetic basis. The "fast" growers attain sexual maturity in less than one year and the "slow" growers in 1-2 1/2 years (Bradshaw 1971). Each growth type has certain "benefits" as well as "costs". Where retreats are limited in number, juvenile "slow-growers" appear to be forced into more marginal shelters where they are susceptible to frost during winter (Baverstock 1978). However, "slow growers" are able to tolerate better than "fast growers" the salt loads in the body fluids that build up during summer drought. The overall variability in growth rate within a population seems to correlate with the unpredictability of the environment: the more unpredictable the local environment, the more variable the growth rate (Bradshaw 1971). Originally these differences were thought to be a dichotomy, but now they are understood to be more of a broad continuum (Bradshaw 1981).

Second, there is a large proportion of the population, primarily consisting of juveniles and young adults, that occurs in habitats marginal to the preferred one; in the case of *C. ornatus* this is at the edge of the main rock outcrop or in small groups of boulders adjacent to it. These individuals are apparently forced into these fringe areas by other, better established conspecifics, and can only get back into the preferred area when established individuals disappear or they themselves can displace such an individual. These lizards may be the ones that actually disperse across unfavourable habitat to new rock outcrops.

Ctenophorus ornatus lays 2-5 eggs in a clutch and has 1-3 clutches per season; both clutch size and clutch frequency show a tendency to increase with age (Bradshaw 1981). Females with oviducal eggs occur from September into March, although the earliest date represents the most northern population. What the fate of the few eggs laid in March might be is an unanswered question.

The second group of rock-associated dragons is the *Ctenophorus decresii* group. It comprises three species (Houston 1974): *C. decresii* (Ehmann 1976a-b; Armstrong 1979), *C. fionni* (Smyth 1971 [as *Amphibolurus* sp.]; White 1976; Robinson 1980) and *C. vadnappa*. They are moderate-sized dragons which are distributed in southeastern South Australia and a small part of adjacent New South Wales. The group provides an excellent example of how behavioural features may be as useful as more traditional morphological features in elucidating phylogenetic relationships, and this is true both at the species group and species level. For example, at the species group level, the *C. decresii* group has been thoroughly studied using morphometric methods (Houston 1974) but none of these features can, as yet, be surely said to be indicative of the close relationships of the three species. However, behavioural analysis (Gibbons 1977, 1979) clearly reveals that the three species possess two features that are unique in dragons and which, therefore, are almost surely indicative that they were inherited from a common ancestor not shared with any other living dragon. These features are the coiling of the tail, slightly watchspring fashion, behind the body, and a series of

push-ups with the hind legs. These behaviours come in the middle of a sequence of displays given in social interactions between adults of both sexes but which are most intense between males. The other components of the total display are widespread in other dragons.

At the species level, morphometric analysis (Houston 1974) revealed differences between the three species, but the differences were restricted to the bright colour patterns of the adult males. Differences between the more drably patterned adult females and juveniles were very difficult if not impossible to detect. However, behavioural analysis (Gibbons 1977, 1979) revealed that the species also differ in components of their displays and these differences occur in adult females as well as adult males (juveniles so far remain difficult to distinguish between the species). For example, *C. vadnappa* coils the tail vertically while the other two species coil it horizontally; *C. fionni* never circumducts the foreleg between hindleg pushups whereas *C. decresii* and *C. vadnappa* do. There are also species differences in the intensity and/or frequency of other display components.

The *Ctenophorus decresii* group also provides an example of another interesting biological principle — substrate colour matching, that is, the lizard's dorsal pattern, especially that of the highly conspicuous male, blends well with the particular rock type it inhabits. This is not a matter of the lizard changing its colour and pattern to suit the rock type, rather, as choice experiments have shown, a matter of the lizard choosing the appropriate rock type. The phenomenon occurs both amongst and within species. For example, in one area where they occur together, male *C. decresii* have a mottled grey and yellow ground colour and a dark lateral band and occurs primarily on rocks which are pinkish yellow, whereas male *C. vadnappa* have a dark brown ground colour with vertical reddish-orange bars on the sides and are found on rocks which are dark reddish brown; similar correspondences occur amongst populations of *C. fionni*. Presumably their genetically based colour and pattern differences are due to selection by visually oriented, colour-perceiving, flying predators, i.e., birds, because they apply primarily to the dorsal colour patterns and are most evident in the males which, because of their territorial habits, often position themselves in conspicuous perches (Gibbons and Lillywhite 1981).

The third group of rock dwellers consists of *Ctenophorus caudicinctus*, a widespread species in the northwestern quarter of Australia. It consists of several races, each associated with separate major rocky areas (Storr 1967b). The distinctiveness of the races is presumed to derive, at least in part, from the geographical isolation of the major habitats. These races would lend themselves well to a biochemical assessment of their degree of differentiation.

Little is known of the biology of any race of *Ctenophorus caudicinctus*. The Arnhem Land population is not only the most tropical of all the *C. caudicinctus* races but also of all the rock-dwelling species. Therefore inferences as to the seasonality of its reproduction may be of interest even though the available data are limited. Eight adults from the late mid-wet (late January) to late dry (late September) were reproductively quiescent; an early mid-wet (December) male may have been reproductively active, and a late mid-wet (late January) specimen is probably a juvenile (pers. obs. on Australian Museum specimens). These data are consistent with reproduction in the early wet, hatching in the late mid-wet and quiescence in the late wet to late dry. Further south, the Pilbara population breeds in February-March, which is an unusual season compared to other dragons; it has been suggested that reproduction coincides with the monsoonal rains of the area (Bradshaw 1981).

Not all *Ctenophorus caudicinctus* populations occur on rock; a few occur on soil, especially hard soil (pers. obs. in vicinity of Sand Fire Flat, W.A.). Whether this is the retention of a primitive ecological association or a secondarily derived one is unknown, but it does show that dispersal between major rock habitats may not be as difficult as once thought.

In comparison to the preceding rock-dwelling species which are fairly well known, the last two members of this ecological group, *Ctenophorus rufescens* and *C. yinnietharra*, are very poorly known largely for reasons of remote distribution. *C. rufescens* is restricted to the granite ranges of far northwestern South Australia (Houston 1978) and *C. yinnietharra* to one set of hills in central Western Australia (Storr 1981a). About all that is known of their biology is that *C. rufescens* matches well the colour of its red rock substrate.

Five species of *Ctenophorus* spend much of their active time on the open ground of sand plains or sand dunes covered with hummock grasses and rather than flee to burrows, which they seem incapable of digging other than for nesting purposes, flee instead to the grass hummocks. The species are *C. femoralis*, *C. fordi*, *C. isolepis*, *C. maculatus* and *C. rubens*, collectively sometimes called the *maculatus* group (Storr 1965). All these species are relatively long-legged reflecting their cursorial habits and well colour-matched to their predominant substrate, usually red or beige sand. All but one species tend to restrict their activity to sand flats and the consolidated base of dunes, but *C. femoralis*, the smallest species in the group, appears to stay largely on the loosely consolidated crests of dunes (pers. obs.). So typical are these species of habitats featuring sandy substrates with hummock grasses, that a map of these habitats, including even the large relict outliers in the eastern states, would come close to mapping the group's distribution. For the purposes of discussion, we may also add to this group a sixth species, *C. maculosus*, which, like the other members of the group, is characteristic of open habitats and seems incapable of burrowing other than for nesting purposes.

Ecological studies are available for two species, *Ctenophorus isolepis* in the west (Pianka 1971c) and *C. fordi* in the east (Cogger 1974, 1978; Baverstock 1979), and most of what we know of the ecology of the group derives from this work. Two features seem peculiar to the group. First, hummock grasses are very important to these lizards, primarily as cover during the night and as a refuge from predators and high temperatures during the day. Second, there is both indirect (seasonal distributions of size classes) and direct (mark and recapture) data to suggest that few individuals live longer than one year, i.e., the populations are largely annual. However, this may be a feature of small-sized dragon species in general and needs looking at in other groups.

It would be difficult to think of a more inhospitable habitat for any vertebrate than the surface of Lake Eyre in northeastern South Australia. The usually dry salt lake lies in the most arid part of one of the most arid countries in the world; solar radiation under the clear skies and on the highly reflective salt crust is intense, and any standing water is sure to be brine. Yet the surfaces of Lake Eyre and a few surrounding saline lakes are the habitat of one of Australia's most unusual dragons, *Ctenophorus maculosus* (Madigan 1930, 1936; Mitchell 1948, 1965a, 1973; Mincham 1966).

Ctenophorus maculosus is a moderate sized dragon with a maximum snout-vent length of 70 cm. It is white to very pale grey above with a row of black blotches on either side of the dorsal midline. The colour matches the dark-spotted light surface of the lake surface so well that the species seems to have lost its fear of predators, because it ignores both low flying hawks and walking people. As a result, it is easily approached and observed, providing unique opportunities for study, and the person who took advantage of this opportunity and provided us with virtually all we know of this species was the late F. J. Mitchell (1973) of the South Australian Museum.

The ability of *Ctenophorus maculosus* to survive on the salt lakes is heavily dependent on two factors: the nature of the salt crust and the substrate below, and a species of ant which also lives on the lake beds. The lizard is found only on those parts of the lake with a cracking crust overlying sediments of sand or clay that are dry above but damp below. The sediments provide a place of both short and long term refuge. However, the sediments can only be entered where the overlying salt crust is cracked, as this allows access to the sediments below and provides something to push against in getting below the surface. The ant, a species of *Melophorus*, is important to the lizard because it forms a large part of its prey and the "collars" it builds around the entrance of its nests on the lake provide observation posts for the lizards and also some small bit of shade.

The annual cycle of *Ctenophorus maculosus* runs approximately as follows. Dominant males of the previous year emerge first from hiberation and establish territories in late winter and early spring (August-September). They are followed by young males and females. If a young male cannot displace an older male it passes the mating season leading a furtive, low profile existence in a weakly defended territory well away from any dominant male.

Mating occurs in mid-spring to early summer (October-December) and egg-laying approximately 20-25 days after fertilization. The eggs are usually laid along the shore-line in from the area of greatest activity. Incubation takes about 70 days and young appear from mid-summer to mid-autumn (January-April).

After the reproductive period individuals devote themselves to feeding (on ants and other arthropods) and building up the fat stores that will take them through the winter and the early stages of the activity period the following year. Hibernation under the salt crust begins in mid to late autumn (April-May).

Although much of the behaviour of *Ctenophorus maculosus* is fairly standard in the context of dragon behaviour in general, it does comprise two unusual features. The first is a method used by subdominant males and non-receptive females to inhibit the attention of dominant males; this involves simply rolling over on the back and lying belly up. As far as is known this behaviour is unique among lizards. The second unusual feature is the assumption of a stronger defensive behaviour on the part of mated females toward males to inhibit their now unneeded attentions. Concomitant with the assumption of this behaviour is the development of colour also typical of highly assertive males:

orange along the lower jaw, between the forelegs and along the ventrolateral part of the body. Similar changes in behaviour and colour pattern on the part of already-mated females are known in a few iguanid lizards (Cooper 1986 and references therein) but in no other dragons.

Eight species of *Ctenophorus* dig burrows in which they shelter during the night and during extremes of weather and season, and to which they flee from predators: *C. clayi, C. gibba, C. cristatus, C. nuchalis, C. pictus, C. salinarum, C. scutulatus* and *C. reticulatus*. To this group we may also add *C. mckenziei* a recently described, but little known dragon, very similar to *C. scutulatus* (Storr 1981a). These are the only Australian dragons known to use burrows regularly in their daily activities, and if *C. nuchalis* and *C. reticulatus* are any indication, they may have more than one burrow, they know where these burrows are at all times, and they are reluctant to enter a strange burrow (White 1949; Heatwole 1970). At least some of the species, e.g., *C. clayi* (pers. obs.), *C. nuchalis* (Rankin 1977), *C. pictus* (H. Ehmann in Rankin 1977) and *C. salinarum* (H. Ehmann, pers. comm.) back-plug the burrow when they are inside.

The best known species in this group is *Ctenophorus nuchalis*, or as it is also sometimes known — *C. inermis*, a lizard whose wide distribution, moderately large size, conspicuous habits, and ease of capture have made it the object of several studies (Bradshaw and Main 1968; Heatwole 1970; Carpenter *et al.* 1970; Pianka 1971a; Bradshaw 1975; Rice and Bradshaw 1980; Bradshaw 1981). Indeed, in spring, one can often drive along the dirt tracks in the arid and semi-arid zones and almost never be out of sight of an individual perched on a stick, shrub or clod of earth. The lizards are easily caught by a slow approach and quick grab, or by digging when they jump down and flee to their nearby short burrows. The species is usually found in open habitats of either a permanent (Heatwole 1970) or transient (Bradshaw 1981) status.

Adults seem to be rather sendentary, confining their activities to a home range in which they dig one or more short burrows. They also seem to be rather solitary, and males at least are known to fight; rarely is more than one individual found in a burrow. The sedentary habits of the lizards are reflected partly by their leg lengths which are shorter, relative to head and body length, than in any other dragon (Storr 1966a; Table 2).

Ctenophorus nuchalis reproduces during spring-summer, e.g., September-February. Clutch size ranges 2-6, and 2-3 clutches may be laid in a season (Bradshaw 1981; Klages 1982). At least in some populations sexual maturity is reached within one year and rarely do individuals live more than two years (Bradshaw 1975); such rapid growth to maturity and high turn-over is intriguing in such a large lizard.

Ctenophorus nuchalis has the lowest rate of evaporative water loss yet measured for any dragon, being nearly 60% lower than *C. ornatus* and 40% lower than *C. caudicinctus* (see also Warburg 1965a, as *Amphibolurus reticulatus inermis*). The low rate is evident in both the respiratory and cutaneous components (Bradshaw 1981) and seems to derive from a reduction in metabolic rate and a more impervious skin (Bradshaw 1977b).

Maximum running speed amongst individuals of *Ctenophorus nuchalis* of similar size seems unrelated to season, sex or body shape (e.g., relative leg length or mass). Furthermore, an animal's maximum speed is highly repeatable over time. This suggests that maximum running speed is probably genetically based, but it leaves unexplained its physiological/morphological basis (Garland 1985).

Ctenophorus reticulatus has long been considered a close relative of *C. nuchalis* (Storr 1966a), the two even having been treated as subspecies of a single species until relatively recently. The morphological basis for the relationship is a similarity in form and colour pattern between the adult males: rounded head, stocky body, relatively short limbs, reticulate dorsal colour pattern, and patternless venter (as opposed to black chest markings as in many of the *Ctenophorus*). The two species are also similar in behaviour, sitting upright on low projections and fleeing to nearby burrows[1]. Ecologically, the two species are among the most herbivorous Australian dragons known, especially for their size (Pianka 1971a). Because juvenile and young female *C. reticulatus* closely resemble other *Ctenophorus*, e.g., *C. fordi*, *C. maculatus* and *C. salinarum*, one might argue that within the lineage, if it is that, the distinctive adult male colour pattern arose first in this sex/age class and then spread "backward" into younger males in *C. reticulatus* and "laterally" into females in *C. nuchalis*.

The only detailed observations on the natural history of *Ctenophorus reticulatus* were made by a school teacher in Western Australia, S. R. White (1949). His observations suggest that males hold territories and that females enter them (but from

where?) for mating. Mating probably occurs in spring (October) and hatching in summer (December). Clutch size ranges 2-8.

Ever since its description, *Ctenophorus clayi* has been allied with *C. nuchalis* primarily on the basis of its similar colour pattern and peculiar, slightly W-shaped arrangement of preanal/femoral pores. However, *C. clayi* also shares two characters with the *Tympanocryptis lineatus* subgroup: (p. 31) the loss of a phalange in the fifth toe of the rear foot and a very high number of postsacral vertebrae with diapophyses (pers. obs.). The possibility exists, therefore, that the species is more closely related to a group of *Tympanocryptis* than to *Ctenophorus nuchalis*.

Ctenophorus clayi inhabits rolling sand plains covered with low open shrubs in central and far western Australia. Although it burrows readily and has been taken from burrows, when pursued it often prefers to dash from shrub to shrub hiding at the base. When the ground is very hot, it may dash to a shrub and climb it to a height of about 30 cm. In doing so it is very reminiscent of miniature *C. nuchalis*. The species is often "out" during mid-day temperatures that cause other dragons to take shelter, which indicates its thermal tolerances may be high. Little is known of the species' reproduction other than egg production occurs during mid to late spring (October and November), and the clutch size is 2-4. Although the species is small, it is often locally common and quite approachable (pers. obs.). It would be an easy dragon to observe closely in the field.

The species *Ctenophorus pictus* and *C. salinarum* are thought to be closely related on the basis of overall similarity, especially of the females which often have the unusual dorsal colour pattern of white spots in transverse series. Indeed, for many years they were simply considered geographic races of one species. *C. pictus* occurs in the south-east and south-central part of the continent and *C. salinarum* in the south-west. Both species are terrestrial and generally associated with low, open shrublands on sandy soils. Their burrows often honeycomb the consolidated soil at the base of dense aggregations of shrubs (R. Tedford, in Mayhew 1963; Gillam *et al.* 1978).

Nest digging by gravid *C. salinarum* has been observed in early spring (19 September, pers. obs.; Fig. 13) and hatchlings have been seen at the same locality (Newman Rocks, WA) in late summer (15 February, G. Shea, pers. comm.). Egg laying occurs in *C. pictus* in both mid-spring (30 October, Antenor 1974) and mid- and late summer (15 December, 10 January, D. Kent, pers. comm.). Two clutches in one season have been recorded for a captive *C. pictus* (D. Kent, pers. comm.). Recorded clutch size is 3 in *C. salinarum* and 2-6 in *C. pictus*.

---

[1]These two species also share, uniquely in *Ctenophorus*, the primitive amphiboluroid pore type, i.e., the pore entirely within the confines of the individual scale. If this pore is truly a retained primitive feature, these two species should probably be relegated to a genus of their own, for which a new name would be needed, and transferred to the assemblage consisting of *Amphibolurus*, *Caimanops*, *Chlamydosaurus*, *Cryptagama* and *Diporiphora*. However, if their pore type represents a reversal to the primitive type, i.e., a secondarily derived character, it would be a further point of similarity between the two species within the group consisting of *Ctenophorus*, *Moloch*, *Pogona*, *Rankinia* and *Tympanocryptis*.

Fig. 13. Gravid *Ctenophorus salinarum* at her nesting burrow; Newman Rocks, W.A.; 19 September 1981.

Three other species that are also thought to be closely related, but again unfortunately only under the unsatisfactory criterion of overall similarity, are *Ctenophorus cristatus, C. mckenziei* and *C. scutulatus*, all from south-west Australia. These are moderately large dragons which forage on the ground in shrubland habitats. *C. cristatus* is said to sit quietly on a vantage point until it sees prey which it then pursues at high speed (Houston 1978). Some basic natural history information is available for *C. cristatus* and *C. scutulatus* (Pianka 1971d; Table 4-5), but the recently described *C. mckenziei* remains almost totally unknown (Storr 1981a).

Ctenophorus gibba is probably the most poorly-known member of the burrowing group of *Ctenophorus*. The sum total of our knowledge of the species is that it lives in the gibber plains of north-east South Australia and one specimen was seen to run into a burrow (Houston 1974) — hence its association here with the burrowers!

The bearded dragons have long been recognized as a distinctive group (called the *Amphibolurus barbatus* group — Caughley 1971; Badham 1976; Houston 1978) but only recently have they been given formal taxonomic recognition with the generic name *Pogona*, which is Latin for "beard" (Storr 1982c). As both the common and scientific name implies, these dragons have a rough-scaled, often dark-coloured, throat which gives the appearance of a beard when distended. In most species the throat is distended in both fright and aggression, perhaps as means of intimidation, but one of the small species, presumably *P. minor* , is said "not [to] display by distending the throat pouch in the characteristic manner of the eastern species" (White 1949), an observation that needs confirmation. Another distinctive feature of *Pogona* is the slightly depressed body which is

depressed even further in display (e.g., *P. barbata* — Brattstrom 1971a; *P. minima* — W. Schevill in Loveridge 1934); other dragons usually compress the body in display. The group is also characterized by larger than average size and a series of spines in species-distinct patterns on the head and neck (Badham 1976). One specimen of *Pogona barbata*, has been reported to possess a caecum (Mackenzie and Owen 1923), but this appears to occur in all dragons (D. Kent, pers. obs.)

Pogona appears to retain, as a primitive skeletal feature, a modal number of 24 presacral vertebrae. In this it is unique among the amphiboluroids with the presacral/femoral pores situated between two or more scales (Table 3).

The group consists of eight species, *Pogona barbata, P. henrylawsoni, P. microlepidota, P. minima, P. minor, P. mitchelli, P. nullarbor* and *P. vitticeps*, which together range over all of Australia except the extreme north and far south-east (including Tasmania). The group was also apparently present on Kangaroo Island 16 000-10 000 years ago but is now extinct there (Smith 1982). As yet there are few clues as to relationships within the group except that *P. barbata, P. henrylawsoni*, the southwestern population of *P. minor*, and *P. minima* share yellow or orange mouth colour, and that *P. barbata* and *P. nullarbor* both lack the second ceratobranchial of the hyoid apparatus (Badham 1976).

Pogona species spend much of their time on raised objects, usually vegetation, apparently both in territorial and thermoregulatory behaviour. Indeed, in the eastern part of the country during spring, it is common to see individuals of the large eastern species, *P. barbata* and *P. vitticeps*, perched singly on stumps, posts, telephone poles and termite mounds. The adult males with their glossy black throats and chests are especially conspicuous. These two species also like to bask on warm bitumen roads and often pay dearly for the pleasure. The lizards can often be approached and grasped by hand without their showing any reaction other than a slight pressing down onto the ground, and should they happen to flee, it may be no further than a few metres to a clump of grass or shrubs where they press down again.

The larger species of *Pogona*, at least, are among the most tameable of Australia's lizards. With little effort they can be taught to become totally accustomed to people, allowing themselves to be handled and carried with no sign of fear. They even tolerate being fitted with loosely attached "devices" (toy party hats in the case of specimens maintained in the Ornithology Section of the Australian Museum).

Despite the large size, abundance, ease of capture and handleability of many of the species of *Pogona*, we have only a patchy understanding of their overall biology. They are often used in laboratory studies of anatomy, behaviour and physiology but to date are almost totally neglected in field studies. Thus, whereas there are detailed studies for the common eastern species, *P. barbata*, on the movements around the joints of the skull (Throckmorton and Clarke 1981), the ultrastructure of the skin sensory receptors (Maclean 1980) and the behaviour in captivity (Carpenter *et al.* 1970; Brattstrom 1971a), there are only incidental and anecdotal accounts of its feeding habits (Rose 1974; Swanson 1976; McPhee 1979 and Kennerson and Cochrane 1981), courtship (C. S. Sullivan in Kinghorn 1931; Pickworth 1981), reproduction (see references, Table 5) and population structure and dynamics. If ever an Australian lizard species lent itself to close field study, one of the large *Pogona* species would be it.

As the beard of *Pogona* is sometimes called a frill and the dragons themselves "frilled lizards", it may be useful to make clear the differences between the beard in *Pogona* and the frill of the "real" frilled lizard, *Chlamydosaurus*. The beard in *Pogona* is simply a more extendable and ornate (with shaggy scales) version of the gular region of other dragons. In all dragons this region is underlain by the hyoid apparatus and when this throat skeleton is depressed and its first branchial arch rotated down and out, a gular pouch is displayed (Throckmorton *et al.* 1985). In *Pogona* the first ceratobranchial arch is longer and its pivotal point more forward than in other dragons; hence it makes a larger pouch when extended (Throckmorton *et al.* 1985). The effect is also enhanced by the rough scales of the throat which stand out when the throat is extended.

In *Chlamydosaurus* the frill is simply a thin but extensive fold of skin surrounding the throat and not a section of the entire throat itself as in the case of the gular pouch. However, the frill may have had its ultimate origin in the gular pouch, because it is invested by the main skeletal element of the pouch, the first branchial arch, and it is probably raised by basically the same muscles acting in the same manner on this arch (Beddard 1905).

A novel method of drinking has been recorded in a single specimen of *Pogona vitticeps* which should be further investigated. During rain the lizard was observed to collect water on its elevated body which then gravity fed to the lower head where it was lapped up from around the mouth (Fitzgerald 1983b). As the animal was able to drink in the normal fashion from a container, the significance of this particular method is obscure.

The genus *Rankinia* comprises three species which share three distinct morphological features: a relatively small body size, a series of tubercular scales in linear series on each side of the base of the tail, and a blue mouth lining. The group also lacks bright hues in the skin. The three species also share a southern distribution, *R. adelaidensis* occurring along the coast of the south-west (but not the far south-west itself), *R. chapmani* along the south-central part of the continent (Storr 1977), and *R. diemensis* in the southeastern highlands and Tasmania; the latter is the world's southernmost agamid. Little is known of the species' biology except that they all occur in low shrublands or heaths, i.e., basically open habitats. *R. adelaidensis* is said to flee to burrows (White 1949) of its own construction (H. Ehmann, pers. comm.) and to have "the habit of 'settling' into the sand with a rapid side to side movement, leaving only the eyes and snout protruding" (True and Reidy 1981). This latter behaviour, which sounds much like the sand-shimmying burrowing behaviour of certain overseas arid area agamids, is unrecorded in other Australian agamids. Nothing is known of food habits or reproduction except that *R. diemensis* is oviparous laying 2-7 eggs in a clutch (James 1983; Kent 1987). Gravid females of this latter species have been seen basking on the horizontal trunks of downed dead trees in late afternoon (pers. obs.).

The genus *Tympanocryptis* is one of the more firmly established groups of agamids because its members share two unusual derived character states: a scaled over tympanum and the loss of a phalange in the fifth toe of the rear foot (Figs 14, 20). Six species are currently recognized in the genus but there are residual taxonomic problems with several forms. The group occurs throughout Australia except for the mesic far east and north coasts.

There appear to be two subgroups within the genus: the *T. parviceps* group and the *T. lineata* group. The former is characterized by the absence of preanal/femoral pores in females and the latter by a complete covering of the tympanic area under the skin by the muscle that opens the lower jaw (as opposed to an incomplete covering in the *T. parviceps* group), a greatly reduced number of preanal/femoral pores (4 versus 7-17) and a high number of postsacral diapophyses (≥ 24). The two subgroups also differ ecologically, the *T. parviceps* group occurring on coastal sand dunes and the *T. lineata* group on generally hard packed or stony substrates in the interior. The

Fig. 14. *Tympanocryptis lineata* from just south-west of Tibooburra, New South Wales showing the scaly epidermis covering the ear region.

T. parviceps subgroup consists of *T. butleri* and *T. parviceps* (Storr 1977; Witten 1982) and the *T. lineata* group of *T. cephala, T. intima, T. lineata* (with several subspecies) and *T. uniformis* (Mitchell 1948; Storr 1964a, 1982b, 1984c).

All the species of *Tympanocryptis* are confined to relatively open habitats in the drier parts of the continent and all are exclusively terrestrial. Substrate matching is close in the genus. For example, members of the *T. parviceps* sub-group are basically very light in colour to match the light sand substrate of their habitat and those species of the *T. lineata* subgroup that occur on bare gibber (i.e., stony) plains have rounded heads and bodies, relatively smooth scales and reddish colour which cause them to closely resemble the reddish-brown, wind polished stones of these inhospitable areas. (Fig. 15).

Colour hues appear to be unusual in the genus, having been recorded in only two species to date: *Tympanocryptis butleri* and the far eastern population of *T. lineata*. In the former case it involves a yellow colour on the throat in both sexes and in the latter a yellow throat and pink pelvic region in breeding males (Jenkins and Bartell 1980). A display sequence, slightly different from other dragons, has been recorded in a captive *T. lineata* (Houston 1978).

Unfortunately, nothing substantive is known of the diet or population biology of *Tympanocryptis*.

Tympanocryptis is oviparous and in the *T. lineata* subgroup clutch sizes are large relative to female size (Table 5); indeed, when dissecting gravid females of this subgroup, it is astonishing to see how completely the eggs fill the body

cavity (Table 6). *T. lineata* with oviducal eggs have been found in late September 1980 (N=5) in central far eastern Western Australia which suggests breeding can occur in spring. Gravid *T. intima* with oviducal eggs have been found in mid to late October (N=3) and in early March (N=3) in northeastern South Australia and northwestern Queensland which suggests breeding in at least spring and early autumn. The latter date is somewhat late in the season and its significance should be investigated further.

In contrast to most dragons, gravid *Tympanocryptis lineata* held in captivity appear to retain their eggs for inordinately long periods. For example, of eight gravid females collected from a single area of northeastern New South Wales on 3 October and held in the laboratory pending their laying, only one laid while the others were finally sacrificed on 12, 17 December and 12, 18 (N=4) January. This is a range of 70-107 days. Given that two of four females collected at the same time had oviducal eggs upon collection (the other two had enlarged yolked follicles), some of the retained females must have also

Fig. 15. Two species of *Tympanocryptis* to show the extremes in the range of body form and colour. *T. lineata* (top) and *T. cephala* (bottom).

been carrying oviducal eggs at capture. Hence the 70-107 day retention time may well apply to oviducal eggs. During their time in captivity these females were supplied with separate enclosures, deep red sand, food and water and lengthy daily basking opportunities. They ate, drank and basked, but with the one exception (which laid in the interval of 75-91 days) they did not lay. It would appear, therefore, that some critical trigger or clue necessary for egg laying was missing. An embryo dissected from an egg retained for the maximum period (i.e., 107 days) showed no further development than the stage at which many other egg-laying lizards lay their eggs (i.e., ca stage 30 of Dufaure and Hubert 1961). Hence despite the elevated temperatures of the basking females, further development seemed dependent upon laying.

Some species of *Tympanocryptis* show an interesting change of colour and pattern in mature females. In *T. intima* these females are lighter and have suppressed partially or entirely the light longitudinal stripes and dark cross-bars of juveniles and adult males. A similar change is also evident in some populations of *T. lineata* (Fig. 16) and perhaps in *T. cephala*. Its significance is unknown.

Studies on the loss of water through evaporation at different temperatures produced an interesting result in *Tympanocryptis* (Warburg 1966). Over moderate to high temperatures, i.e., 30-40°C, *T. lineata* from around Mt Olga proved to be 5-80 times more resistant to water loss than *T. lineata* from the Lake Eyre Basin. Furthermore the two populations were the most resistant and the least resistant, respectively, of five dragons tested under similar circumstances, the other three being, in decreasing order of resistance, *Diporiphora winneckei* (as *D. australis*), *Ctenophorus maculatus* (as *C. isolepis*), and *C. fionni*.

Fig. 16. Male (top) and female (bottom) *Tympanocryptis lineata* from just SW of Tibooburra, New South Wales showing the strong sexual dimorphism in mature individuals.

Fig. 17. *Tympanocryptis lineata* in extreme thermoregulatory posture. Photo: B. Miller.

If these results are correct, they indicate a high degree of interpopulation variation in one species in a very basic physiological parameter.

The far eastern population of *Tympanocryptis lineata* is said to utilize burrows, at least sometimes and perhaps always, of another animal's (e.g., a spider) making (Jenkins and Bartell 1980). Burrow use has not been reported in other forms. The animals from northeastern New South Wales which I have seen all fled only a few metres at most and then hid behind rocks, under small shrubs or in earth crevices.

Two species of *Tympanocryptis* have been reported sitting upright for long periods on the hind limbs and tail with the front limbs held firmly along the sides (e.g., *T. intima* — Swanson 1976, as *T. cephala* and *T. lineata* — Fig. 17). Just what this behaviour is all about is difficult to tell; one observer suggested it was to gain a better view of the surroundings, but in the only species in which similar behaviour has been reported, i.e., *Ctenophorus nuchalis* (Bradshaw 1986), it appears to be thermoregulatory. This explanation may also apply to *Tympanocryptis* because *T. lineata* heat-stressed in the laboratory behave in a similar fashion (pers. obs.).

A possible close relative of *Tympanocryptis* is Australia's strangest and perhaps most widely publicised lizard, *Moloch*. The word most commonly associated with *Moloch* is "bizarre", a much over-worked word in zoology but in this case apt, for even the most jaded biologist is usually amazed when he sees his first specimen, be it in a jar of preservative in a museum or alive in the field (Fig. 18). The head, body and tail are covered with hard spiny scales which may render the animal unswallowable to snakes; the head emerges from a collar-like fold of skin around the neck, and the neck itself carries a spiny hump which looks like a false head when

Fig. 18. Moloch horridus from the Northern Territory. Note the many large spines, lack of an external ear opening, and the spiny hump on the neck.

the animal lowers its real head when touched. Furthermore, the tail is very short with only 25-27 vertebrae and most of the toes have lost some phalanges so that the primitive digital formula for the genus is 2.3.4.4.3/2.3.4.4.3 instead of the more usual, and primitive for dragons, 2.3.4.5.3/2.3.4.5.4. A further, less obvious, feature is the loss of all preanal/femoral pores.

The phalangeal formula reported for *Moloch* above, 2.3.4.4.3/2.3.4.4.3, is considerably less reduced than the one quoted in the literature (Siebenrock 1895; Stephenson 1960): 2.2.3.3.2/2.2.3.3.2. (Figs 19-20). The reason for this is that the quoted figure is typical of the animals from throughout most of the range, and hence purely by chance, happened to be the one first studied. However, it appears that in a small area of the far western part of the range, roughly in the area from Exmouth to the Murchinson River, the population is characterized by the higher, more primitive count (pers. obs.). Whether this difference in phalangeal formula has taxonomic significance, i.e., represents a second species of *Moloch*, remains to be determined by further study of the zone of contact between the two populations.

The phalangeal formula for the interior population, i.e., 2.2.3.3.2/2.2.3.3.2, is the most reduced for any agamid species in the world. In most squamates this degree of "limb reduction" would be associated with a pronounced degree of body elongation as evidenced by an increased number of presacral vertebrae. However, in *Moloch*, the body, if anything, has become shorter, with a presacral count of 20-22, the lowest count for any squamate. The 2.2.3.3.2/2.2.3.3.2 condition also represents an extreme evening-up through loss across the entire spread of digits. A similar truncation is seen in other reptiles only in turtles and *Tiliqua rugosa* — 2.3.3.3.2/2.2.3.3.2.

As bizarre as *Moloch* appears to be, even to the tutored eye, many of its distinctive features can be readily interpreted as rather extreme elaborations of features already evident in *Tympanocryptis*. For example, this latter genus has scattered enlarged dorsal scales, a short tail, a greatly reduced number of preanal/femoral pores and has lost a phalange in the fifth digit of the rear foot. Furthermore, like *Moloch*, it is completely terrestrial.

Moloch occurs throughout most of the arid zone but only where there are sand substrates; it is conspicuously absent from regions of hard stony soils such as the Pilbara and Nullarbor Plain. It is often locally common, and in spring can often be seen in numbers on the roads in sand-plain country. One of the species' common names is mountain devil, but on the basis of its known habitat, the first part of the name is as much a misnomer as the latter.

Other aspects of *Moloch*'s biology are as unusual as its morphology. For example, it eats nothing but small ants, mainly of the genus *Iridomyrmex*, and these are consumed in large quantities (up to approximately 2 500 in two cases) usually from ant columns (references in Table 4). Second, although it will drink water both from a container and as droplets, it can also drink by immersing a part of its body and letting the water be conducted into the corners of its mouth by the capillary attraction of the spaces between its small scales (Davey 1923; Bentley and Blumer 1962; Gans et al. 1982). In the laboratory we have often seen *Moloch* standing calmly with one front leg in the water dish, apparently drinking. Another unusual feature we have noticed is that individuals will inflate themselves by means of a rapid series of short intakes of air. These intakes are accompanied by a throat pulse, a slight movement of the appendages and a barely perceptible swelling of the body and give the overall impression of a little *Moloch*-shaped balloon being inflated by a small pump located in the throat. The purpose of the behaviour is unknown.

There are a number of distinctive features in the locomotion of *Moloch*. For example, the ordinary locomotion of the lizards is slower than in other dragons, except *Chelosania*, both in walking and running. Furthermore, they often walk with a jerky motion; this occurs nowhere else in agamids but does occur in chamaeleonids. Its function is unknown. Another peculiarity is "freezing" in place while walking, sometimes even with one foot off the ground. This may serve to help conceal a *Moloch* caught out in the open. Finally there is the female's method of rejecting the unwanted embrace of an amorous male; when he seeks to straddle her back, she may execute a single, very rapid 360° rotation of her body along the long axis, throwing him off (pers. obs.). For a species that otherwise moves with what seems to be only a slow and deliberate pace, the rapidity of this action is remarkable.

Given the general slowing down of most locomotion in *Moloch*, it would be interesting to know how the repertoire of social behaviour may have been affected, because much of this behaviour is performed at high speed in other dragons. To date only head bobbing and an alteration in the orientation of the body have been recorded in a social context (Houston 1978; Johnston 1981).

There is some very circumstantial evidence that *Moloch* can store sperm for at least several weeks if not longer. For example, one female laid a clutch of fertile eggs two months after her last possible contact with a male (Philipp 1979), and there are reports that the species sometimes mates during early autumn (Sporn 1965), well after the females have laid their eggs for the season. There is evidence that females may also "get a jump" on the next breeding season, in that some females carry ovarian eggs at the end of autumn (May) that are of a size similar to those in spring in other dragons (Pianka and Pianka 1970).

Moloch played a small but significant role in one of paleontology's more infamous cases of mistaken identity. The story began when the famous British paleontologist Richard Owen (1858) described a "gigantic land lizard" on the basis of three fossil vertebrae from eastern Australia. He named the lizard *Megalania* and compared it with the then known large Australian varanids, noting that aside from some differences in proportions, the correspondence was close between the extinct and existing forms.

In the following years Owen received other large reptile bones from eastern Australia which he attributed to *Megalania*: in 1862 a sacral vertebra and a scapula, in 1866 a caudal vertebra and the back part of the skull, and in 1880 a trunk vertebra and the front part of the skull. The front part of the skull was the only unusual part. It was short-snouted, without teeth and had horn cores and looked as if it had come from a horned turtle. This skull caused Owen to immediately write another paper, not only to describe the new fossils but also to expand his overall description of *Megalania* — as a giant, toothless, horned lizard. Unfortunately, however, what Owen had in this peculiar skull was an example not of *Megalania* but of the giant horned turtle of Australia and Lord Howe

Island. Why Owen confused the two reptiles is not clear other than it was perhaps the most parsimonious way to reconcile all the reptile fragments of obviously large size from the same general area and, in some cases, from the same general facies. In addition, there may have been a desire on Owen's part to justify having already given the original bits of bone, all clearly from the giant varanid, a new generic name when they were so remarkably similar in form, at least, to living varanids (Owen 1858). Certainly a lizard with a varanid body but a horned, turtle-like head merited a new genus!

Be that as it may, Owen may never have persisted in his audacious concept of *Megalania* had it not been for *Moloch*, the little horned lizard described from Australia 40 years previously, for in *Moloch* Owen saw a living lizard similar in form, if not in size, to his extinct beast. (Owen mistakenly thought *Moloch* lacked premaxillary teeth, hence giving it a tendency toward toothlessness).

Shortly after the 1880 paper was published fate handed Owen an even bigger surprise than his horned skull: tail fragments from the very same specimen and giving every indication that it too was horned or spiked like the head. Owen's astonishment is almost palpable in the paper (Owen 1881) which quickly followed receipt of these specimens; he wrote: "if, indeeed, the last received fossils had first come to hand a conclusion that they formed part of some huge Armadillo might have been condoned". However, Owen again found comfort in *Moloch*, for after making comparisons with other lizards (as well as with armadillos) he concluded that the spiny caudal armature of *Moloch* was more similar to that of the giant extinct Australian lizard than was that of any other living lizard.

Owen maintained his concept of *Megalania*'s morphology and relationships for the next six years, because in 1887 he called *Moloch* "the now nearest known ally of *Megalania*" (Owen 1887a) and shortly thereafter he described the megalanians (see below) as saurians of the "many-horned, toothless kind" (Owen 1887b). In this latter paper Owen described fossils from Lord Howe Island which he regarded as closely allied to *Megalania* but sufficiently distinct to name as a new genus, *Meiolania*.

Clearly Owen's concept could not last, and the denouement came in a matter of weeks in the form of a brief paper by T. H. Huxley (1887). Huxley re-examined much of Owen's Lord Howe Island material and while disclaiming any idea of what *Megalania* might be, correctly

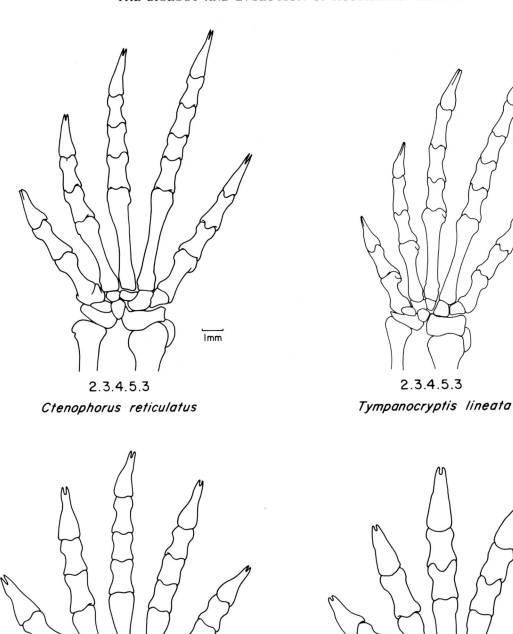

2.3.4.5.3
Ctenophorus reticulatus

2.3.4.5.3
Tympanocryptis lineata

2.3.4.4.3
Moloch horridus

2.2.3.3.2
Moloch horridus

Fig. 19. The loss of phalanges in the front foot of dragons: *Ctenophorus reticulatus* and *Tympanocryptis lineata* with the primitive formula: 2.3.4.5.3; *Moloch horridus* (far southwestern population) with a phalange loss in the fourth toe: 2.3.4.4.3, and *Moloch horridus* (central populations) with yet further losses in the second through fifth toes: 2.2.3.3.2.

Fig. 20. The loss of phalanges in the rear foot of dragons: *Ctenophorus reticulatus* with the primitive formula: 2.3.4.5.4; *Tympanocryptis lineata* with a loss of one phalange of the fifth toe: 2.3.4.5.3; *Moloch horridus* (far southwestern population) with an additional loss in fourth toe: 2.3.4.4.3, and *Moloch horridus* (central populations) with yet further losses in the second through fifth toes: 2.2.3.3.2.

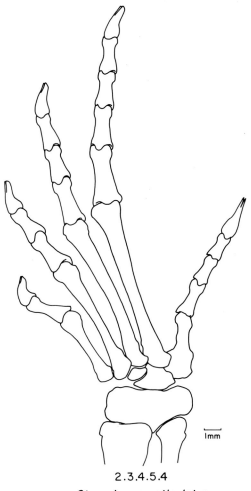

2.3.4.5.4

Ctenophorus reticulatus

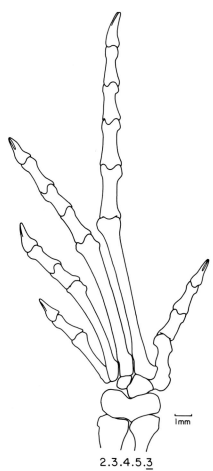

2.3.4.5.<u>3</u>

Tympanocryptis lineata

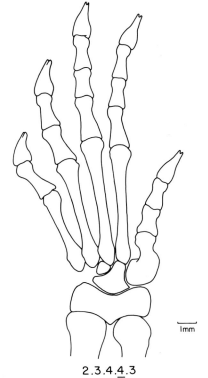

2.3.4.<u>4</u>.<u>3</u>

Moloch horridus

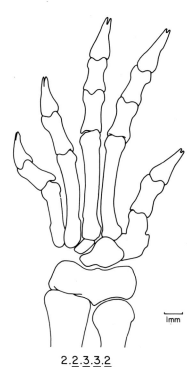

2.<u>2</u>.<u>3</u>.<u>3</u>.<u>2</u>

Moloch horridus

identified *Meiolania* as a turtle, renaming the genus, unnecessarily, *Ceratochelys*.

Owen published a final paper on the megalanians in 1888; in this paper he retained the view of *Megalania* and *Meiolania* as close relatives but gave up the idea of their close relationship with any living group, considering them to be the sole representatives of a new suborder of reptiles, the Ceratosauria. Needless to say there was no further mention of *Moloch*. Strangely, Owen made no mention of Huxley's paper which first established the distinction between *Megalania* and *Meiolania*, but he did use the prefix of Huxley's name for the genus in his new subordinal name. Stranger still was the timing of the two men's papers: Huxley's was received by the society that was to publish it on the 24th of March 1887, Owen's on the 29th of March 1887.

There remains one Australian dragon genus to discuss, and it has been left to last because its relationships are very uncertain. This is the monotypic genus *Chelosania* from northern Australia. This dragon's small lacrimal bone tentatively associates it with the amphiboluroids. However, in that it has not yet been karyotyped, it is not known if it has the other hallmark of this group, a diploid chromosome number of 32 instead of the more primitive 36. Unfortunately, preanal/femoral pores are no help because *Chelosania* lacks them, being the only Australian genus, besides *Gonocephalus* and *Moloch*, to consistently lack them. If *Chelosania* is an amphiboluroid, it is a rather primitive one on the criterion of presacral vertebrae, for it retains 24 instead of the more usual 22 or 23.

There is, of course, a possibility that *Chelosania* is not an amphiboluroid; it could be a third non-amphiboluroid group in Australia, along with *Gonocephalus* and *Physignathus*. In fact, it has been suggested that *Chelosania* is most closely related to *Calotes cristatellus* from Timor and New Guinea, a species with which it is said to share posterolaterally oriented mid-dorsal scales, a feature not seen in other Australian dragons (Witten 1982).

Whatever the relationships of *Chelosania* may be it is a peculiar beast about which little is known. Its remote distribution plus cryptozoic habits make it one of Australia's least well-known large dragons. It is predominantly arboreal and apparently capable of only slow, deliberate movement (Husband 1979a). It is oviparous with a clutch size of 4-9 eggs (Table 5). Egg laying has been recorded during the dry season, but it is not yet known if the species also lays during the wet (Anonymous 1973b; Pengilley 1982; James 1983; James and Shine 1985). So far *Chelosania* appears to be the only tropical dragon that breeds throughout much of the dry (cf. *Chlamydosaurus, Diporiphora* and *Amphibolurus* — Harcourt 1986; James 1983; James and Shine 1985; pers. obs.). *Chelosania* may also have a slightly shorter incubation period than other Australian dragons for which there is information, i.e., 59-61 days vs 67 or more days (Table 8).

Another remarkable feature of *Chelosania* may be its high relative clutch mass. If the mass of the eggs actually weighed (N=5) is indicative, then the entire clutch (N=9) of the only female observed to date (Pengilley 1982) would give an RCM(1) of .94 (Table 6). This is the highest RCM(1) known to date for any Australian lizard.

From this brief review of agamid lizards in Australia, it should be evident that the single most pervasive factor in their biology is their diurnal, out-in-open habit. As this habit is also characteristic of all other agamids, it is probably indicative of the family's ancestral behaviour.

Much of agamid biology seems to follow directly from this basic habit. It "explains" their highly visually oriented and heliothermic behaviour, their physiological resistance to high temperatures (Table 11), dehydration and salt loading, and perhaps even their persistance in their "conservative" body plan of relatively large limbs and robust bodies.

The only other squamates that also seem to be primitively active during the day in open spaces are chamaelonids, iguanids, some lacertids, some teiids and varanids. Because the first two groups are thought to be related to agamids and because this combined group is thought to be relatively primitive within squamates, there is a possibility that the first squamates also possessed similar habits. However, against this must be set the fact that most other squamate groups and the only living relative of squamates, *Sphenodon*, are basically cryptozoic and/or nocturnal and/or inhabitants of confined spaces. This might suggest that the open, diurnal habits of agamids and their relatives are derived within squamates. Be that as it may, such habits are a basic aspect of agamid biology and may be a useful primary consideration when attempting to interpret any aspect, old or new, of their biology.

Table 1. Derived character states shown by the Agamidae in comparison to the hypothetical common ancestor of all lizards and snakes.

Skull and hyoid

> Premaxillae fused
> Frontals fused
> Parietals fused
> Maxillary and dentary teeth mostly acrodont
> Vomerine, palatine and pterygoid teeth absent
> Stapes imperforate
> Second epibranchial absent

Postcranial skeleton

> Vertebrae procoelous, anotochordal
> Intercentra absent over mid and posterior presacral vertebrae
> Inscriptional ribs separated posterior to mesosternum
> Caudal autotomy planes absent
> Distal end of terminal phalanges vertically notched

Table 2. Data on snout-vent length (SVL, mm), hind limb length (HLL, as % of SVL), tail length (TL, as % of SVL) and sex of largest specimen recorded for a population and/or the species in Australian dragon lizards. Values are ranges in all cases. Data are primarily from the taxonomic literature.

Taxon	SVL	HLL	TL	Sex of Largest Specimen	
				Population	Species
Amphibolurus					
burnsi	34-125				♂
gilberti	28-128	76-87	213-324		
longirostris	27-114	80-95	244-368		
muricatus	39-125		170-230		♂
nobbi	25-80	77-82	236-285		♀
norrisi	31-117				♀
temporalis	37-130		261-318		
Caimanops					
amphiboluroides	45-94		147-181		
Chelosania					
brunnea	46-118		135-169		
Chlamydosaurus					
kingii	50-258		182-222		
Cryptagama					
aurita	29-46	61-63	69-77		
Ctenophorus					
Rock dwellers					
caudicinctus					
caudicinctus	23-89		154-258		♂
graafi	-79				♂
infans	-67				♂
mensarum	-81				♂
slateri	-83				♂
decresii	-82	68-95	160-220		♂
fioni	-96	70-92	140-220		♂
ornatus	30-93		163-245	♂	
rufescens	44-97	87-101	184-249		
vadnappa	-82	77-100	180-230		♂
yinnietherra	34-87		178-204		
Open spaces					
femoralis	32-57		247-317	♂	♂
fordi	23-53	78-100	187-271		♀
isolepis	28-70		200-260		♀
maculatus	28-59		213-288		♀
rubens	41-83		211-264		♂
maculosus	27-69	61-74	101-153		♂
Burrowers					
clayi	24-54	56-72	127-173		♂
gibba	-82	75-90	126-154		
cristatus	35-110	80-104	152-273		
mckenziei	-66		220-238		
nuchalis	31-115	52-62	110-152		♂
pictus	35-65	71-81	162-206		
salinarum	31-70		129-184		♀
scutulatus	31-115	74-90	194-273		
reticulatus	27-108	62-76	126-175		♂
Diporiphora					
albilabris	27-55		164-232		
arnhemica	27-63		185-254		
australis					
bennetti	23-80		189-288		
bilineata	36-64				
convergens	34		204		
lalliae	35-76		183-265		
linga	-61	58-86	165-271		♀
magna	27-87		208-304		
pindan	31-61		198-274		
reginae	29-72		183-265		
superba	61-93		329-413		
valens	32-66		206-268		
winneckei	21-65	61-76	219-304		
Gonocephalus					
boydii	53-173	90-98	181-202		
spinipes	44-131	79-94	175-227		♀

Table 2. (continued)

Taxon	SVL	HLL	TL	Sex of Largest Specimen	
				Population	Species
Moloch					
horridus	34-110		62-86		♀
Physignathus					
lesueurii	49-245			♂	
Pogona					
barbata	38-250	48-56	117-132		
microlepidota	93-180	59-71	181-212		
minima	39-115	60-77	175-246		
minor	34-149	47-72	138-229		
mitchelli	37-163	49-64	143-205		
nullarbor	34-141	51-62	122-160		
vitticeps	39-250	46-58	104-131		
Rankinia					
adelaidensis	21-52	54-76	118-189	♀	♀
chapmani	25-53	51-70	113-160		
diemensis	25-82		167-211		♀
Tympanocryptis					
butleri	23-43	67-82	117-154		
cephala	22-57	62-76	102-160		
intima	-61	55-73	100-142		
lineata					
centralis	23-61	52-69	137-185		♀
houstoni	22-68	56-80	134-196		♀
lineata	30-72	66-88	124-192	♂	♂
macra	36-64	64-81	137-185		
parviceps	23-46	68-83	118-160		♀

Table 3. Summary of counts for presacral and postsacral vertebrae and postsacral diapophyses for Australian dragon lizards. Numbers for presacral vertebrae counts are range, mode and sample size, and for postsacral vertebrae and diapophyses counts, range and sample size.

Taxon	Vertebrae		Postsacral diapophyses
	Presacral	Postsacral	
Amphibolurus			
burnsi	22-23, 23, 9	61-63, 5	14-20, 9
gilberti	22-23, 23, 8	69-70, 3	14-16, 8
longirostris	24-25, 24, 18	73-79, 11	17-20, 9
muricàtus	22-23, 23, 12	52-57, 7	16-19, 5
nobbi	23-24, 23, 16	52-58, 11	13-17, 11
temporalis	24, 24, 15	61-69+, 7	12-14, 14
Caimanops			
amphiboluroides	23, 23, 1	—	14, 1
Chelosania			
brunnea	24, 24, 1	55+	9, 1
Chlamydosaurus			
kingii	22-23, 23, 14	55-59, 7	17-19, 4
Cryptagama			
aurita	23, 23, 1	32, 1	21, 1
Ctenophorus			
caudicinctus			
caudicinctus	23, 23, 5	49-53, 2	16-17, 5
infans	22-23, 22, 5	51-54, 2	15-20, 4
macropus	22-24, 23, 12	55-61, 7	17-21, 4
slateri	23, 23, 4	—	16-18, 4
clayi	22-24, 23, 15	39-44, 12	23-32, 9
decresii	23, 23, 8	51-56, 6	14-17, 12
femoralis	22-24, 23, 36	52-59.5, 31	12-15½, 35
gibba	22, 22, 1	46, 46, 1	29, 1
maculatus	22-23, 23, 12	53-56, 8	14-16, 8
nuchalis	23, 23, 12	44-49, 9	18-24, 11
ornatus	23, 23, 13	50-57, 6	14-19, 9
pictus	23-24, 23, 15	49-52, 9	16-21, 17
reticulatus	22-24, 23, 13	47-54, 6	17½-22, 13
rubens	22-24, 23, 16	49-53, 9	13-17, 17
rufescens	23, 23, 4	50-53+, 3	17½-22, 13
salinarum	22-23, 23, 13	42-47, 12	19-24, 13
scutulatus	22-23, 22/23, 4	54, 1	22, 4
vadnappa	22-23, 23, 3	52-57, 2	14-15, 3
Diporiphora			
australis	22-23, 23, 8	52-60, 16	13-16, 9
magna	23-24, 23, 5	60-65, 4	14-17, 6
pindan	22-24, 23, 11	55-61, 9	15-22, 11
Gonocephalus			
spinipes	23, 23, 5	53-55, 2	11-19, 4
Moloch			
horridus	20-21, 21, 17	22-27, 14	9-14, 13
Physignathus			
lesueurii	23-24, 24, 19	64-70, 13	11-14, 17
Pogona			
barbata	24-25½, 25, 27	45-52, 14	18-22½, 25
minor	24-25, 24, 9	53-56, 6	15-20, 7
mitchelli	24, 24, 4	53-54, 3	15-19, 4
nullarbor	24, 24, 2	47-48, 2	19-20, 2
vitticeps	24-25, 24, 11	46-51, 11	19-25, 10
Rankinia			
adelaidensis	21-22, 21/22, 4	38-42, 3	14-17, 4
chapmani	20-22, 22, 3	37-41, 2	16-19, 2
diemensis	22-23, 23, 49	40-48, 20	11-16, 35
Tympanocryptis			
butleri	21.5-22, 22, 7	38-41, 6	16.5-22, 6
cephala	22-23, 23, 12	37-43, 12	25-34, 10
intima	23-23½, 23, 10	43-45, 5	27-40, 11
lineata	22-24, 23, 8	45-50, 6	24-37, 6
parviceps	22-23, 22, 13	38-42, 11	18-22, 5

Table 4. References to food eaten by Australian dragon lizards both in the wild (W) and in captivity (C).

Taxon	Reference
Amphibolurus	
gilberti	James *et al.* 1984 (W); Weigel 1984 (?)
longirostris	Scott 1962 (W); Baehr 1976 (W); Pianka 1986 (W)
muricatus	Groom 1973a-b (C); Peters 1974, 1975, 1983 (C)
nobbi	Bustard 1968e (W, as *A. muricatus*)
norrisi	Witten and Coventry 1984 (W)
temporalis	Cogger and Lindner 1974 (W)
Caimanops	
amphiboluroides	Pianka 1986 (W)
Chlamydosaurus	
kingii	Saville-Kent 1895, 1897 (W, C); Worrell 1963 (?); Peters 1974, 1975, 1983 (C); Barnett 1977c (C); Harcourt 1986 (C)
Ctenophorus	
clayi	Pianka 1986 (W)
cristatus	Pianka 1971d (W); Chapman and Dell 1978 (W); Dell and Chapman 1978 (W)
decresii	Peters 1975, 1983 (C)
fionni	Smyth 1971 (W)
fordi	Baverstock 1979 (W)
isolepis	Loveridge 1938 (W, as *Amphibolurus maculatus*); Pianka 1971c (W); Pianka 1986 (W)
maculatus	Chapman and Dell 1979 (W)
maculosus	Madigan 1944 (W); Mincham 1966 (C); Mitchell 1973 (W)
nuchalis	Bradshaw 1965 (W, as quoted by Pianka 1971a); Heatwole 1970 (W); Pianka 1971a (W, as *Amphibolurus inermis*); Klages 1982 (W, C); Dell and Chapman 1981b (W, as *A. inermis*); Pianka 1986 (W, as *A. inermis*); Bradshaw 1986 (W, C)
ornatus	Dell and Harold 1979 (W)
pictus	Mayhew 1963 (C)
reticulatus	Chapman and Dell 1980a (W); Pianka 1986 (W)
scutulatus	Pianka 1971d (W); Dell and Chapman 1979a (W); Chapman and Dell 1979 (W); Pianka 1986 (W)
Diporiphora	
winneckei	Houston 1978 (W); Pianka 1986 (W)
Gonocephalus	
boydii	Zwinenberg 1974 (?); Barnett 1981a (C)
spinipes	Longley 1943 (C, as *G. cristipes*)
Moloch	
horridus	Lucas and Frost 1896; Saville-Kent 1897; Davey 1923; White 1948; Sporn 1955; Paton 1965 (C); Pianka and Pianka 1970 (W); Peters 1974, 1975, 1983 (C); Chapman and Dell 1980a (W); Pianka 1986 (W)
Physignathus	
lesueurii	Mattingley, in Barrett 1931 (W); Longley 1947b (C); Mackay 1959 (W); Rose 1974 (W); Anonymous 1976 (W, C); Davey 1970 (C); Peters 1974, 1975, 1983 (C); Hardy and Hardy 1977 (W, C); Clifford and Hamley 1982 (W)
Pogona	
barbata	Coleman 1944 (C); Lee and Badham 1963 (C); Rose 1974 (W); Schafer 1979 (C)
minor	Bradshaw 1965 (?), as quoted in Davidge 1979; Smith 1976 (W); Chapman and Dell 1978 (W); Davidge 1979 (W); Chapman and Dell 1980a (W); Pianka 1986 (W)
vitticeps	Lee and Badham 1963 (C, as *Amphibolurus barbatus*, part); Smith 1974 (C); Kennerson and Cochrane 1981 (W)
Rankinia	
adelaidensis	Davidge 1979 (W)
diemensis	Webb 1983 (W)
Tympanocryptis	
lineata	Robson 1968 (C)

Table 5. Clutch and female size in Australian dragon lizards. Figures are ranges and means (latter in parentheses).

Taxon	Female size	Clutch size	N/n	Reference
Amphibolurus *burnsi*	—	8	1	Bustard 1968e (as *Amphibolurus* sp.)
gilberti	?-? (82.1)	4-8 (5.5)	17	James 1983
	—	5	1	Smith 1976
longirostris	—	— (3.9)	7	Pianka 1986
	81-88 (84.5)	5-6 (5.3)	2/3	Pers. obs.
muricatus	—	3-7	?	Weekes 1934
	—	8	1	Meredith and Cann 1952
	—	7-8	?	Groom 1973a
	—	6	1	Lawton 1982
	78-104 (91.9)	4-8 (5.9)	14	James 1983
nobbi	—	6	1	Bustard 1968e (as *A. muricatus*)
norrisi	—	3-7 (4.8)	17	Witten and Coventry 1984
temporalis	75	6	1	James 1983
	83-95 (87.0)	4-5 (4.5)	2	Pers. obs.
Chelosania *brunnea*	—	4	1	Anonymous 1973b
	—	9	1	Pengilley 1982
	—	4-6 (5.0)	2	Sadlier 1981 (as quoted by James 1983)
	80	5	1	James 1983
Chlamydosaurus *kingii*	—	13	1	Barnett 1977c
	—	7-14 (10.0)	3	Harcourt 1986
	—	12	1	T. Boylan, pers. comm.
Ctenophorus *caudicinctus*	—	4-8 (6.0)	?	Bradshaw 1981
	—	— (5.4)	18	Pianka 1986
clayi	—	— (1.9)	7	Pianka 1986
	42-48 (45.0)	2-4 (3.0)	2	Pers. obs.
cristatus	95	5	1	Pianka 1971d
	95	4	1	Chapman and Dell 1977
	85	2	1	Chapman and Dell 1978
	—	5	1	Dell and Chapman 1978
	85-108 (97.7)	3-9 (6.1)	7	Pers. obs.
decresii	61-85 (73.0)	3-7 (4.7)	4	Pers. obs.
femoralis	44-52 (46.8)	2 (2.0)	10	Pers. obs.
fordi	—	2-3 (2.3)	?	Cogger 1969 (as quoted in Bradshaw 1981)
	—	— (2.4)	15	Pianka 1986
isolepis	—	4-5	?	Lucas and Frost 1896
	51-67 (58.9)	1-6 (3.3)	106	Pianka 1971c
maculatus	49	3	1	Dell and Chapman 1977
	49-57 (51.3)	2-3 (2.2)	6	Pers. obs.
	62 (62.0)	3 (3.0))	2	Chapman and Dell 1977
	50	4	1	Dell and Chapman 1979b
	—	2 (2.0)	2	Dell and Chapman 1981b
	51	2	1	Dell and Chapman 1981b
maculosus	—	2-4	?	Mitchell 1973
nuchalis	—	2-6 (3.4)	96	Pianka 1971a (as *Amphibolurus inermis*; including 72 records from Bradshaw 1965)
	—	6 (6.0)	1/3	Klages 1982
	—	— (4.0)	27	Pianka 1986 (as *A. inermis*)
ornatus	—	2-3 (2.5)	2	Chapman and Dell 1975
	91	3	1	Dell and Chapman 1978
	77-86 (82.3)	2-3 (2.7)	3	Chapman and Dell 1980b
	—	2-5 (3.7)	?	Bradshaw 1981
pictus	—	about 8	?	Lucas and Frost 1896
	—	4	1	Antenor 1974
	—	4	1	D. Kent, pers. comm.
	52-73 (59.5)	2-6 (3.3)	11	Pers. obs.

Table 5 (continued)

Taxon	Female Size	Clutch Size	N/n	Reference
Ctenophorus reticulatus	—	6-8	?	Lucas and Frost 1896
	73	4	1	Chapman and Dell 1980a
	—	— (4.0)	7	Pianka 1986
	62-79 (69.3)	2-8 (4.6)	20	Pers. obs.
salinarum	60-64 (62.0)	3 (3.0)	2	Pers. obs.
scutulatus	93-105 (98.4)	5-10 (6.6)	7	Pianka 1971d (includes Loveridge's 1938 record)
	97	6	1	Dell and Chapman 1979a
	90	4	1	Chapman and Dell 1979
Diporiphora albilabris	57-59 (58.0)	1-8 (3.7)	4	Pers. obs.
bennetti	50-56 (54.3)	4-5 (4.3)	4	G. Shea, pers. comm.
bilineata	?-? (54.5)	4-8 (6.0)	13	James 1983
winneckei	—	1-3	?	Houston 1978
Gonocephalus boydii	175	3	?	Wells 1972
Moloch horridus	—	about 8	?	Lucas and Frost 1896
	—	6	1	LeSouef and McFadyen 1937
	—	7-10 (8.5)	4	White 1948
	—	5-9 (6.3)	8	Sporn 1955, 1965
	82-110 (94.9)	3-10 (7.3)	37	Pianka and Pianka 1970
	—	8	1	Hudson 1977
	98	4	1	Chapman and Dell 1979
Physignathus lesueurii	—	6	1	Mattingley, in Barrett 1931
	—	8-17	?	Hay 1972
	—	6-12	1/5	P. Harlow, in Anonymous 1976
	—	18	1	M. Maddocks, in Anonymous 1976
	—	10	1	Smith 1979
	170	8	?	Giddings 1980
	212	13	1	Giddings 1983b
Pogona barbata	—	28	1	Meredith and Cann 1952
	185	31	1	Bustard 1966b
	—	19	1	Bustard 1966b
	—	6	1	Wells 1971
	—	18-23 (20.5)	2	Badham 1971
	—	24	1	Schafer 1979
	—	15-35	6	Smith and Schwaner 1981 (but includes Smith's [1974, 1984] data which is for *P. vitticeps* [pers. comm., H. Ehmann])
minor	100	6	1	Smith 1976
	—	2-6 (4.5)	4	Dell and Chapman 1977
	107	5	1	Chapman and Dell 1977
	103-108 (104.7)	6-9 (7.3)	3	Chapman and Dell 1978
	—	7	?	Dell and Chapman 1978
	—	7	1	Dell and Chapman 1979a
	110-126 (118.0)	6	2	Chapman and Dell 1980a, b
	96-103 (99.5)	8-9 (8.5)	2	Dell and Chapman 1981b
	—	3-8	?	Davidge 1980
	—	5-19 (8.2)	?	Bradshaw 1981
	—	5-9	?	Bush 1981
	113	5+	1	Browne-Cooper 1985
	—	— (7.6)	73	Pianka 1986
nullarbor	135-140 (137.5)	14 (14.0)	2	Smith and Schwaner 1981
vitticeps	221	22	1	Loveridge 1934 (as *Amphibolurus barbatus*)
	—	15	1	Williams 1964
	—	13	?	Kennerson and Cochrane 1981
	—	11-16 (13.5)	2	Johnston 1979
	—	24-35 (28.0)	2/3	Smith 1974, 1984 (as this species *fide* H. Ehmann)

Table 6. Relative clutch mass in Australian dragon lizards.

Taxon	RCM (1)			RCM (2)			RCM (3)			RCM (4)			Reference
	x̄	R	SD	x̄	R	SD	x̄	R	SD	x̄	R	SD	
Amphibolurus													
gilberti	—			—			—			.22		.06	James 1983 (N=6)
longirostris	—			—			15.1			—			Pianka 1986 (N=4)
muricatus	—			—			—			.18		.07	James 1983 (N=7)
temporalis	—			—			—			.27			James 1983 (N=1)
Chelosania													
brunnea	.94			.49			—			—			Pengilley 1982 (N=1)
	—			.31			—			—			James 1983 (N=1)
Ctenophorus													
caudicinctus	—			—			16.6			—			Pianka 1986 (N=10)
clayi	—			—			12.9			—			Pianka 1986 (N=6)
fordi	—			—			11.8			—			Pianka 1986 (N=5)
isolepis	—			—			12.9			—			Pianka 1986 (N=76)
nuchalis[1]	—			—			15.1			—			Pianka 1986 (N=8)
reticulatus	.54			.35			—			—			Pers. obs. (N=1)
	—			—			19.2			—			Pianka 1986 (N=3)
scutulatus	—			—			11.8			—			Pianka 1986 (N=2)
Diporiphora													
bilineata	—			—			—			.24		.02	James 1983 (N=6)
Moloch													
horridus	—			—			14.9			—			Pianka 1986 (N=34)
Pogona													
barbata	.36			.26			—			—			Pers. obs. (N=1)
minor	—			—			19.5			—			Pianka 1986 (N=27
Rankinia													
diemensis	—			—			—			.23		.03	James 1983 (N=3)
Tympanocryptis													
lineata	.66	.76-.86	.13	—			—			—			Pers. obs. (N=7; eggs dissected from females)

[1]As *Ctenophorus inermis*

Table 7. Correlation coefficients for female size *vs* clutch size in Australian dragon lizards.

Taxon	Correlation Coefficient	N	Reference
Amphibolurus			
gilberti	.56**	17	James 1983
muricatus	.55**	14	James 1983
Ctenophorus			
cristatus	.73*	10	Pers. obs. (data from Table 5)
decresii	.97*	4	Pers. obs.
femoralis	.00^{NS}	10	Pers. obs.
isolepis	.46***	106	Pianka 1971c (data from this work)
maculatus	.36^{NS}	11	Pers. obs. (data from Table 5)
nuchalis	.75***	24	Pianka 1971a (data from this work; as *Amphibolurus inermis*)
pictus	.91**	10	Pers. obs.
reticulatus	.71***	20	Pers. obs.
scutulatus	.22^{NS}	9	Pers. obs. (data from Table 5)
Diporiphora			
bilineata	.15^{NS}	13	James 1983
Moloch			
horridus	.18^{NS}	37	Pianka and Pianka 1970
Pogona			
minor	—.22^{NS}	7	Pers. obs. (data from authors in Table 5)
Rankinia			
diemensis	.71**	18	Kent 1987
Tympanocryptis			
cephala	.81^{NS}	5	Pers. obs.
intima	.26^{NS}	16	Pers. obs.
lineata	.78***	14	Pers. obs.

Table 8. Incubation times for Australian dragon lizard eggs. Means in parentheses.

Taxon	Incubation Time (days)	Incubation Temperature (°C)	Reference
Chelosania			
brunnea	59-61	—	Anonymous 1973b
Chlamydosaurus			
kingii	88-92	30	Barnett 1977c (see also Harcourt 1986)
	95	30	T. Boylan, pers. comm.
	67-70	30	Harcourt 1986
	75	30	Harcourt 1986
Ctenophorus			
maculosus	about 70	—	Mitchell 1973
nuchalis	77-79	27	Klages 1982
pictus	109	19-37(27)	Antenor 1974
Physignathus			
lesueurii	98-100	—	Smith 1979
Pogona			
barbata	75-84	25	Bustard 1966b
	63-70	—	Schafer 1979
vitticeps	89-96	26	Johnston 1979
	137	27-29	Smith 1974 (as this species *fide* H. Ehmann)

Table 9. Body temperatures (°C; FBT) of Australian dragon lizards active in the field.

Taxon	Mean	Range	SD or SE	N	Reference
Amphibolurus					
longirostris	37.0	34.2-39.0	1.48	26	Licht *et al.* 1966a
	34.1	—	3.9	69	Pianka 1986
muricatus	34.8	29.5-38.6	±0.34	32	Heatwole *et al.* 1973
Caimanops					
amphiboluroides	36.6	—	2.6	13	Pianka 1986
Ctenophorus					
caudicinctus	39.0	34.8-41.0	1.84	15	Licht *et al.* 1966a
	36.5	—	±0.87	17	Bradshaw and Main 1968
clayi	36.7	—	2.6	23	Pianka 1986
fordi	36.9	—	2.2	104	Pianka and Pianka 1970
	36.9	—	±0.16	82	Cogger 1974
nuchalis[1]	36.1	—	—	40	Bradshaw 1965 (as quoted in Pianka 1971a)
	39.3	34.5-43.0	1.24	4	Licht *et al.* 1966a
	36.7	34.0-40.3	±0.53	12	Bradshaw and Main 1968
	39.4-41.0	—	—	?	Heatwole 1970 (range of means for different subgroups)
	36.1	—	3.8	129	Pianka and Pianka 1970; Pianka 1971a
isolepis	37.7	—	2.3	511	Pianka and Pianka 1970; Pianka 1971c
	37.3	28.0-40.1	3.18	20	Pers. obs.
ornatus	36.6	33.0-39.6	±0.23	48	Bradshaw and Main 1968
reticulatus	37.0	35.0-40.6	2.20	6	Licht *et al.* 1966a
	34.4	—	3.5	30	Pianka 1986
scutulatus	38.9	33.5-44.0	1.87	81	Pianka and Pianka 1970; Pianka 1971d
vadnappa	37.2	—	—	1	Pers. obs.
Diporiphora					
winneckei	33.5	—	3.1	31	Pianka and Pianka 1970
Moloch					
horridus	34.1	31.5-39.9	3.38	5	Licht *et al.* 1966a
	33.3	27.3-40.2	2.9	89	Pianka and Pianka 1970
Pogona					
barbata	34.8	—	±0.37	58	Lee and Badham 1963
minima	32.8	28.4-36.7	2.13	36	Licht *et al.* 1966a
	34.1	32.6-34.8	±0.31	6	Bradshaw and Main 1968
minor	34.8	—	3.6	63	Pianka and Pianka 1970
	35.0	31.8-38.4	2.92	13	Licht *et al.* 1966a
	34.0	—	±0.57	17	Bradshaw and Main 1968
vitticeps	34.3	24.0-39.5	3.75	31	Pers. obs.
Tympanocryptis					
lineata	37.2	—	—	1	Pers. obs.

[1] As *Amphibolurus inermis* in all studies.

Table 10. Preferred body temperatures (°C; PBT) of Australian dragon lizards in a laboratory thermal gradient.

Taxon	Mean	Range	SD or SE	N/n	Reference
Amphibolurus					
muricatus	34.6	24.5-42.4	±0.56	16/110	Heatwole *et al.* 1973
	36.0	—	1.22	6	Licht *et al.* 1966a
Ctenophorus					
caudicinctus	37.7	—	1.45	8	Licht *et al.* 1966a
maculatus	37.0	—	1.28	6	Licht *et al.* 1966a
nuchalis	36.4	—	2.39	15	Licht *et al.* 1966a (as *Amphibolurus inermis*)
ornatus	36.6	33.0-39.6	±0.25	22	Bradshaw and Main 1968
	36.3	—	±0.67	6	Baverstock 1975 ("fast growers")
	35.5	—	±0.91	8	Baverstock 1975 ("slow growers")
pictus	39.0	36.6-40.4	—	1/?	Mayhew 1963
reticulatus	36.9	—	0.93	5	Licht *et al.* 1966a
scutulatus	38.2	—	1.06	7	Licht *et al.* 1966a
Moloch					
horridus	36.7	—	1.17	9	Licht *et al.* 1966a
Physignathus					
lesueurii	30.1	23.7-36.0	1.6	10/1000	Wilson 1974
Pogona					
barbata	35.7	—	0.80	6	Licht *et al.* 1966a
minor	35.8	—	0.68	7	Licht *et al.* 1966a
minima	36.3	—	1.11	11	Licht *et al.* 1966a

Table 11. Critical thermal maximum temperatures (°C; CTMax) of Australian dragon lizards.

Taxon	Mean	Range	SD or SE	N	Reference
Amphibolurus					
gilberti	41.8	41.6-42.0	—	2	Spellerberg 1972a
muricatus	42.3	42.1-42.4	—	2	Spellerberg 1972a
Ctenophorus					
cristatus	43	—	—	?	Warburg 1965a
decresii	42	—	—	2	Warburg 1965a
	43.8	43.7-43.9	—	2	Spellerberg 1972a
fordi	—	44-46	—	?	Cogger 1969, 1974
maculosus	48.9	—	—	5	Mitchell 1973
nuchalis	46.5	—	—	?	Warburg 1965b (as *Amphibolurus reticulatus inermis*)
	48.5	47.7-49.5	?	8	Heatwole 1970 (as *A. inermis*)
pictus	45	—	—	?	Warburg 1965a
	46.5	—	—	?	Mitchell 1973
Moloch					
horridus	42.5	—	—	1	Spellerberg 1972a
Rankinia					
diemensis	44.8	44.0-45.6	—	2	Spellerberg 1972a

CHAPTER 3

Gekkonidae — Geckos

GECKOS are the second largest family of lizards after skinks. They are widespread throughout the world and in mainland Australia occur in most areas except the south-east. There are approximately 800 species in approximately 90 genera worldwide and 87 species (Cogger 1986) in 18 genera in Australia. A recent trend in gecko taxonomy in Australia is the recognition of undescribed species initially on the basis of differences in their chromosomes and then subsequently on subtle morphological criteria (King and Rofe 1976; King and King 1977; King 1977b, 1979, 1981a, 1982a-b, 1983a-b, 1984a; Moritz 1983).

In general, geckos are small to large lizards with well-developed limbs, large eyes and uniformly small, bead-like scales which give the skin a soft, velvety feel (Fig. 21). In many species the skin is so thin, especially on the belly, that the internal organs and eggs can be seen through it. Morphologically, gecko genera differ most from one another in the digits, tail and colour pattern. Indeed, so important are the digits (Fig. 25) that the Latin word for these structures, *dactylus*, when appropriately prefixed, forms the basis for many gecko generic names (Table 1).

Geckos occur in a wide variety of habitats in Australia from the rainforests of the north-east to the arid plains and dune fields of the centre. They are absent generally only in the cool moist areas of the south-east. They may be terrestrial, saxicolous or arboreal. In addition, several species live in and around human habitations.

Most geckos, including all Australian species, are nocturnal in their foraging times, generally becoming active just after sunset and remaining abroad until they eat their fill or it becomes too cool due to falling temperatures or too light due to a rising moon or the dawn. One consequence of foraging at night, of course, is that the temperatures in all parts of the environment are dropping and hence becoming increasingly inimical to physiological processes. This leads to the question of how low a temperature foraging geckos will tolerate and whether there are any species differences at these low levels. The answer to the first question appears to be about 16°C (several species, Table 10; see also Bustard 1968f) and the answer to the second is probably "yes" but we lack sufficiently comparable information to be very precise (one example: *Nephrurus* species seem to be active at generally lower body temperatures than other desert geckos — Pianka and Pianka 1976). However, more work, preferably on a large local gecko fauna, is needed to provide more complete answers to both questions.

The few studies on feeding in Australian geckos (Table 5) suggest that they are primarily opportunistic predators on arthropods, that is, they will take whatever prey in this group that comes their way, with the exception of ants. A few species, e.g., *Diplodactylus conspicillatus, D. pulcher* and *Rhynchoedura ornata*, may be termite specialists (Pianka and Pianka 1976; Dell and Chapman 1978, 1979a; Pianka 1986) but additional observations would be desirable to confirm this. The larger geckos such as *Oedura* (Cogger 1957), *Nephrurus* (Schmida 1973b) and *Underwoodisaurus* (Davey 1945) may also prey on other lizards including other geckos. As might be expected, larger gecko species, as measured by their head size, take larger prey (Pianka and Pianka 1976; Pianka 1981, 1986). In addition to their normal prey, many geckos are also known to eat their sloughed skin (Bustard and Maderson 1965); this may be a way of recouping protein resources that would otherwise be lost if the slough were left in the environment.

Geckos have peg-like teeth but compared to other lizards with similarly shaped teeth, e.g., many skinks and some pygopodids, gecko teeth

are thinner and more numerous (Figs 5, 22). The functional significance of these differences is unclear.

One peculiar gecko feeding habit is to lick liquids or semi-liquid substances, either natural ones like sap or nectar (Dell 1985) or man-made ones like jam, honey, and semi-dissolved sugar, e.g., *Christinus marmoratus* — T. F. Houston in Dell 1985; *Heteronotia binoei* — Peters 1968, and *Lepidodactylus lugubris* — Schnee 1901; McCoy 1980. Some species will also eat sugar grains, e.g., *Lepidodactylus lugubris* — Sabath 1981; *Christinus guentheri* — Cogger *et al.* 1983 with the latter species even breaking into sugar bags for their contents. All these substances are presumably energy rich and hence highly attractive. Whether geckos exploit these foods more often than other lizards simply because they are arboreal and more likely to encounter them or whether they have a special predisposition to them is unknown.

It has been said that "unlike the majority of other reptiles, geckos do not drink from an open body of water but obtain their water . . . by lapping from moist surfaces such as leaves and bark wet with rain or dew, or water-soaked rocks" (Cogger 1958). However, other than this account, observations on drinking in Australian geckos are almost nonexistent, and as other, overseas geckos clearly drink from open water, e.g., *Eublepharis* (pers. obs.), further specific observations would be desirable.

Little is known of the specific mating times of geckos, but at least two southern species are known or suspected to mate in the autumn after egg-laying has finished for the season. The sperm are stored in the female's reproductive tract until the following spring when the eggs are ovulated and fertilized. The advantage of this autumn mating, which also occurs in many southern skinks, is thought to derive from the head start it gives the reproductive process and ultimately the young in an environment where the activity season is relatively short. To date, autumn mating is thought to occur in *Christinus marmoratus* (King 1977a) and *Phyllurus platurus* (Green 1973b).

Little is also known about the actual mating behaviour of Australian geckos. Our only information is that the male grasps the female by the skin of the neck in *Hemidactylus frenatus*, *Lepidodactylus lugubris* (Werner 1980), *Nephrurus levis* (Ehmann 1976c) and *Strophurus williamsi* (Bustard 1969c). This is apparently the primitive mating grip for lizards (Böhme and Bischoff 1976).

Most geckos, including all the Australian species are egg layers; the only known live-bearers are the nine New Zealand endemics and one New Caledonian species. As a group, geckos are characterized by a constant clutch size: two in most species and probably primitively, but reduced to one in a few species. Amongst Australian geckos only the members of the *Gehyra variegatus* species group lay only one egg per clutch; all others lay two[1].

An interesting direct consequence of the evolution of a constant clutch size in geckos, has

[1]This constancy of clutch size in geckos may not be as regular as widely believed; see the more than two counts for *Christinus guentheri* and *Strophurus spinigerus* (pp. 72 and 84 respectively).

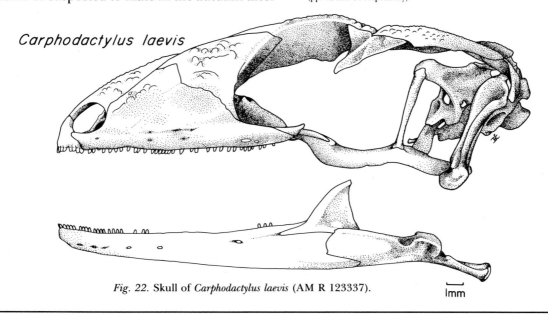

Carphodactylus laevis

Fig. 22. Skull of *Carphodactylus laevis* (AM R 123337).

1mm

◁ *Fig. 21.* Eight diplodactyline gecko species to show the variation in form within the Australian members of the family. Note especially the tail. First row (left to right): *Crenadactylus ocellatus* and *Strophurus spinigerus;* second row: *Diplodactylus ornatus* and *Underwoodisaurus milii;* third row: *Oedura castelnaui* and *Nephrurus wheeleri;* fourth row: *Diplodactylus conspicillatus* and *Phyllurus platurus*. Photos in fourth row: P. R. Rankin.

been the loss of one of the two ways of varying the number of young per season. Lizards with variable clutch sizes can vary both the number of young per clutch (often as a function of female size) and the number of clutches per season. Geckos have lost the former capacity and retain only the latter. Why geckos have "given up" this component of fecundity is not clear. However, other lizards with constant clutch sizes — also one or two — are strongly associated with the tropics, and while this may not as yet help explain why a constant clutch size evolves, it may point to the ecological associations of the first geckos.

Despite its seeming importance, there is relatively little information on clutch frequency in Australian geckos. Some temperate species or populations lay two clutches (e.g., *Gehyra variegata* — Bustard 1968a, 1969b; Dell 1985) whereas others appear to usually lay only one (e.g., southern *Gehyra dubia* — Bustard 1969c [as *G. australis*], *Heteronotia binoei* — Bustard 1968g and *Christinus marmoratus* — King 1977a). *Strophurus* species for which there are adequate sample sizes, often carry yolking follicles in addition to shelled oviducal eggs, indicating that at least two clutches per season may be laid (How *et al.* 1986). The only completely tropical species for which there is information also lays at least two clutches (i.e., northern *Gehyra dubia* — Bustard 1969c, as *G. australis*).

The seasonality of reproduction is fairly well-studied in southern, i.e., temperate, geckos in Australia but much less well-known in northern forms. Southern forms generally reproduce during spring and summer, i.e., during the mid-part of the activity season (e.g., *Gehyra dubia* — as *G. australis*, *G. variegata* — Bustard 1968a, 1969a,c; *Heteronotia binoei* — Bustard 1968g; *Oedura reticulata* — How and Kitchener 1983; *Christinus marmoratus* — King 1977a). No complete cycle is known for northern species, but many, if not most, appear to reproduce at least during the winter and early spring, i.e., the dry season. This is in contrast to most other lizards, which appear to be reproductively quiescent during the winter dry (James and Shine 1985). It has been suggested that geckos with hard, calcareous shelled eggs (p. 59), have diminished hatching success in conditions of high humidity and moisture (Bustard 1968a, 1969b).

Several gecko species reproduce parthenogenetically, that is, the populations are entirely female and reproduction occurs without males. Amongst Australian species this asexual reproductive mode is known in many populations of the composite endemic species *Heteronotia binoei* and in *Lepidodactylus lugubris* (Moritz and King 1985) and in some overseas populations of *Hemidactylus frenatus* and *Nactus pelagicus*. The evolutionary advantages of parthenogenesis are thought to derive from the ability of a single animal to colonize and then rapidly fill disturbed or transient habitats, i.e., those that last only a short time and then give way to a more permanent habitat. In parthenogenesis, it takes only one individual in any reproductive condition to invade a new habitat and ultimately establish a line, but in bisexuals it takes a fertilized female or a pair to invade a habitat with any chance of establishing a population. Furthermore, a parthenogenetic population can grow twice as fast as a bisexual population in which half the individuals (males) are incapable of producing young themselves. This is the benefit of parthenogenesis; the "loss" is a theoretical reduction in genetic variability, the basis of evolutionary change.

Many geckos lay communally, that is, many females lay their eggs in one spot. Communal nests may consist of the eggs of more than one species, and in some species, the nests may be used over several years. To date communal nesting has been reported in the following Australian species: *Christinus guentheri* (Cogger 1971; Cogger *et al.* 1983), *C. marmoratus* (Tubb 1938; Ehmann 1976a; Robinson 1980; Hudson 1981), *Lepidodactylus lugubris* (Schnee 1902; Cagle 1946; McCoy 1980; Gibbons and Zug 1987), *Oedura tryoni* (Milton 1980b) and *Phyllurus platurus* (Green 1973b). The adaptive significance of the behaviour is unknown, and it is not known if the females are attracted to the eggs of preceding females or to the site itself.

In the Australian gecko species that have been studied to date, the age to maturity is generally two or three years (Table 6; actually measured as season cycles, hence the absolute time is less than full years). For animals of comparable size, this appears to be a longer time than faced by certain other lizards such as dragons where the age to maturity is often only one year (p. 13). The reason for this difference is unclear as yet but may reflect the generally overall low temperature regimes under which geckos lead their lives.

Sexual dimorphism in geckos is most obvious in structures located at the base of the tail. Males have a general enlargement of this area, larger cloacal spurs (enlarged knobs on the lateral base of the tail) and one or two pairs of postcloacal bones internally (Fig. 27; some non-Australian geckos lack postcloacal bones). Sexual dimorphism in maximum size shows virtually all the possibilities. Both sexes appear to be approximately equal in size in *Christinus guentheri* on Philip Island (Cogger *et al.* 1983), *Gehyra variegata* (Bustard 1968a), *Hemidactylus frenatus* (Sabath 1981), and

Heteronotia binoei (Bustard 1968g); males are larger than females in *Christinus guentheri* on Nepean I. (Cogger *et al.* 1983), and females are larger than males in *Oedura monilis* (Bustard 1971, as *O. ocellata*), *O. reticulata* How and Kitchener 1983, and *Strophurus* (Bustard 1969c; How *et al.* 1986). Sexual dichromatism does not appear to occur in Australian geckos.

Morphologically distinguishable sex chromosomes are relatively rare in geckos and amongst Australian species have only been reported in *Gehyra purpurascens* (Moritz 1984b) and one population of *Christinus marmoratus* (King and Rofe 1976).

Although most geckos are nocturnal and hence are often active at lower temperatures than most other lizards, e.g., 14-21°C (Table 10), when tested in laboratory thermal gradients, they seem to prefer higher temperatures, sometimes even as high as those in the range preferred by diurnal lizards, i.e., 30-38°C (Table 11). It has also been shown that some geckos thermoregulate in their day retreats to achieve similarly high temperatures. For example, during the day *Gehyra variegata* moves around under the exfoliating bark of dead trees taking advantage of the natural gradient of temperatures arising from having one side of the tree in sun and the other in shadow (Bustard 1967c). Some geckos even bask, e.g., *Strophurus* (p. 84). Geckos also have very high critical thermal maximum temperatures, often as high as many other lizards that are active in full sun. The range of the mean CTMaxs for the six Australian species examined to date is 40.4-43.8°C (Table 12). The large difference between the temperature at which geckos will remain voluntarily active and what they can stand under extreme stress is one of their most remarkable physiological features.

When geckos approach critically high temperatures they open their mouths and pant, like many other lizards. However, they also have an additional behaviour, not seen in other lizards, a rapid vibration or fluttering of the gular area (Heatwole *et al.* 1975). The physiological significance of this behaviour is unclear, but it has systematic importance in apparently being a derived character state amongst lizards.

Geckos shelter by day and many terrestrial species are known to shelter in burrows, specifically: *Diplodactylus conspicillatus* (Bustard 1970d; Pianka and Pianka 1976) *D. maini* (Chapman and Dell 1985), *D. pulcher* (Chapman and Dell 1985); *D. squarrosus* (Kluge 1962b), *D. stenodactylus* (Pianka and Pianka 1976) *Heteronotia binoei* (Pianka and Pianka 1976), *Lucasium damaeum* (Cogger 1969; Bustard 1970d; Peters 1970c; Schmida 1973b; Johnston 1982), *Nephrurus*

laevissimus, N. levis (Pianka and Pianka 1976), *N. stellatus* (Galliford 1978; H. Ehmann, pers. comm.) and *N. vertebralis*, and *Rhynchoedura ornata* (Pianka and Pianka 1976; Storr and Smith 1981); *Underwoodisaurus milii* (Hudson, *et al.* 1981; Sadlier pers. comm.). Usually these burrows have been made by other animals such as spiders or lizards. Only *Lucasium* (Schmida 1973b) and *Nephrurus* (Bustard 1970d; Schmida 1973b; Galliford 1978) appear to dig their own burrows, although there is as yet no first hand account of this for any species. Some of the burrow-inhabiting species are said to plug the burrows or tunnels either with their tails, e.g., *Diplodactylus conspicillatus* (Bustard 1970d) or with dirt, e.g., *Nephrurus levis* (Schmida 1973b) and *N. stellatus* (Galliford 1978; H. Ehmann, pers. comm.). Sealing the burrow of course, would both inhibit the entry of predators and help curtail evaporative water loss. It would be interesting to know if these burrow-inhabiting geckos return to the same burrow every night.

Other geckos seem to shelter or, indeed, to pass most of their lives in hummock grass. The best known species in this regard is *Strophurus elderi* (Bustard 1970d; Storr and Smith 1981) but it is probably also the case with *S. mcmillani, S. michaelseni* (Storr and Harold 1980; Smith and Johnstone 1981), *S. taeniata* (Storr and Hanlon 1980) and northern populations of *Crenadactylus ocellatus* (Storr and Harold 1980; Storr and Hanlon 1980; Smith and Johnstone 1981).

Geckos that live on rocks and trees usually shelter in crevices, but only one arboreal species, *Pseudothecadactylus australis*, appears to inhabit regularly holes in trees (L. Cameron, pers. comm.; Cogger 1975).

Geckos have the best-developed capacity among lizards for producing sound; indeed the name of the family derives from the repeated cry of "gecko" given by a large Asian species often found in houses. Gecko calls are usually given in social encounters or in predator defense, but the nuances of calling are known only for one species in Australia *Hemidactylus frenatus* (Marcellini 1974) which happens to be introduced.

Most Australian geckos that have been observed closely enough to make a fair test have been found to possess a voice. In most species a vocalization is given primarily, if not exclusively, in situations of extreme stress, such as during agressive displays (usually actual fighting, e.g., *Diplodactylus* — Bustard 1965c; *Gehyra* — Bustard 1969a; Frankenberg and Werner 1984) or confrontation with a predator (e.g., *Christinus* — Daniels *et al.* 1986; *Nephrurus asper* — Bustard 1967a; pers. obs.; *Phyllurus platurus* — pers.

obs.). Only a few species appear to actually call in a more "relaxed", social context (e.g., *Hemidactylus* — Marcellini 1971, 1972, 1974, 1977, 1978; pers. obs. and perhaps *Strophurus* — Bustard 1969c). The audiographic structure of the distress call has been studied in *Gehyra variegata* (Frankenberg and Werner 1984), *Hemidactylus frenatus* (Marcellini 1974; Brown 1985), *Oedura lesueurii*, *Phyllurus platurus* and *Underwoodisaurus milii* (Brown 1985) and the social call in *Hemidactylus frenatus* (Marcellini 1974).

Geckos have a characteristic defensive posture which usually involves extending the throat, raising the body on extended limbs, arching the back, and raising the tail and waving it slowly from side to side, e.g., *Heteronotia binoei* (Bustard 1968g; Armstrong 1978); *Nephrurus* spp. (Galliford 1978); *Oedura tryoni*; *Phyllurus platurus* (H. Ehmann, pers. comm.); *Strophurus taeniatus* (Schmida 1973a). Some species also do slow "push ups" from this posture, e.g., *Nephrurus asper* (Longman 1918; Bustard 1967a), *N. levis* (Shea, pers. comm.), *N. stellatus* (H. Ehmann, pers. comm.) and *Phyllurus platurus* (Mebs 1973). Many species also open the mouth and vocalize during the display.

Although a few geckos retain the moveable scaly lower eyelid that is primitive for squamates, most species, including all the Australian ones, have the lower eyelid transformed into a large clear spectacle and fixed in the up position by fusion to the underside of the rudimentary upper eyelid. The function of this spectacle is unclear but if the clues available from other lizards (p. 120) are any indication, it may have evolved to retard evaporative water loss from the moist cornea.

In all Australian geckos the pupil is vertical with "scallops" caused by the multiply-lobed inner edge of the iris. When the iris closes under conditions of bright light, the scallops overlap, leaving a vertical series of pin-hole openings through which the light passes (Fig. 28 and p. 60).

Eye size, as measured by the size of the spectacle, is relatively larger in terrestrial geckos than in climbing geckos. This relationship was first established in overseas geckos (Werner 1969) but has subsequently been found to hold for Australian species (Pianka and Pianka 1976). The relationship is thought to arise from the more limited visual field of an animal living on the ground being compensated for by larger eyes.

A very characteristic and peculiar gecko trait is licking the face, including the eye, with the tongue (Mitchell 1958; Bustard 1963b; Fig. 23).

Fig. 23. Facial tongue wiping in geckos: *Oedura monilis*. Compare with Fig. 44.

This is often done in a very systematic and deliberate way, the upper side of the mouth being licked with the dorsal surface of the tongue and the lower side with the ventral surface. The behaviour usually occurs after some strenuous activity like feeding or being handled. The standard explanation for the trait is cleaning, but why geckos should be so fastidious is not obvious. Another explanation might be to highten the sensitivity of the microscopic hair-like papillae which are abundant on the face through either making them moister or more erect. Although again why geckos should have special requirements in this regard is unclear, as these sensors seem to occur in most, if not all, lizards. Tellingly, this licking or tongue wiping trait is also seen in geckos' closest living relatives, the pygopodids, and in a group sometimes thought to be related to these two families taken together, the New World xantusiids (Greer 1985b).

One of the features in which gecko lineages vary most noticeably amongst themselves is in the morphology of the scales on the underside of the digits (Figs 24, 25). The primitive condition appears to be a single row of narrow scales, often called lamellae, along the length of the digit. Amongst Australian geckos such a digit occurs in *Carphodactylus, Cyrtodactylus, Heteronotia, Nactus, Phyllurus* and *Underwoodisaurus* (Table 1). From this condition various gecko lineages have evolved highly modified scales, often called scansors, integrated into groups called pads. The scales on these pads are transversely enlarged and carry thousands of microscopic, terminally branching, hair-like outgrowths of the scale's surface, called setae (Hill, 1976)[1]. These setae engage microscopic irregularities

---

[1]The setal morphology is inferred from studies done on *Hemidactylus frenatus* and overseas species. A survey of the actual distribution and structure of the setae in other Australian geckos has yet to be carried out.

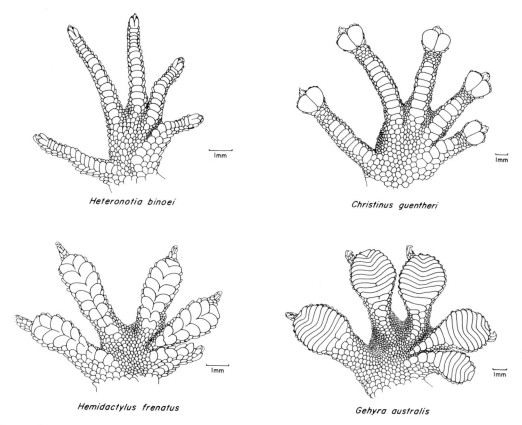

Fig. 24. The rear foot in ventral view in various Australian gekkonine species to show the range of variation in the subfamily and the convergence with the diplodactylines (compare Fig. 25).

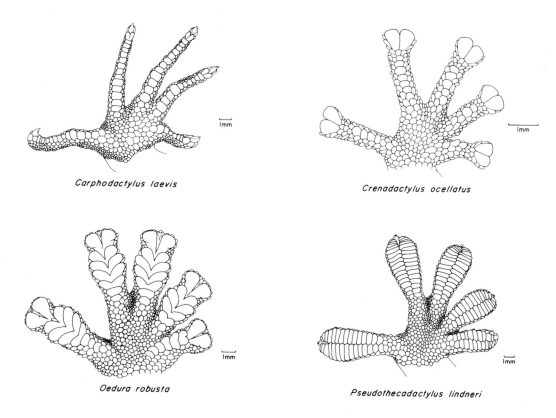

Fig. 25. The rear foot in ventral view in various Australian diplodactyline species to show the range of variation in the subfamily and the convergence with the gekkonines (compare Fig. 24).

upon whatever surface the gecko may be climbing and enable the animals to negotiate not only vertical surfaces but even the undersides of structures, such as ceilings.

The pads themselves may develop over only the basal part of the digit, over only the terminal part, over both the basal and terminal parts separately, or over both basal and terminal parts continually. amongst Australian geckos a basal pad only occurs in *Gehyra*, *Hemidactylus* and *Lepidodactylus*; a terminal pad only occurs in *Christinus*, *Crenadactylus*, *Diplodactylus* and *Rhynchoedura*; a combination basal/terminal pad occurs in *Oedura* and *Strophurus*, and a continous pad in *Pseudothecadactylus* (Table 1).

The species in the *Diplodactylus-Lucasium* group show a range of terminal pad types from well-padded to padless, and it is thought that in this case, but contrary to what is thought for the family as a whole, the evolutionary sequence has been from well-padded to padless with *Lucasium* representing an extreme reduction in scale size to small spines (Kluge 1967b; Russell 1979 (Fig. 38)).

Padless geckos are usually thought of as terrestrial and padded geckos as climbers either on trees or rocks. In general this is true but there are many exceptions. For example, amongst the padless forms some *Cyrtodactylus* (e.g., *C. louisiadensis*), some *Nactus* (e.g., *N. galgajuga*) and all *Phyllurus* are excellent climbers, while amongst the padded forms most *Strophurus* are at least partially terrestrial and even the most well-padded *Diplodactylus* are almost exclusively ground-dwelling.

The astounding adhesive properties of gecko toe pads — or rather their microscopic setae — pose two difficult questions. First, if geckos' toe pads are so good at catching in the small rugosities of the substrate, how can a gecko disengage its toes quickly enough to run, and second, why don't the pads constantly pick up pieces of sand and grit? The answer to the first question lies, at least partly, in the way a gecko disengages its toes. It does this not by carrying through in the same direction as the pad was applied but, instead, by reversing and removing it in the direction opposite to that in which it was initially applied. This involves a distinct curling back of the top part of the ends of the digits in a way that would be quite painful if not impossible in most vertebrates. This curling back is called hyperextension and can be seen clearly when a gecko with toe pads is walking slowly. Geckos without toe pads do not hyperextend; this is true both of species that are primitively and secondarily (e.g., *Lucasium* pers. obs.) padless. Parenthetically it may be noted that some gecko species are also able to curl the

digits around perches of very small diameter, e.g., the hummock grass inhabiting *Strophurus*, but whether this flexion is extraordinary (i.e., hyper) is unknown. The answer to the second question remains a mystery; we don't know why geckos with pads fail to pick up pieces of the substrate in their pads.

The similar foot structures (pad types) and functions (hyperextension) in only distantly related gecko lineages, provides one of the most remarkable cases of parallel evolution in lizards (Russell 1976, 1979). Preliminary comparisons have been carried out on Australian species (Russell 1979) but additional morphological and functional analysis amongst Australian species would be useful.

Geckos' tails, like their feet, are also variable in size, shape (Fig. 21) and internal morphology and have a correspondingly wide variety of functions. Predator escape is probably the most general function. Most geckos wave the tail slowly back and forth in a slightly raised s-shaped position when agitated, and this may serve to distract a predator's attention to the tail and away from the more vital head and body. Most geckos, like many other lizards, are also able to "drop" or autotomize the tail when attacked by a predator.

Like other biological processes the ease of tail loss is largely temperature dependent: it occurs more readily at high temperatures and less readily at low temperatures. However, it has been discovered in several, mainly Australian, gecko species that at very low temperatures, 4-10°C, the tail is again readily lost. This effect, called cold-enhanced autotomy and has been observed in *Christinus marmoratus* — Daniels 1984, *Gehyra variegata*, *Heteronotia binoei*, and *Oedura monilis* (as *O. ocellata*) — Bustard 1967e, 1968d). The adaptive significance of the effect is thought to be a kind of "last chance" mechanism to deflect a predator's attention at a time when the gecko is virtually incapacitated by cold. In *Gehyra variegata* the effect is further enhanced by virtually the whole tail being shed instead of just a part as is the case at higher temperatures. The most detailed observations yet available on cold-enhanced autotomy are for the Australian gecko, *Christinus marmoratus*, and are discussed further under that species (p. 72).

It is sometimes asked whether geckos can drop their tails spontaneously, that is, without being grasped. There is no better answer to this question than the following observation for *Gehyra variegata* of H. M. Hale quoted in Waite (1929). "Some years ago, in company with Mr. Tindale, I was camped in a deserted tin-roofed hut at Owieandana, in the northern part of the Flinders Ranges. Numbers of little geckoes of this species were walking about the inside of the roof, but were difficult to capture, for as we approached them they retreated into crevices. One of us suggested that if a blank shot-gun cartridge were fired in the hut the shock would no doubt cause the lizards to drop to the ground. The gun was accordingly fired and a sparse shower of small bodies came tumbling down. 'That's done the trick,' said my companion, but when we came to collect the spoil we found that the fallen objects were merely fat little tails, which the geckoes had thrown off in their alarm! We noticed later that most of the geckoes in the hut were tailless after this disturbance." Cold *G. variegata* have also been observed to drop the tail without its having been touched (Bustard 1967e).

Another very distinctive defensive function of the tail occurs in *Strophurus*. Species of this genus eject a sticky noxious substance from the dorsal surface of the tail which probably acts as a repellant (Bourne 1932; p. 82). Yet other tails, notably the leaf-tails of certain *Phyllurus*, may act as disruptive camouflage.

The knob tail of the genus *Nephrurus* is one of the most unusual tail types amongst geckos, and its function is one of herpetology's enduring mysteries. The tails have the shape of a short, flat carrot, ending in a small ball or knob. Like many other lizards, when agitated, knob-tails rapidily vibrate the tail tip in the substrate (pers. obs.) and if done in dry litter this can make a buzzing sound which could act to distract predators. Perhaps the knob tail makes an especially efficient kind of predator-distracting vibrator.

Tails are also used in climbing by certain species. The two most arboreal groups, i.e., *Strophurus* and *Pseudothecadactylus*, have prehensile tails, and one of these, *Pseudothecadactylus*, has a toe-pad like structure on the tip of its tail.

Another general function of gecko tails is energy storage. Most geckos store fat in the tail and in some the capacity for storage and hence the shape of the tail (fat vs thin) is remarkable. For example, a well-fed *Lucasium damaeum* which has a tapering tail presumably with little capacity for storage will lose condition in three weeks if not fed (Bustard 1965c, 1967b), whereas well-fed *Oedura lesueurii*, *O. marmorata* and *O. monilis* which have fat tails will live for 9-12 months if they have water (Cogger 1957).

Tail shape can vary greatly even amongst closely related species. For example, some species of *Nephrurus*, *Oedura* and *Phyllurus* have thin, gradually tapering tails even when well-fed, e.g., *Nephrurus asper*, *Oedura gracilis* (King 1984c) and some populations of *Phyllurus*

caudiannulatus (Covacevich 1975), while their congeners have thick, rapidly tapering tails. Could this variation be related to the variability of the environment or the habitat within which the various species live? Do thin-tailed species live in situations where food is regular and predictable and hence there is no need to store energy against hard times; conversely, do fat-tailed species inhabit unpredictable environments in which energy needs to be stored against hard times (Cogger 1957)?

Geckos are renowned for their ability to change colour in response to temperature and mood, but not all species have the same capacity. In general, geckos lighten when warm and undisturbed, and darken when cool or frightened (e.g., *Gehyra variegata* — Rieppel 1971). Amongst Australian species, members of the *Gehyra australis* species group and *Hemidactylus frenatus* show the greatest range, from dark grey or brown to an almost ghost-like off-white.

THE LINEAGES OF AUSTRALIAN GECKOS

The family Gekkonidae can be divided into three subfamilies (Fig. 26). Two of these occur in Australia: the Gekkoninae (in my view including the New World Sphaerodactylinae of other authors), which has almost the same distribution as the family as a whole and the Diplodactylinae which is restricted to Australia, New Caledonia and New Zealand. The third subfamily is the relatively primitive Eublepharinae which is scattered disjunctly around the world generally north of the equator (Kluge 1967a) and approaches Australia in the form of the Indonesian genus *Aleuroscalabotes*, a group superficially similar to the Australian diplodactyline genus *Carphodactylus*.

The Gekkoninae are a well-diagnosed group. For example, they have the most anterior bones of the skull, the premaxillae, fused throughout life, in contrast to the more primitive condition, seen in the Diplodactylinae, of the premaxillae paired. The Gekkoninae also have enlarged endolymphatic sacs (Fig. 27), basically post-cranial extensions of a calcium-filled system that is usually confined to the skull of most other lower vertebrates. The sacs are paired and when full, are often noticeable as two whitish bulges just below the skin on either side of the neck. The function of these sacs is unknown, but because they are larger in females than males, they are thought to have something to do with egg shell production (Bustard 1968b).

The gekkonines also have a hard, calcareous egg shell which is relatively impermeable (Bustard 1968b; Dunson 1982). The egg is soft and pliable when laid but dries hard in a matter of minutes or hours (Cagle 1946). Most other lizards,

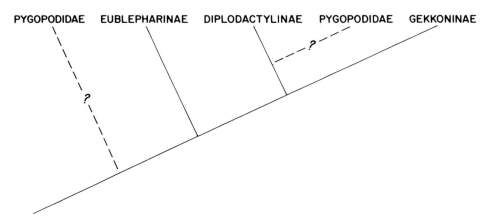

PYGOPODIDAE   EUBLEPHARINAE   DIPLODACTYLINAE   PYGOPODIDAE   GEKKONINAE

Fig. 26. Schematic diagram of the probable relationships of the three subfamilies of the family Gekkonidae and of the family Pygopodidae. The gecko subfamily Gekkoninae is taken here to include also a fourth subfamily of geckos often recognized, the Sphaerodactylinae.

including the diplodactylines, have a relatively pliable and permeable egg shell which has as one of its functions, the ready uptake of water from moist substrates but also has as one of its hazards the equally rapid loss of water in dry situations. Because the egg shells of gekkonines are relatively impermeable, they can withstand exposure to both dry air and salt water (Brown and Alcala 1957). The first fact explains why gekkonine eggs are often found under only the most superficial cover (a rock or piece of bark or in a shallow hollow or crevice) and the second, perhaps, why gekkonines have been so successful in colonizing the Pacific. Gekkonines also appear to have a more spherical egg than other geckos (Table 8), a shape that would help reduce water loss (low surface area to volume relationship).

In contrast to the well-diagnosed Gekkoninae, the Diplodactylinae are characterized by only one special feature: more than the usual number of scleral ossicles, the thin overlapping bones that surround and help support the eyeball in certain fishes, reptiles and birds. Diplodactylines have 21-40 ossicles whereas gekkonines have, at least primitively, only 14, the probable primitive number for lizards in general.

Within Australia the two subfamilies are also distinguished by additional features and characteristics, the phylogenetic significance of which for the subfamilies as a whole have yet to be determined. For example, the gekkonine pupil usually has only three scallops or lobes on each edge and these overlap upon closure of the pupil to give four pin holes; in contrast, the diplodactyline pupil usually has four or more (usually many more) scallops which give five or more pin holes upon closure (pers. obs.; Fig. 28).

A second difference between the two subfamilies relates to chromosome number: most gekkonines have a relative high number while diplodactylines have a low number. All the Australian gekkonine genera, except *Christinus*, have at least one form with 40 or more chromosomes whereas no diplodactylines recorded to date has more than 38 (Table 4). Gekkonines also seem to be characterized by a very high frequency of inter and intraspecific karyotypic variablity and diplodactylines by a very low frequency (Table 4). This could be fortuitous based on small samples, but if real, it could mean that in gekkonines karyotypic change is intimately related to the speciation process itself[1].

A third difference between the subfamilies relates to where they break the tail. Both gekkonines and diplodactylines have the majority of their tail breaks concentrated in the base of the tail, but gekkonines have more breaks in the rest of the tail than do diplodactylines (Table 3). The reasons for these different distributions are unknown but may relate to the all or nothing usefulness of the highly specialized tails of the diplodactylines.

There is also an intriguing possibility that the relative clutch mass is lower in gekkonines than in diplodactylines. For example, according to the most complete available data set, RCM(3) in two Australian gekkonine species ranged 9.9-10.6 ($\bar{X}$ = 10.3; the value for *Gehyra variegata* which lays only one egg has been doubled to bring it in line with the other species which lay two eggs) whereas nine diplodactyline species ranged 10.9-19.6 ($\bar{X}$ = 14.9) (Pianka 1986: appendix F; see also Table 7)[2]. The significance

[1] In addition to detailing chromosome differences between different populations and species in Australian geckos, much effort has gone into trying to reconstruct the possible evolution of the karyotype in different groups, especially in the most variable and best-studied genus, *Gehyra*. However, most of these attempts are based on either a very weak primary assumption about the most primitive karyotype, i.e., the most common karyotype within the group is the primitive one — King 1979, 1982a, 1983a, 1985, or when outgroups are used to determine the primitive karyotype, the choice is based on geographic instead of phylogenetic criteria — Moritz 1986. Furthermore, in some cases such as *Gehyra*, we can not even be sure that we are dealing with a lineage as there is no available diagnosis for the group.

[2] The differences in RCM between gekkonines and diplodactylines may be more general than just for the Australian species. The same date set, which also includes five African gekkonines, shows that the reproductive effort in all seven gekkonines ranges 8.2-17.6 ($\bar{X}$ = 11.5) versus the nine diplodactylines at 10.9-19.6 ($\bar{X}$ = 14.9; t = 2.81*).

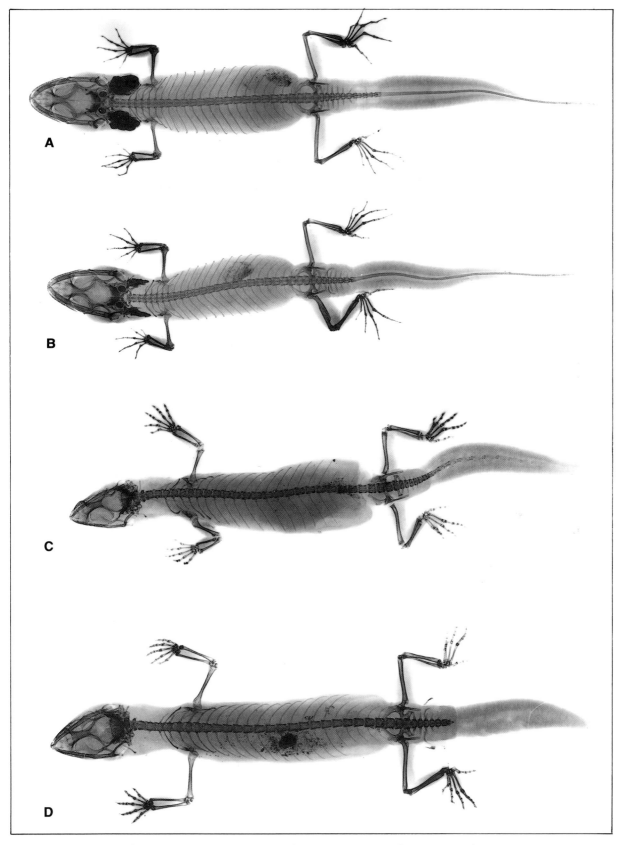

Fig. 27. Prints of X-rays of a gekkonine (A-B: *Heteronotia binoei* — AM R 113081, ♀ and AM R 113083, ♂) and a diplodactyline (C-D: *Rhynchoedura ornata* — AM R 90893, ♀ and AM R 114111, ♂) gecko to show the presence of the endolymphatic sacs in the former and their absence in the latter. Note that the sacs are more well-developed in the female (A) than in the male (B) gekkonine. Note also the presence of a distinct cartilaginous rod in the regenerated tail of the gekkonine but not in the diplodactyline (male only with regenerated tail) and, in the males, a single pair of postcloacal bones in the gekkonine but two pairs in the diplodactyline.

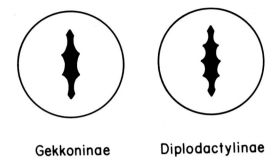

Gekkoninae   Diplodactylinae

Fig. 28. The two basic pupil types in Australian geckos: with three lobes closing down to four pinholes which is typical of gekkonines and the diplodactyline *Crenadactylus,* and with four or more lobes closing down to five or more pinholes which is typical of diplodactylines (except *Crenadactylus*).

of this difference, if it is real, and its possible relationship to the differences in egg shape and shell texture between the two families remains to be investigated.

Within the Gekkoninae there is a large lineage diagnosed on the basis of a single derived character state in the hyoid apparatus: the loss of the second ceratobranchial (Figs 29, 30). This lineage, called the tribe Gekkonini, comprises a large number of the genera and has a distribution almost as large as the subfamily itself (Kluge 1983). Five of the seven Australian gekkonine genera belong to this lineage: *Cyrtodactylus, Gehyra, Hemidactylus, Heteronotia* and *Lepido-*

dactylus. Unfortunately relationships within this group are unclear, except possibly for *Lepidodactylus* (p. 69). There is also a slightly disquieting discrepancy in the distribution of the diagnostic character state for the Gekkonini in *Gehyra.* Some individuals of this reputed member of the tribe have well-developed second ceratobranchials (Mitchell 1965b). The "left over gekkonines", an assemblage called the Ptyodactylini, include only two Australian genera: *Christinus* and *Nactus.* Only the relationships of the former genus seem clear (p. 71).

The two most primitive Australian genera in the Gekkonini are *Heteronotia* and *Cyrtodactylus.* Both genera retain, in the skull, paired vomers and a perforate stapes (Fig. 31), and, in the foot, padless toes and the terminal phalange on the first toe of both front and rear feet differentiated as a claw.

Heteronotia is one of the most widespread and abundant gecko genera in Australia. It occurs throughout the continent except for the far southwestern and the southeastern corners and can be incredibly abundant, especially in disturbed habitats such as rubbish tips. Indeed, when collecting in tips, one sometimes gets the impression that there is at least one *Heteronotia* under every piece of ground cover, and it is axiomatic amongst reptile collectors that when nothing else is stirring one can always count on uncovering *Heteronotia.*

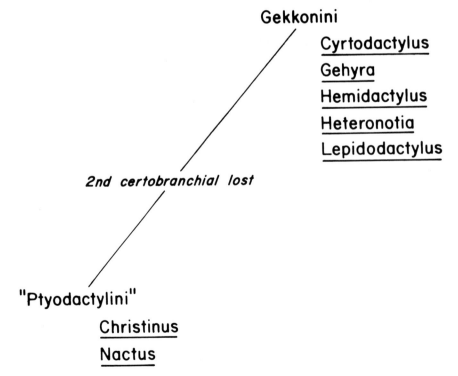

Gekkonini

Cyrtodactylus

Gehyra

Hemidactylus

Heteronotia

Lepidodactylus

2nd certobranchial lost

"Ptyodactylini"

Christinus

Nactus

Fig. 29. The relationships of the two groups in the subfamily Gekkonidae, one of which, the Gekkonini, is a lineage and the other of which, the "Ptyodactylini", is an assemblage. The Australian genera are listed for each group. For details see Kluge 1983.

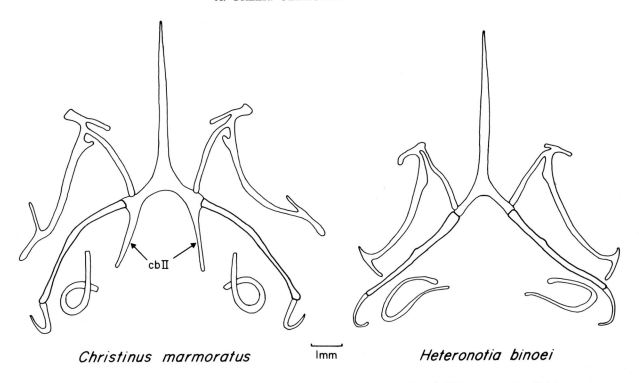

Christinus marmoratus   Imm   *Heteronotia binoei*

Fig. 30. Hyoid apparatus in the two groups of the subfamily Gekkoninae: the "Ptyodactylini" (represented by *Christinus marmoratus* — AM R 111052) which retain the second certobranchial (cb II) and the Gekkonini (represented by *Heteronotia binoei* — R 122999) which have lost it.

Traditionally only two species of *Heteronotia* have been recognized; the distinctly banded *H. spelea*, said to be associated with caves and crevices in the rocky country of north-west Australia and the variably patterned *H. binoei* thought to be widespread throughout most of arid, semiarid and seasonally dry Australia. There was little to disturb this tidy picture other than the occasional individual of *H. binoei* that approached *H. spelea* in its degree of banding or the occasional "out of range" *H. spelea*, which amounts to the same thing stated differently. However, this picture has recently changed into something slightly more complicated, and therefore more interesting (Moritz 1984a; Moritz and King 1985).

In general the current understanding is as follows. There are two chromosomally distinct, sexually reproducing "forms" of "*H. binoei*". Each form is widely distributed and itself consists of geographical delimited, chromosomally recognizable populations. Both forms overlap broadly in range, and in at least one area they have been found together without evidence of interbreeding, hence they are almost certainly distinct species. Within and near one area of the overlapping ranges of these two species there are several all female, parthenogenetically reproducing populations which, although now reproductively distinct from the two parental species, appear to have each arisen from them through hybridization. These populations are triploids, carrying two sets of chromosomes from one parental

species and one set from the other. In most, but not all, cases the chromosome sets can be related to one of the chromosomally distinct populations within each parent species. Furthermore, within these asexually producing populations there appear to be very local populations, also asexually reproducing, of course, which are identifiable on the basis of what appear to be specific, locally derived chromosome mutations. The biological significance of all this cytogenetic variation remains to be revealed. But one thing is certain, taxonomically it is a waking nightmare.

One of the widespread sexually reproducing forms of *Heteronotia binoei* has been extensively studied by Bustard (1968g) at a study site in the Pilliga Scrub, and virtually all our knowledge of the natural history of the "species" comes from the population at this one site. However, before describing Bustard's results, it may be useful to describe briefly the site itself for not only did Bustard study *Heteronotia* there but four other gecko species (*Diplodactylus williamsi*, *Gehyra dubia* [as *G. australis*], *G. variegata* and *Oedura monilis* [as *O. ocellata*] and one skink species (*Egernia striolata*) as well.

The Pilliga Scrub is a vast flat sandy area in northern New South Wales, covered by an open forest comprised of cypress pine, eucalypts and she-oaks with a shrub understorey of acacias and a groundcover of grass. The climate is continental with generally cool nights, even during

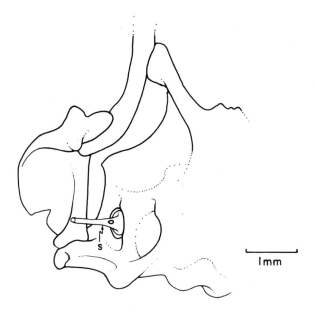

Hemidactylus frenatus

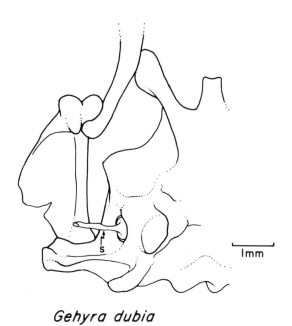

Gehyra dubia

Fig. 31. The two types of ear bones (stapes) in geckos: the primitive perforate condition (*Hemidactylus frenatus* — AM R 123990) and the derived imperforate condition (*Gehyra dubia* — AM R 123989). In the former the facial artery passes through the foramen and in the latter it passes anteriorly.

most of the summer, and warm to hot days. Rainfall is highly variable (range 305-1090 mm) but averages about 580 mm.

At the time of Bustard's studies (years 1963-1965) forestry activities had resulted in an "unnaturally" large number of lizard micro-habitats. For example, the exploitation of the cypress pines had left many dead stumps half a meter to a meter in height and several ring-barked whole trees. In addition a few trees had

also been killed by lightning. Dead cypress pine weathers initially by the bark separating slightly from the bole, and the resulting crevice between bark and bole (a semi-stable condition that may last for several years) is ideal arboreal lizard microhabitat. Furthermore the practice of felling, barking and cutting up eucalypts on the site had left many scattered piles of bark and board which provided ideal ground lizard micro-habitat. There were thus many more lizard micro-habitats than would be found in a virgin section of forest.

With regard to *Heteronotia binoei* specifically, Bustard found that the species occurred only under or amongst piles of bark and board on the ground or at the base of cypress pine stumps. Eggs and individuals of all ages occurred in and under the bark and boards while only about two-fifths of the adult population occurred at the base of stumps. For this reason the latter micro-habitat was considered to be only marginally suitable.

The piles of bark and board usually contained only one male — unless the pile was large, several usually well-spaced females and a number of young. This suggests that adult males are very intolerant of each other (a supposition supported by the laboratory observation that only sexually mature males display to each other and fight) but tolerant of females; adult females are relatively tolerant of each other and of males, and adults of both sexes are very tolerant of young. Adult tolerance of young appears to be widespread in lizards, and although it is easy to understand its adaptive significance, it remains to be learned how the adults discriminate between the young of their species and similar-sized prey.

During summer, activity begins shortly after dusk and lasts for two to three hours, and thermoregulation occurs during the day by changes of position within the microhabitat. Some activity may occur during the winter, as the only specimen collected during winter (June) had been feeding.

Prey consists of beetles, spiders, grasshoppers, cockroaches, scorpions, true bugs and lepidopterans, but there was no indication of specific prey selection.

Reproduction occurs in late spring and summer. Gravid females occur between October and December, but actual egg laying is confined to late November through December. Two eggs comprise a clutch. A female usually lays only one clutch per season but sometimes two.

All eggs are laid under the bark/board piles which means that the females living on stumps (nearly 40 per cent of the adult female population) must leave their retreats and move to the piles to lay. Clutches are usually laid singly, occasionally together with another clutch or beside the eggs of *Gehyra variegata* (they differ from the *Gehyra* eggs in being slightly smaller).

The two egg clutch of a north Queensland *Heteronotia binoei* comprised 30% of the female's weight (RCM 1) and the two eggs of another north Queensland female took 47-48 days to hatch at 25°C (comparable data were unavailable for Pilliga Scrub animals).

The eggs hatch between the end of February and late March. Growth is variable depending on when the eggs hatch and the amount of warm weather available. Most animals breed in the second or third season after hatching. The hatchlings are about 18 mm SVL, most breeding females 44-47 mm and the largest individuals for both sexes 49 mm.

The only other padless gecko amongst the Australian Gekkonini is *Cyrtodactylus louisiadensis*, a large, spectacularly cross-banded species that is widespread in New Guinea and the Solomon Islands and occurs in Australia only in eastern Cape York Peninsula as far south as the Atherton Tableland. In contrast to *Heteronotia, Cyrtodactylus louisiadensis* is a strong climber. Little is known of the species, other than in Australia at least, it is often found under bridges and in large culverts. In the Solomon Islands it is said to prey on insects, spiders and small geckos (McCoy 1980). Two Australian females collected in December contained oviducal eggs, indicating that the species is reproductive at least during the first part of the wet season.

Cyrtodactylus is a large and widespread genus and is almost certainly an assemblage of groups more closely related to other geckos than to one another (Kluge 1983). The relationships of *C. louisiadensis* itself are uncertain but may lie with some of the New Guinea *Cyrtodactylus* (Brown and Parker 1973; Brown and McCoy 1980). The possibility should also be considered that *C. louisiadensis* is not too distantly related to *Heteronotia*.

Like *Heteronotia* and *Cyrtodactylus, Hemidactylus* also retains the primitive features of paired vomers, perforated stapes and claws on the first toes of the front and rear feet, but in contrast to these genera, it has developed toe pads of the basal type. The genus is also characterized by very small, antepenultimate (next to the last but one) phalange in digits 3-4 of the front foot (Fig. 32) and digits 3-5 of the rear foot. The phalange is so small in fact that it was overlooked in the early literature (Steiner and Anders 1946; Russell 1977).

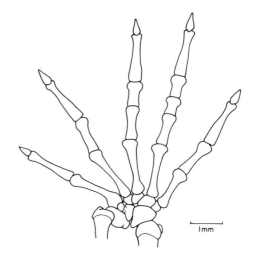

Christinus marmoratus

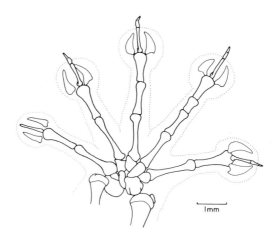

Gehyra punctata

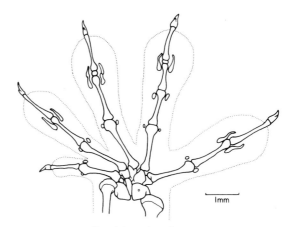

Hemidactylus frenatus

Fig. 32. Right manus of *Christinus marmoratus* (AM R 111052), *Gehyra punctata* (AM R 117608) and *Hemidactylus frenatus* (AM R 117107) to show the generally primitive gekkonine condition of the first in comparison to the derived condition in the last two. Note in *Gehyra* the dedifferentiated, elongate first phalange in the first digit (a claw in other forms), and the thinned pentultimate phalange in the other digits; in *Hemidactylus* the much reduced antepentultimate phalange in digits 3-4, and in both genera the cartilaginous pad supports of each digit.

Hemidactylus is a large and widespread genus. It comprises approximately 70 species and ranges from Africa to southern Asia, thence east into the Indonesian Archipelago and the islands of the south-west Pacific and west into tropical America. Only one species, *H. frenatus,* occurs in Australia and then only in and around active human settlements in the north and north-east (Fyfe 1981a; Wright 1982). There is some geographic variation in the karyotype of the species, the taxonomic significance of which has yet to be determined: Australian and Vietnamese populations are karyotypically similar, but both differ substantially from two Formosan populations (King 1978; Darevsky *et al.* 1984).

Within the areas of human settlements both in Australia and elsewhere, *Hemidactylus frenatus* appears to be a formidable colonizer having recently expanded its range into such areas as the Solomon Islands, Hawaii, Mexico, Florida, Texas, the Mediterranean, Cocos (Keeling) and Christmas Island, in some places displacing other human-associated gecko species, e.g., *Gehyra australis* and *Oedura rhombifer* in Australia and *Hemidactylus garnoti* and *Lepidodactylus lugubris* in Hawaii (Oliver and Shaw 1953; Hunsaker and Breese 1967; McCoy and Busack 1970). The basis of *H. frenatus'* ability to replace other house geckos is unknown. In Australia the species' large numbers, i.e., up to 4-6 around one light, and characteristic "geck, geck" call repeated several times (pers. obs.) make it a conspicuous part of the night life in certain settlements of the tropical north.

Certain aspects of the biology of *Hemidactylus frenatus* are well known, but mostly from study of overseas populations. For example, at a locality in Mexico, the number of geckos leaving their day-time retreats and appearing on the dimly lighted walls of buildings increases during spring, in a pattern closely following the variable, but steadily increasing, temperatures. On a daily basis, these same geckos appeared rapidly as the day ended, maintained their numbers throughout the night and rapidly disappeared as day broke (Marcellini 1971). A similar survey in the Philippines showed that the geckos (along with a second species from which it could not be distinguished) followed a similar diel cycle appearing shortly before sunset, the numbers peaking in the early morning hours and declining before sunrise (Feder and Feder 1981). Observations on a population in southern India found that activity began around dusk, was most evident between 2000-2400 hrs and declined after midnight (Sahi 1980).

A laboratory study of activity in New Guinea *Hemidactylus frenatus* showed that actual physical activity, that is, how much the geckos moved, increased rapidly in the late afternnon (i.e., prior to their emergence time if they maintained the schedule of the Mexican and Philippine populations), peaked just after sunset (when they would have just emerged from their retreats) and then declined steadily throughout the rest of the night (Bustard 1970c). Around Port Moresby, *Hemidactylus frenatus* is said to be the only gecko species which is active during the day as well as the night, but no details are given of this diurnal activity (Pernetta and Black 1983).

While out on the walls, the Mexican population showed a very close correspondence between body temperatures and both air and substrate temperatures. The animals also seemed to prefer the warmer and calmer parts of the buildings, i.e., ceilings and upper parts of walls, and stairwells and corners (Marcellini 1976).

Vocal behaviour was also studied in the Mexican population (Marcellini 1972, 1974, 1977, 1978). The species has three calls: a multiple chirp call, consisting of a series of 5-15 "gecks", given at a rate of 3.1-5.8/second, the entire call lasting 1.0-3.7 seconds; a single chirp call, consisting of a single "geck"; and a churr call, "an extremely rapid series of short chirps similar in sound to the rattle of a high speed teletype machine".

The multiple chirp call is by far the most common (this is the one commonly heard around settlements) and seems to be given by animals of either sex larger than 45 mm, but most frequently by "aggressive males". The call has been noted after animals first emerge from their retreats and before they begin feeding, before and after aggressive encounters between males, by males before courtship and after mating, after feeding, after defecation — in sum, before or after almost any actual or anticipated change of state. The call can also be given for no apparent reason. The only functional effect demonstrated so far for this call is that, in the laboratory, males tend to move away from the source of a played-back, taped call, whereas females show no consistent response. The rate of the multiple chirp call, i.e., the "gecks" per second, increases directly with temperature, as one expects with most biological processes.

The single chirp call is given under stress, e.g., when a gecko of either sex is grasped by a predator or one male is grasped by another. It is given by adults, subadults and large juveniles of both sexes.

The churr call is given only by adult males and only during high intensity aggressive encounters, often just prior to an actual attack.

The eggs of *Hemidactylus frenatus* when laid on vertical surfaces are very sticky (Petzold 1965) whereas those laid on horizontal surfaces are not (Church 1962); this implies the female has some control over the "glue" responsible for the stickiness. In the Darwin area, egg-laying has been recorded in mid-March (Husband 1980b), and females with enlarged ovarian eggs have been taken in June and early September (pers. obs.). These dates represent the period from the end of the wet season to the late dry. The population on the north-east Indian Ocean island Cocos Keeling, which is about the same latitude as Darwin, is reproductive in May (Cogger *et al.* 1983) and the population at Bandung, western Java, about 5½° closer to the equator than Darwin is reproductively active year around (Church 1962). It would be interesting to know the length of the breeding season in northern Australia.

The remaining two genera of the Gekkonini in Australia, *Gehyra* and *Lepidodactylus*, are members of a larger group of basically south-west Pacific genera which is characterized by the loss of the stapedial foramen (Fig. 31) and by the dedifferentiation of the last phalange in the first digit of each foot into a long thin rod of bone instead of a claw as in most other geckos (Fig. 32). Most of the genera in the group, e.g., *Gehyra, Gekko, Lepidodactylus* and *Pseudogekko*, are widespread on the islands of the south-west Pacific and hence, presumably excellent water crossers.

Gehyra is centered over south-east Asia, the islands of the south-west Pacific and Australia, but one of its species, *G. mutilata*, has spread across the Pacific as a stowaway in human cargo. There are 29 species in the genus, 15 of which occur in Australia, 14 exclusively. All members of this genus live off the ground, on trees, rocks, termite mounds or buildings. Although the genus *Gehyra* has long been recognized as a taxonomic entity, there is as yet no diagnosis for it. One of its most distinctive features — a long thin ultimate (digit 1) or pentultimate (digits 2-5) phalange (Fig. 32) — occurs in certain other gekkonines, where it may be indicative of a lineage, and it also even occurs in one diplodactyline, *Pseudothecadactylus*.

The Australian *Gehyra* species can be divided into two groups on the basis of osteology and clutch size: the *G. australis* group with only two ribs in contact with the sternum and a clutch size of two and the *G. variegata* group with three sternal ribs and a clutch size of one (Mitchell 1965b). Also, the species in the former group generally attain a larger size than those in the latter and generally have at least some if not all the subdigital lamellae single instead of all completely divided.

The *Gehyra australis* group comprises six described species in Australia and an as yet undetermined number throughout the rest of the genus' distribution. However, several undescribed species are known in Australia, having been recently detected chromosomally. The named species are *G. australis, G. borroloola, G. catenata, G. dubia, G. occidentalis, G. pamela, G. robusta* and *G. xenopus*. Members of this group are almost invariably found on either trees or rocks and are generally quite specific for one perch or another (King 1984a). For example, *G. catenata* and *G. dubia* have only been found on trees and *G. booroloola, G. pamela, G. robusta* and *G. xenopus* on rocks. *G. australis* shows geographic variation in its choice of substrate; in the Top End of the Northern Territory it occurs only on trees, but further west it occurs on both trees and rocks. The reason for this variation is unknown but possibly relates to some sort of interaction with the saxicolous *G. pamela* in the Northern Territory. Two species of this group, *G. australis* and *G. dubia*, have also moved on to buildings and can often be seen on the walls of houses after dark (Limpus 1982). When thus seen they are usually in their extreme pale colour form.

The *Gehyra variegata* group comprises eight described species in Australia and an undetermined number throughout the rest of the genus' distribution, and, as in the case of the *G. australis* group, several additional Australian forms recently detected chromosomally await taxonomic treatment. The described Australian species are *G. minuta, G. montium, G. nana, G. pilbara, G. punctata, G. purpurascens* and *G. variegata*. Most members of the *variegata* group show the same species-specific fidelity to substrate type as do members of the *G. australis* group (Bustard 1965a-b; King 1981b, 1984a-b). For example, *G. purpurascens* is a tree dweller and *G. minuta, G. montium, G. nana, G. punctata*, rock dwellers. Two species, *G. pilbara* and *G. variegata*, are said to occur on both trees and rocks, but given the substrate specificities of the known species and the tradition of sibling species in *Gehyra*, one has to wonder if the two generalists may not themselves still be composite species.

One of the best-studied geckos in the world is *Gehyra variegata* and most of what we know comes from Bustard's work in the Pilliga scrub (p. 63). Bustard (1969a) found that most *Gehyra variegata* occurred on the same cypress pine stump or tree during his study period. This observation suggests that once established on a home site adults rarely move. It was also found that males only very rarely occurred together on a home stump or tree, but a male with two or three females was not uncommon. This indicates that males may defend their areas against other males but tolerate females, and that females themselves are relatively, but not completely, tolerant of each other (Bustard 1970b). Juveniles occur commonly on stumps occupied by adults and hence must be tolerated by both sexes.

During the warmer months (October to April) the animals become active and start feeding shortly after dusk. They remain active for two or three hours or until the temperature drops to about 18°C. As digestion is temperature dependent and the lizards have no way to raise their temperature during the night, they must wait until morning when the sun begins to heat their stump or tree to find a suitable warm spot beneath the bark to raise the temperature and begin digesting the previous night's meal. As the day progresses and the sun's rays strike different parts of the tree, the geckos can regulate their body temperatures by moving in the crevice between the bark and bole of their stump or tree. During the winter (May to September) activity is much curtailed and no feeding occurs; during very cold periods the geckos presumably retreat deep within their crevices, although a true hibernation does not seem to occur.

The prey of *Gehyra variegata* consists mainly of beetles (adults), spiders, termites, grasshoppers, cockroaches and ants. All these animals except the grasshoppers occur in the same crevices as the geckos. The grasshoppers may be taken at night outside the crevices while they are sleeping. Like many other geckos, the *Gehyra* also eat their shed skin, a habit that presumably serves to recycle the protein of which the slough is largely comprised. Although *G. variegata* feeds throughout the warmer months, it is most active in the spring and early summer. Presumably this is to replenish reserves run down over winter and to prepare for reproduction.

Reproduction occurs in late spring and summer. Gravid females occur between October and mid-January and may be readily identified due to their translucent belly skin which allows the single egg to show through. Although only one egg is laid at a time, two eggs are laid each season, the first in late November or early December and the second in the first half of January. The single egg of *Gehyra variegata* weighs 0.38-0.58 g ($\bar{X}$ = 0.48 g, N = 20) and comprises between 20 and 30% of the female's weight (RCM 1). To judge from the seasonal variation in the state of the gonads and accessory structures in males, mating and fertilization probably occur during the laying season of the females, i.e., late November to early January.

Females only rarely lay their eggs in their home stump or tree. Instead, they deposit them on the ground beneath piles of bark which may be 15-20 meters from the home site. The females obviously run considerable risk to lay in these sites and the obvious question is why. Part of the answer may be the more equable temperatures under the piles. Microenvironmental temperatures under the bark at the home stump or tree can vary as much as 20°C, but they vary only 7°C under the bark piles.

The females also often lay their eggs together with two to seven other females and often in close association with the eggs of the common ground gecko *Heteronotia binoei*.

Hatchlings from the first laying begin appearing in February, those from the second in late March or later. At 25°C incubation in the laboratory takes 61-79 days. There is evidence from both the field and the laboratory that damp or wet conditions can kill developing eggs and reduce hatching success.

Growth in *Gehyra variegata* like most other physiological processes in reptiles, is temperature dependent, hence fast in warm seasons and slow or nonexistent in cool periods. Overall it appears to take nearly two years to reach the size of sexual maturity, but because this is achieved in the late summer or autumn, reproduction does not actually occur until spring-summer of the following year when the animals are approaching three years old. The geckos are about 23-25 mm SVL at hatching, approximately 45 mm at sexual maturity (smallest gravid female) and average 50-55 mm adult size, both sexes being nearly the same size. The average age of an adult in the population is difficult to compute but appears to be between four and five years. Given the relative stability of the population and the long survivorship of adults, it is fair to surmise that mortality amongst hatchlings and juveniles is high.

Heteronotia and the *Gehyra variegata* species group are the only two gekkonine taxa that occur widely throughout the interior of Australia. Indeed, in many arid and semi-arid areas they occur together abundantly, with *Heteronotia* being the ground gecko and *Gehyra* being the tree/rock gecko; both taxa, of course, are nocturnal. Given their widespread co-occurrence, one might expect the two taxa to have similar physiological parameters. However, such seems not to be the case, as *Heteronotia* is much less able to withstand high temperatures than is *Gehyra* (Licht *et al.* 1966b: table 3; Table 12). Why this should be so is not clear, as the ground microhabitat of *Heteronotia* would appear to be no less thermally extreme than the tree or rock microhabitat of *Gehyra*. However, precise measurements of the microhabitats actually occupied by the two taxa have yet to be made.

Bustard (1969c) also studied *Gehyra dubia* (as *G. australis*) at his Pilliga Scrub site and this work provides some interesting contrasts with his work on sympatric *G. variegata* from the other subgroup of the genus. *G. dubia* is a larger (max. SVL = 70 mm vs 55) and rarer species than *G. variegata*. It occurs under the exfoliating bark of stumps like *G. variegata* but, perhaps due to its larger size, prefers larger lightning struck stumps. Generally only a male and female are found together indicating that both sexes, not just males as in *G. variegata*, are strongly exclusive of other members of their own sex. In the laboratory, females display to one another in a fashion similar to males whereas female-female display was never observed in *G. variegata*[1]. Two eggs are laid in a clutch, not one as in *G. variegata* and only one clutch is laid a season not two. The RCM(1) is similar in both species (Table 7) but as clutch size is larger in *G. dubia*, its young are probably relatively smaller. The eggs of *G. dubia* are laid beneath the bark of the homesite and not on the ground under cover away from the homesite. The eggs are also bigger and take longer to hatch (75-101 days, mean = 87 days, N = 8 vs 58-79 days, mean = 65 days, N = 12). The extent to which these differences may extend to other species in the respective species groups is an intriguing question.

Gehyra dubia is widely distributed in northeastern Australia and the Pilliga Scrub is near the southern limit of its range. As Bustard (1965a, 1969c) also had some information on central and northern Queensland populations, some north-south comparisons could be made. For example, whereas the southern populations have the traditional spring-summer breeding season, northern (Cape York) populations were breeding during the winter (dry season). Furthermore, southern animals lay only one clutch per season but northern animals lay several clutches. Finally, and perhaps most interesting of all, eggs from Pilliga Scrub *G. dubia* are larger than those from more northern populations (northern Queensland) (average egg weight: .71 gm vs .52 gm), but incubation times are much lower under identical conditions (75-101 days, mean = 87, N = 8 vs 139-148 days, mean =143, N = 12). Bustard speculated that there might be a physiological adaptation to development at cooler temperatures in the more southern populations. If so this would be another clear exception, along with ease of autotomy at cool temperatures

to the general rule that physiological processes in reptiles are regulated solely by temperature.

The winter (dry) season reproduction of northern *Gehyra dubia* mentioned above may also be characteristic of other northern members of the group. For example, four of five adult (SVL ≥ 69 mm) female *G. australis* collected in the Kimberleys in the period 22-29 August were gravid with either enlarged ovarian or shelled oviducal eggs (pers. obs.). This is the height of the dry season in the Kimberleys.

Lepidodactylus is a genus of approximately 19 species which ranges widely throughout the islands and continental fringes of the tropical Indian and Pacific Oceans but has its centre of abundance in New Guinea (Kluge 1968; Brown and Parker 1977). All species are relatively small in size, and arboreal and nocturnal in habits. The genus is thought to be most closely related to *Pseudogecko* and *Hemiphyllodactylus* which are also centred over the islands and continental fringes of the south-west Pacific (Kluge 1968).

Two species of *Lepidodactylus* occur in Australia: *L. pumilus* which occurs in southern New Guinea and the islands of Torres Strait and *L. lugubris* which has a distribution nearly as wide as the genus and enters Australia at a few coastal localities in northeastern Queensland. The latter species is thought to have been introduced into Hawaii (Oliver and Shaw 1953; Hunsaker and Breese 1967) by Polynesians and into South America by Europeans (Smith and Grant 1961; Fugler 1966; Kluge 1968). In Australia both species are most common in coastal habitats (e.g., for *L. lugubris*, see Mau 1978), and both are also associated with human habitations.

Perhaps the most interesting aspect of the biology of *Lepidodactylus lugubris* is that all the populations investigated to date consist almost entirely of females (Schnee 1901; Smith 1935; Cuellar and Kluge 1972; Cuellar 1984), the few males having been discovered (four out of 673 specimens examined) generally appearing to be sexually impaired (Cuellar and Kluge 1972; Sabath 1981). This plus the fact that many populations occur in areas without closely related bisexual species suggests that the species is capable of reproducing parthenogenetically. This supposition is also supported by the demonstration that females raised from eggs laid in captivity without males can themselves produce viable eggs (Mau 1978).

Furthermore, it has recently been discovered that there are not only diploid populations of *L. lugubris*, that is, with two sets of chromosomes, as has long been known (Cuellar and Kluge 1972) but also triploid populations, i.e., with three sets of chromosomes (Moritz and King 1985).

---

[1]The asocial behaviour of *G. dubia* may even be carried a step further in another member of the *G. australis* subgroup. *G. catenata* was reported to have been found only singly on trees in 21 of 25 cases (Low 1980).

These triploids have two identical sets of chromosomes and one set that, although similar in morphology to the other two sets, stains differently. This suggests that triploids may have arisen by a mating between a diploid parthenogenetic female and a male of a sexually reproducing diploid species or population (Moritz and King 1985). Parenthetically, one of the triploid populations was originally described as a subspecies of *Gehyra variegata*, and this led to the latter species being mistakenly listed as triploid in some later reviews of lizard chromosomes (Kluge 1982).

Theoretically, parthenogenetic lineages should be less variable genetically than sexually reproducing lineages, because their only source of new genes is mutation of their own germ line instead of introduction of new genes from another germ line through sex. One consequence of this genetic similarity is that in a parthenogenetic species the immune system of one individual is less likely to recognize the tissue of another individual as "foreign" than is the case in a sexually reproducing species. Hence skin grafts from one parthenogenetic individual to another are likely to "take". In fact, it is thought that the relative degree of rejection of skin grafts between different sublineages of a parthenogenetic species may reflect the number of accumulated mutations between them, and if mutation rate is stochastically constant, then reflect also the relative, if not the absolute, time of divergence between them. This theoretical notion appears to be supported by *Lepidodactylus lugubris* in the Hawaiian Islands and Tahiti where there seems to be at least a loose positive association between the degree of geographic isolation, and hence presumed phylogenetic distance, and the rejection frequency of reciprocal skin grafts (Cuellar 1980).

Some *Lepidodactylus lugubris*, even though they are females, show certain male behaviour, specifically the characteristic male mating grip. This behaviour has been suggested to be a form of dominance display (Werner 1980), but in other parthenogenetic lizard species it has been shown that post-ovulatory females play a male role with pre-ovulatory females in order to provide the stimulus necessary for them to ovulate (Crews 1987).

The exclusively female reproduction of *Lepidodactylus lugubris* plus its association with both coastal habitats and human habitations may have facilitated its transport, both natural and human, throughout a wide distribution. This is because only one individual, a single female, instead of two, a pair, is adequate to start a population, and its coastal habitats and close association with man both increase its chance of being transported over water either naturally on a log or as a stowaway on a boat.

The diel (24 hour) variation in oxygen consumption was studied in *Lepidodactylus lugubris* in the Philippines. A distinct cycle was evident with maximum values in the early evening. This is the period when the geckos first emerge from their day-time retreats, and it is tempting to attribute the increase in oxygen consumption to the increased level of activity. However, a second species (*Cosymbotus platyurus*) with approximately the same emergence time had a different pattern. This shows that a more detailed knowledge of exactly what the animals are doing, both behaviourally and physiologically, is required before any convincing explanation can be given for the patterns observed (Feder and Feder 1981).

Finally, we might conclude this discussion of *Lepidodactylus* by acknowledging just how much can be learned about an animal when a dedicated and astute observer is given even a limited opportunity. In 1974 Klaus-Georg Mau, a German school teacher, captured two live *Lepidodactylus lugubris* on Green Island off Cairns and took them home for observation. From his observations (Mau 1978), based mainly on one animal and her offspring (the second specimen died the following year), he was able to establish the following facts, in addition to demonstrating for the first time that females can produce viable eggs without males: age to maturity can be as brief as 8½ months; the eggs are laid well away from the female's place of retreat; a second clutch of eggs can be produced about 39-41 days after the first; the incubation period is 93 days at 22°C (half day); individuals give very low intensity calls in social interactions, and longevity is at least 2¾ years. All of this from an original two animals! One wonders if Herr Mau would be able to get the necessary permits to catch and export two animals if he were in Australia today.

Of the two Australian genera of the "Ptyodactylini" assemblage, *Nactus* is, on the basis of its padless toes and terrestrial habits, more primitive than *Christinus* with its padded toes and at least partially arboreal habits. The relationships of *Nactus* are unknown, but those of *Christinus* appear to lie with a South African species of *Phyllodactylus* (see below).

The genus *Nactus* has recently been proposed to accommodate four species of the large and widespread genus *Cyrtodactylus* (Kluge 1983; Zug 1985). The four species include the well-known New Guinea and south-west Pacific and Australian species *Nactus pelagicus* and three more locally distributed species: one on two small islands north of Mauritius in the western Indian Ocean, one in a small area on the north coast of New Guinea and one, *N. galgajuga*, on a single, small mountain range in Queensland. The fact that *N. pelagicus* has a wide distribution covering two of the other three species and is clearly able to cross water gaps of a kind that would bring it within range of the third, makes it possible that this species has given rise to each of the others.

Nactus pelagicus itself is variable both cytogenetically and morphologically in ways which suggest that more than one species is involved. For example, populations in the western part of the distribution i.e., New Guinea, Australia and Vanuatu appear to be bisexual whereas those from the Pacific Islands to the east appear to be parthenogenetic. Furthermore, the western populations vary greatly both within and amongst themselves in karyotype, mostly due it seems to fusion between chromosomes (Moritz and King 1985). In Australia only one chromosome form has been identified, and it has not been found elsewhere (Moritz and King 1985). However, morphologically there appear to be two allopatric populations: a larger (max. SVL = 54 mm) northern one with 5-11 preanal pores (males only) and a smaller (max. SVL = 48 mm) southern one with 0-3 preanal pores. The dividing line between the two appears to be in the vicinity of Princess Charlotte Bay (pers. obs.).

Little is known of the ecology of the two Australian *Nactus* species. *N. pelagicus* is generally found under the exfoliating bark of fallen dead trees, under rocks or piles of surface debris. In size, general appearance (light brown with darker crossbands and tuberculate) and habits it is very similar to *Heteronotia binoei* and the two could be competitors on Cape York. With this in mind it would be interesting to establish their microgeographic distributions and relative abundances in areas where their ranges overlap.

Nactus galgajuga occurs in one of the most restricted and distinctive terrestrial habitats in Australia: the bare black boulders of the Trevethan Range just south of Cooktown (Ingram 1978). Here huge, round boulders are piled up on one another like giant marbles, with the interstices and exfoliations making excellent retreats for the gecko and an endemic skink (*Carlia scirtetis*) — and absolutely terrifying conditions for observers.

The far northern distribution of *Nactus* raises the question of the seasonality of reproduction in the two species. Nothing is known about this in *N. galgajuga* but the available specimens of *N. pelagicus*, mostly from the dry season, indicate that at least some females are reproductive throughout the dry (mid May-end of August; pers. obs.).

Two Australian species, long-placed in the otherwise American/African genus *Phyllodactylus* but also long-suspected of being distinct from this geographically distant group, have recently been recognized as a distinct lineage and given generic rank under the name *Christinus* (Wells and Wellington 1984). Unfortunately, however, the lineage was not rigorously diagnosed when it was proposed so it is difficult to evaluate. In the spirit of future testing, I can provide the following diagnosis.

The genus *Christinus* differs from all other gekkonines in the following combination of characters: nasal bones fused; elements of atlantal arch fused; presacral vertebrae ≥ 27 (Table 2); preanal and femoral pores absent; terminal subdigital scales expanded into fan-like scansors, and tail of young with an orange-red or salmon hue (Fig. 33). The next-to-last character largely characterized the taxonomically outmoded genus *Phyllodactylus*. *Christinus* shares all its derived characters except for the elevated number of presacral vertebrae with *Phyllodactylus porphyreus*, a southern African form

Fig. 33. Juvenile (above) and adult (below) of *Christinus marmoratus* to show the characteristic red tail of the former.

(Dixon 1964; Bustard 1963a, 1965c; Cornwall 1965 — these last two authors for tail colour). If the closest relatives of these two taxa taken together were to be amongst the "other" southern Africa *Phyllodactylus* — perhaps an uncertain proposition given the taxonomic shakiness of "*Phyllodactylus*", then a logical origin for *Christinus* would be via an African *Phyllodactylus* transported to Australia on a natural raft carried by the west wind drift[1].

Of the two species of *Christinus*, one, *C. marmoratus*, occurs across the southern part of the continent including many of the off-shore islands (Johnston and Ellins 1979) and the other, *C. guentheri*, occurs only on the Lord Howe Island and the Philip Island groups off the coast of New South Wales (Cogger *et al.* 1983). The continental species has several geographically, and in some cases morphologically and ecologically distinct chromosome races (King and Rofe 1976; King and King 1977), the taxonomic status of which is just beginning to be assessed (Storr 1987). One of these races is characterized by a pair of heteromorphic chromosomes in females, one of the few cases of sex chromosomes known in geckos. The island species is nearly twice the size of the mainland species (maximum SVL = 102 mm vs 55 mm).

Both *Christinus* species occur on both rocks (although generally low flat rocks) and trees, as well as on the ground. *C. marmoratus* has been found in south-west Australia on just-felled eucalypts at a point previously 30 meters above the ground (P. J. Darlington in Loveridge 1934). Both species will also occasionally enter buildings (True and Reidy 1981; Cogger *et al.* 1983).

One of the peculiarities of both species of *Christinus* is the prolonged incubation time required by some of the eggs (Table 9). For example, eggs from two South Australian populations of *C. marmoratus* took 70-87 and 85-92 days to hatch at "room temperature" and 25°C, respectively (King 1977a; Bustard 1965c), a period fairly typical for reptile eggs; however, eggs from a Victorian population took 207 days to hatch (Davey 1924). Similarly, while one *C. guentheri* freshly-laid egg from the Philip Island group maintained at 25°C, took 91 days to hatch and five field-collected eggs took 79-87 days (the exact period was not recorded) (Cogger *et al.* 1983), five field collected eggs from the Lord Howe Island group took 210-273 days to complete development and hatch (Cogger 1971).

The reasons for these extended incubation periods in certain populations of both species have yet to be determined.

Clutch size ranges 1-2, usually 2, in *Christinus marmoratus* (Bustard 1965c; King 1977a) and 1-3 (mean = 1.3) in *C. guentheri* (Cogger *et al.* 1983). *C. marmoratus* produces only one clutch per season (King 1977a; Daniels 1983). Mating appears to occur after egg-laying in *C. marmoratus* and sperm are stored in the female reproductive tract until ovulation the following spring. This is the best-known example of post-laying mating in geckos (King 1977a; King and Hayman 1978).

There is seasonal variation in both the kind (terminal vs internal) and frequency of crossing-over between homologous chromosomes in male *Christinus marmoratus*. When mating actually takes place (January-April), the testes are producing sperm showing high frequencies of internal recombination, a recombinant mode thought to lead to the high levels of genetic variablity (King and Hayman 1978).

Observations on *Christinus marmoratus* have provided interesting insights into the details of the structure and function of the tail (Daniels 1983, 1984; Daniels *et al.* 1986). Juveniles have relatively short, narrow tails with little fat, but as they grow, the tail becomes relatively longer and wider and more invested with fat. In adults the complete, original tail comprises nearly a quarter of the entire body weight ($\bar{X}$= 23.9 ± 2.0%), and the tail's stored fat comprises nearly a third of its own weight ($\bar{X}$± 32.4 ± 8.7%) or, by re-calculation, approximately one seventh of the entire body weight. Starved but watered geckos with complete tails, either original or regenerated, survived longer than those without tails (90.4 ± 40.5 and 142.4 ± 99.4 vs 49.2 ± 39.7 days, respectively), and starved geckos used nearly all the fat in their tails. Clearly tail fat is critical to survival during periods of reduced feeding.

At very low temperatures (ca 5°C), the geckos autotomize the tail very readily, but at slightly higher temperatures (ca 10°C) much less so; from this point on there is a steady increase in the readiness with which the tail is lost with increasing temperature until a plateau of very ready loss is reached again at about the average preferred body temperatures (ca 27°C). The ready autotomy at low temperatures is difficult to explain mechanistically, but its survival value seems clear in that at these temperatures the geckos are practically immobile. The steady increase in readinesss to drop the tail with rising temperatures from about 10°C is explained mechanistically as being simply another example of a biological process increasing its

---

[1]Continental drift can be ruled out because there is no other evidence for it elsewhere amongst the lizard faunas of the two continents. Furthermore, water crossing abilities appear to be in the biology of *Christinus*, at least as witnessed by *C. guentheri* on islands in the Tasman Sea.

rate with temperature. Cold enhanced autotomy also occurs in two other gekkonines *Gehyra variegata* and *Heteronotia binoei* and in the diplodactyline *Oedura monilis* (Bustard 1967e, 1968d).

At low temperatures juveniles lose the tail more readily than adults, but at higher temperatures there is no difference between the two groups. Why such a temperature-dependent age difference should occur is unclear.

Geckos starved either experimentally or naturally (after winter) lose the tail less readily than fed animals. Possibly starvation places a higher premium on the retention of remaining energy reserves.

Finally, although it should come as no great surprise considering the weight and frictional "drag" a tail represents, it has been confirmed experimentally that geckos without tails run appreciably faster (nearly twice) than geckos with tails. Despite this, in confined experimental situations with natural predators (e.g., *Antechinus* — a small mammalian carnivore), adults with tails escape more often than either tailless adults or tailed or tailless juveniles (Daniels *et al.* 1986). Apparently the decoy and shielding benefit of a tail outweighs its cost as a drag to running in these encounters with predators.

There is some evidence that *Christinus marmoratus* and members of the *Gehyra variegata* species group are competitors. The general distributions are largely abutting but non-overlapping (see maps in Cogger 1986), and where they do come together, they are often separated on a microgeographic scale (H. Cogger in Bustard 1965c; Ehmann 1976a; King and Rofe 1976). Furthermore, the two taxa are similar in size, shape and colour (so much so that misidentifications are easily made; for example, all reports of *Gehyra variegata* from Kangaroo Island, e.g., Waite 1929 are presumably based on this species) and occupy similar sites in similar habitats. There is even some evidence for a driving force behind the competitive interaction, for when held together in captivity, *Gehyra* predominates in feeding opportunities, indicating a more aggressive behaviour (Bustard 1965c).

On the other hand there is also some evidence that the distribution between the two taxa is influenced at least in part by environmental factors. For example, *Christinus marmoratus* appears to prefer lower body temperatures (Licht *et al.* 1966a; Froudist 1970), to be more photophobic (Froudist 1970), and to be much more susceptible to dehydration and thermal stress (Licht *et al.* 1966b; Warburg 1966) than

members of the *Gehyra variegata* group. Hence it may be limited by heat and/or aridity from extending north into the areas occupied by members of this group. At present the precise nature of the interaction between the two taxa remains to be demonstrated, but the situation is interesting and would lend itself to further study in both the field and laboratory.

The other subfamily of geckos in Australia, the Diplodactylinae, has been divided into two subgroups or tribes: the Carphodactylini and the Diplodactylini (Kluge 1967a). The Carphodactylini are characterized by a single, irregular patch of preanal pores (but in three genera they are presumably secondarily lost). The group comprises 10 genera: *Carphodactylus, Nephrurus, Phyllurus, Pseudothecadactylus* and *Underwoodisaurus* in Australia; *Bavayia, Eurydactylodes* and *Rhacodactylus* in New Caledonia, and *Hoplodactylus* and *Naultinus* in New Zealand (Fig. 34).

The Diplodactylini are characterized by a partial or complete fusion of the premaxillary bones in adults (Kluge 1967a) and terminal toe pads (except in *Lucasium* where they are presumably secondarily lost). The group comprises five genera, all Australian: *Crenadactylus, Diplodactylus, Lucasium, Oedura, Rhynchoedura* and *Strophurus* (Fig. 35).

There is also a possible physiological difference between the two tribes, hinted at in some data on thermal biology. In field body temperatures, preferred temperatures in the laboratory, and various measures of critical thermal limits, the Carphodactylini seem to have consistently lower values than the Diplodactylini (Tables 10-11). This comparison needs to be extended to additional species, but if the differences held up in other members of each tribe, it could help explain, at least in part, why the Carphodactylini are relatively poorly represented in the hot interior Australia, being concentrated instead in the more equable periphery (Kluge 1967b).

Most Diplodactylinae have a diploid karyotype of 38 consisting of 19 pairs of acrocentric chromosomes which grade evenly in size from large to small. This karyotype occurs in the Australian genera *Crenadactylus, Diplodactylus* (some), *Nephrurus, Phyllurus, Oedura, Rhychoedura* and *Underwoodisaurus* and in the New Zealand genus *Hoplodactylus* and has been hypothesized to be the primitive one for diplodactylines because it is widespread in the group (King 1977b). However, this is not a logically robust criterion for inferring character state polarities, and it should be treated with reserve for the subfamily as a whole. Unfortunately we are not yet in a position to apply the more rigorous "out-group" criterion, because it is uncertain

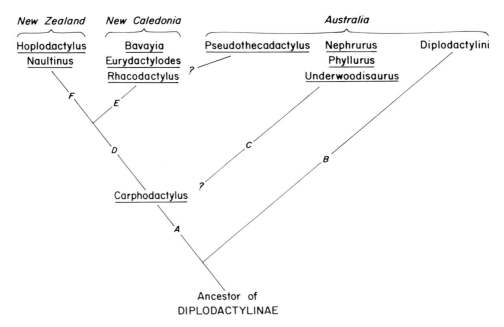

Fig. 34. Schematic diagram of the probable relationships of the genera within the tribe Carphodactylini in the subfamily Diplodactylinae (largely after Kluge 1967b). The diagnostic characters signified by the letters are as follows: A — preanal pores in large patch; B — premaxillary bones fused; terminal digital pads; C — spiny scales; carrot-shaped tail, and preanal pores absent; D — premaxillary bones fused; E — basal digital pad; terminal tail pad; F — ovoviviparous reproduction.

to which group of geckos diplodactylines are most closely related. All other Australian diplodactylines for which there is karyotypic information, i.e., various species of *Diplodactylus* and *Oedura*, have diploid numbers lower than 38, e.g., 36, 34 and 28, but a higher number of metacentric chromosomes (King 1973a, 1977b). Hence the chromosome morphocline in diplodactylines is very neat and tidy; the only thing lacking is a polarity.

The most generally primitive member of the Carphodactylini is thought to be *Carphodactylus* (Kluge 1967b), a monotypic genus from northeastern Queensland. It inhabits rainforest, is semi-arboreal (W. E. Schevill and P. J. Darlington, in Loveridge 1934) and has the normal gecko clutch of two eggs (pers. obs.). Beyond this, nothing is known.

The three Australian genera *Nephrurus*, *Phyllurus* and *Underwoodisaurus* appear to form a distinct lineage within the Carphodactylini as they share spines on the body and tail, a distinctly "carrot-shaped" tail and the loss of preanal pores. The largest and most widely distributed of the three genera is *Nephrurus*. The seven species of this genus are medium to large geckos that are exclusively terrestrial (Storr 1963; Harvey 1983). The genus is represented in virtually all the arid, semi-arid and seasonally dry parts of Australia. Externally, these geckos are characterized by a bulky head and body, relatively long, thin limbs, and a short, fat, but rapidly tapering, tail terminating in a firm spherical

knob. This latter feature gives the group its common name: knob-tailed geckos. The function of this tail is unknown, but when some species such as *N. asper* are agitated, they vibrate the tip rapidly on the substrate and if this includes dry litter, a rapid whirring sound is produced (pers. obs.). The frequency of individuals in natural populations which have lost their tails is very low (0.6-8.6%) compared to other geckos (although about comparable to most "tail squirters" — see below) (Pianka and Pianka 1976), although it is not known if this means these geckos have a higher threshold for tail loss or are at less "risk" than other geckos. At least one species, *N. asper*, has lost the autotomy planes which facilitate tail loss (Holder 1960).

Internally, the genus is characterized by a reduced number of presacral vertebrae (modal number of 25 instead of 26 or more; Table 2), generally three (or more) sacral vertebrae (instead of the usual two; Holder 1960), and, primitively, the loss of one phalange in the fourth toe of the front and rear feet giving a basic phalangeal formula of 2.3.4.4.3/2.3.4.4.4 (Holder 1960; Stephenson 1960; Fig. 36).

Many, perhaps all, species occasionally occupy burrows, either of their own or another animal's making (Pianka and Pianka 1976). Captive *Nephrurus levis* (Schmida 1973b) and *N. stellatus* (Galliford 1978) plug the entrance to their burrows with sand when inside, and *N. levis* is said "to close" the internal entrances to the escape tunnels of the *Egernia* burrows (p. 130) it

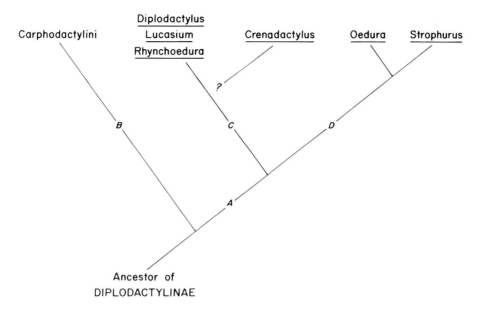

Fig. 35. Schematic diagram of the probable relationships of the genera within the tribe Diplodactylini in the subfamily Diplodactylinae (largely based on personal observations). The diagnostic characters signified by the letters are as follows: A
— terminal digital pads; B — preanal pores in large patch; C — external ear opening relatively small; overall size small;
D — basal digital pads; lateral cloacal bones lost.

inhabits (Bustard 1965c, 1970d), although what
kind of closure is meant is uncertain. In captivity,
a *N. stellatus* fleeing to its burrow customarily
stopped and licked the edge of the entrance
before entering. This was suggested to be a
means of determining whether another animal,
perhaps a predator, had entered the burrow
while the owner was away (Galliford 1978).

Like most geckos, the species of *Nephrurus* are
fairly catholic in the prey they take, but in
addition to the usual variety of invertebrate
prey, they also take lizards, especially geckos and
skinks (Pianka and Pianka 1976; Swanson 1976;
Schmida 1985: 184-85). This may simply be due
to their larger size and hence greater ability to
overcome larger prey.

Nephrurus laevissimus has been reported
"wandering in pairs" during the end of January
(Delean and Harvey 1981), but the significance
of this apparent social behaviour is unclear.

In the southern species *Nephrurus deleani*,
gravid females have been reported from mid-
summer to late autumn (January, April-May) as
well as in spring (October) (Delean and Harvey
1984). Furthermore a gravid female collected in
mid-autumn (24 April) still held her eggs in mid-
winter (18 July; Harvey 1983). It appears, therefore, that eggs may be carried over winter in this
species. This bears further investigation. A *N.
asper* female from Alice Springs area laid two
eggs in late spring (late November; H. Ehmann,
pers. comm.) while another from the Top End
laid at the beginning of the summer wet (5
December; Gow 1979).

A distinct subgroup may be recognizable within
Nephrurus on the basis of a further loss of
phalanges. *N. laevis, N. laevissimus, N. stellatus* and
N. vertebralis have lost additional single phalanges
in the third and fourth toes of the front foot and
in the third, fourth, and fifth toes of the rear
foot to give an overall phalangeal formula of
2.3.3.3.3/2.3.3.3.3. This is in comparison to the
primitive (for the genus) but already reduced
formula of 2.3.4.4.3/2.3.4.4.4 seen in *N. asper*
and *N. wheeleri* (Stephenson 1960; pers. obs.;
Fig. 33). The *Nephrurus* with the further reduced
digits are also relatively more smooth-skinned
than the two less reduced species and lack the
distinct crossbands often seen in these latter two
species. Unfortunately, the phalangeal formula
of *N. deleani*, a relatively smooth-skinned
species, has not yet been determined.

The genus *Phyllurus* comprises four species of
large, leggy geckos with depressed heads and
bodies, rough scales, and long, padless toes. All
species have flat, widened tails — one species, *P.
caudiannulatus*, only slightly, the other three, the
"leaf tails", *P. cornutus, P. platurus* and *P. saleb-
rosus*, spectacularly (Fig. 21). They are arboreal
and saxicolous and their mottled pattern and
rough outline makes them inconspicuous against
the substrate. *P. cornutus* has been seen as high as
9 m up a tree (P. J. Darlington, in Loveridge
1934). This species has also been reported living
on giant stinging trees (*Laportea gigas*) with no
ill effects (Bustard 1965c), but how it manages
this is unknown. The genus ranges along the
east coast of Australia with one species extending
into central Queensland (Covacevich 1975).

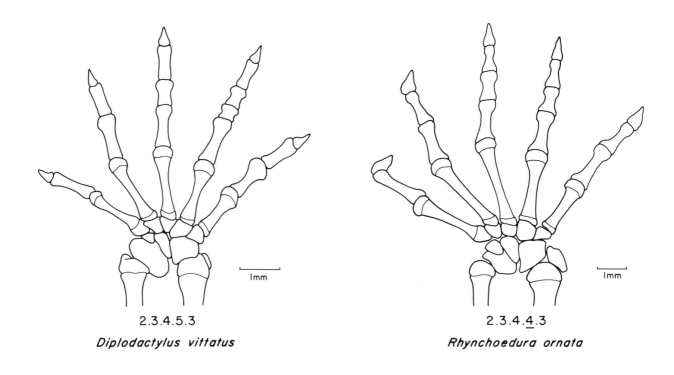

2.3.4.5.3

Diplodactylus vittatus

2.3.4.4.3

Rhynchoedura ornata

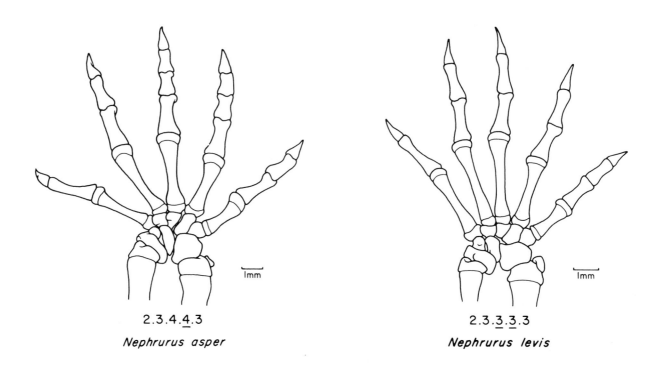

2.3.4.4.3

Nephrurus asper

2.3.3.3.3

Nephrurus levis

Fig. 36. The loss of phalanges in the front foot of diplodactyline geckos: *Diplodactylus vittatus* (AM RR 4) with the primitive formula: 2.3.4.5.3; *Rhynchoedura ornata* (AM R 13243) with loss of one phalange in the fourth toe: 2.3.4.4.3; *Nephrurus asper* (AM R 107165) with the same configuration, and *Nephrurus levis* (AM R 113269) with additional losses in the third and fourth toes: 2.3.3.3.3.

The only well-known species of *Phyllurus* is *P. platurus*, the southern leaf tail, which is centered over the Hawkesbury sandstone, the geological formation over which Sydney is also centred. Here leaf-tails live primarily in and around sandstone cliffs and ledges, but sometimes move into nearby buildings. These geckos may be locally common, several occurring together in close proximity: once, 16 individuals were found in the same crevice. Activity only begins after dark and generally only when the temperature is above 17°C (although H. Ehmann, pers. comm., reports an active specimen at 15.5°C). Prey includes a wide variety of arthropods, probably almost anything that can be overcome and swallowed. Mating has been observed in the autumn (May) which suggests that sperm may be stored over winter by the females. Females are usually heavily gravid in mid-spring (October-November) and the two eggs are probably laid about this time. Hatching occurs in mid-summer (January). The eggs are laid within rock crevices and communal nesting is known to occur (Green 1973b; see also Green 1973a; Harlow and Van der Straaten 1976).

The genus *Underwoodisaurus* comprises two species, both with terrestrial/saxicolous habits. One species is very widespread, the other quite local. The widespread species, *U. milii*, occurs throughout the southern part of the country except for the south-east corner (including Tasmania). This distribution covers many of the major population centres in Australia (Tyler *et al.* 1976; Houston and Tyler 1979), and consequently the species is one of the better known Australian geckos.

The species is often kept in captivity and most of what we know about it comes from such specimens. For example, it is said to eat small insects and their larvae, spiders and small lizards. It sheds four to six times a year (a frequency thought to be high), and lives for at least five years (Davey 1945; Peters 1970b). It is often called the barking gecko in regard for its characteristic call in defensive situations.

In a photothermal gradient *Underwoodisaurus milii* is said to be "strongly photophobic" choosing to assume a wide variety of temperatures (15-34°C) in order to avoid the light (Williams 1965; Froudist 1970).

The second species in the genus, *Underwoodisaurus sphyrurus* is much less well known. It is confined to the granite country in the eastern highlands near the Queensland-New South Wales border. A female in the Australian Museum collected in mid-spring (October) contained two enlarged follicles.

Years ago, on the basis of a chance observation of a captive *Underwoodisaurus milii* feeding on some relatively large skinks, Davey (1945) suggested that these "large-headed", terrestrial geckos were capable of taking larger prey than smaller-headed arboreal species. This made sense because terrestrial prey was likely to be larger, in general, than arboreal prey. This is an interesting suggestion and might be profitably pursued not only with this species but also with other seemingly large-headed species, such as *Nephrurus*. There are two aspects to the problem. First, one would want to determine whether these geckos really do have relatively larger heads than arboreal geckos of similar size (presumably this is the sense of larger that Davey had in mind, not just the proportionally bigger heads to be expected in larger geckos), and second, one would have to determine if the prey taken were really relatively larger.

Some taxonomists have put *Underwoodisaurus* in the genus *Phyllurus* (Kluge 1967b; Russell 1980). However, its relationships have yet to be thoroughly studied (they could conceivably be closer to *Nephrurus* than to *Phyllurus*), and there are several morphological and ecological differences between the two taxa that make it difficult for anyone who knows them well to associate them at the generic level. For example, in contrast to *Phyllurus*, *Underwoodisaurus* is deep-bodied instead of depressed, has smooth instead of sculptured cranial bones, a cylindrical instead of flattened tail, and is terrestrial instead of arboreal or saxicolous (Wermuth 1965; Covacevich 1975).

The closest relatives of the only other Australian genus of the Carphodactylini, *Pseudothecadactylus*, appear to be the New Caledonian genera *Bavayia*, *Eurydactylodes*, and *Rhacodactylus* (Kluge 1967b). These genera share broadly oval toe pads, a long thin penultimate phalange and a toe pad-like structure on the underside of the tail tip. All these features are unique within the diplodactylines, although they do occur elsewhere in gekkonines. *Pseudothecadactylus* differs from its New Caledonian relatives in lacking a claw on the inner toe of each foot.

There are two species of *Pseudothecadactylus*, one, *P. australis*, lives on Cape York where it occurs primarily in paper bark woodland and the other, *P. lindneri*, occurs in sandstone country in Arnhem Land (subspecies *lindneri*) and the Kimberleys (subspecies *cavaticus*). The natural history of neither species is particularly well known. *P. linderi* has been found in caves by day and either on sandstone faces or in fig trees growing out of sandstone by night. Those in the trees have been observed leaping up to 2 m between branches and dropping 6 m to the ground to escape. In Arnhem Land the species is said to be most active in "showery or drizzly weather" (Cogger 1975; Swanson 1979; Smith and Johnstone 1981). *P. australis* shelters in tree hollows by day and will give a "low, prolonged grating call" to passers-by (Cogger 1975).

There is some information on reproduction in *Pseudothecadactylus lindneri*. Specimens from the Northern Territory contained enlarged yolky ovarian eggs at the end of July and middle of November and oviducal eggs at the begining of August and end of September (pers. obs.). These data indicate that the species is reproductively active at least from the middle part of the dry into the early part of the wet.

The tongue and lining is the more usual pink colour in *Pseudothecadactylus lindneri* but a very dark brown to almost black colour in *P. australis*. The function of the distinctive mouth colour in *P. australis* is unknown. Other morphological differences between the two species that are as yet unexplained functionally are in *P. lindneri* the loss of the lateral cloacal bones in males and the loss of a sternal rib (from three ribs to two), and in *P. australis* the increase by one in the number of presacral vertebrae (from 26 to 27).

The tribe Diplodactylini, which occurs only in Australia, can be divided into two groups. One group is characterized by a relatively small external ear opening and relatively small size (maximum SVL of largest species: 65 mm). This group includes *Crenadactylus*, *Diplodactylus*, *Lucasium* and *Rhynchoedura*. These taxa are largely ground dwelling. The other group is characterized by basal toe pads and the loss of the lateral pair of postcloacal bones. This group includes *Oedura* and *Strophurus*. These taxa are largely arboreal/saxicolous.

The largest genus within the Diplodactylini is *Diplodactylus*. It comprises 18 largely terrestrial species and occurs throughout Australia, exclusive of the far south-east (Cogger 1986; King *et al.* 1982). The group is probably not a lineage, at least there are no derived character states to identify it as such as yet; rather, it is probably an assemblage of relatively primitive diplodactyline species. All species of *Diplodactylus* are terrestrial, although *D. pulcher* is said to climb occasionally (Pianka and Pianka 1976).

Within the *Diplodactylus* assemblage, there are two subgroups that are likely to be monophyletic. One of these is the *D. stenodactylus* group. It is characterized primarily by the loss of a phalange in the fourth toe of the front foot giving a phalangeal formula for this foot of 2.3.4.4.3 instead of 2.3.4.5.3. This group includes *D. alboguttatus*, *D. maini*, *D. squarrosus* and *D.*

stenodactylus which have been checked for the critical phalange character and possibly the following species which have not been checked: *D. fulleri*, *D. occultus* and *D. wombeyi*. Only *D. stenodactylus* is well-known. It is completely terrestrial and occurs in a wide variety of habitats from woodland to open sand plain.

A second group in the *Diplodactylus* assemblage that is possibly monophyletic comprises *D. conspicillatus* and *D. savagei* which have the ordinarily large labial scales, i.e., those bordering the mouth, reduced to granules the same size as the neighbouring scales. These two species may in turn be related to *D. tessellatus*, a species with which they share a somewhat short, fat, blunt tail. The karyotype of the latter species differs from that of most diplodactylines (Table 4), i.e., it has 14 pairs of chromosomes of which six are large and eight small, with five of the six large pairs metacentric and the one remaining pair acrocentric. This is the lowest diploid number yet recorded for geckos and the most metacentric karyotype in diplodactylines.

The *Diplodactylus vittatus* group is a third group within the *Diplodactylus* assemblage which has been proposed and widely used (Kluge 1967b; Storr 1979b); unfortunately, however, it does not appear to be based on any special features. Thus its included species — *D. granariensis*, *D. ornatus*, *D. polyophthalmus* and *D. vittatus* — can be combined with the remaining species of the genus *Diplodactylus* — *D. byrnei*, *D. galeatus*, *D. mitchelli*, *D. pulcher* and *D. steindachneri* — in a "left over" assemblage requiring further work to establish their relationships.

The three genera closely related to *Diplodactylus* — *Rhynchoedura*, *Lucasium* and *Crenadactylus* — are all monotypic. *Rhynchoedura* is a small, reddish to beige, terrestrial species with a pinched, beak-like snout (Fig. 27) and a pair of enlarged preanal pore scales (Fig. 37). It is widespread in the arid and semi-arid interior. It is probably related to the *Diplodactylus stenodactylus* group, with which it shares not only size, colour, habits and distribution but also the loss of a phalange in the fourth toe of the front foot. Alternatively, it may be related to *Diplodactylus pucher* with which it shares a narrow snout and a high number of presacral vertebrae ($\geqslant$27).

Rhynchoedura's wide distribution and local abundance have made it a fairly well-studied species. It feeds largely, if not exclusively, on termites (Pianka and Pianka 1976), a habit which might be thought to offer an adaptive explanation for its beak-like snout were it not for the fact that other small, terrestrial diplodactylines without modified snouts also feed heavily on termites. By day, *Rhynchoedura* often shelters in vertical

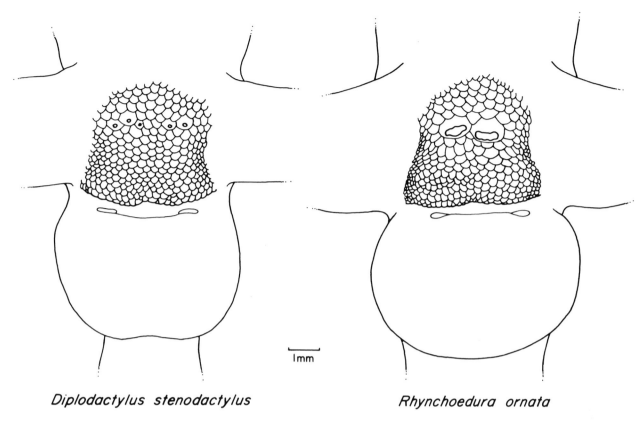

Diplodactylus stenodactylus   *Rhynchoedura ornata*

Fig. 37. Male *Diplodactylus stenodactylus* (AM R 100693) and *Rhynchoedura ornata* (AM R 114102) to show the enlarged preanal pore scales in the latter. The paired postcloacal pouches are also indicated schematically.

spider burrows, a habit not uncommon in other small diplodactylines. Like other members of its tribe, *Rhynchoedura* also appears to be relatively heat resistant (Licht *et al.* 1966a-b).

There is some information from a Kimberleys population of *Rhynchoedura* that provides insight into the seasonality of reproduction in a far northern population of this wide-ranging species. At a locality 7.2 km W of the Barnett River via the Gibb River Road sampled during the late dry season (1 September), the four adult females found were reproductively active with either small to large yolking ovarian follicles or oviducal eggs (pers. obs.).

Compared to other diplodactylines of similar size, including its probable close relatives in the *Diplodactylus stenodactylus* group, *Rhychoedura* retains certain embryological features. For example, the skull and vertebrae are poorly ossified and large fontanels remain in the skull Whether the pointed snout of the taxon is also an embryologic feature awaits further knowledge of gecko embryology. The possible adaptive significance of the retained embryonic features in *Rhynchoedura* is unknown.

Lucasium is also a small, reddish terrestrial species with a wide distribution in the arid and semi-arid interior. Its principal distinguishing feature is the small, spiny scales on the under-

sides of the digits, a feature which in terms of scale size is the extreme end of a trend seen in several species of the genus *Diplodactylus* (Fig. 38).

Lucasium is also a fairly well known species, but there is little in its biology that is exceptional. In some areas (e.g., vicinity of Renmark, South Australia) *Lucasium* is said to shelter exclusively in the burrows of the dragon *Ctenophorus pictus*. Fights between males have been observed in captivity; these conflicts are usually accompanied by a low, continuous chirping call (Bustard 1965c).

Lucasium was previously allied with the *Diplodactylus stenodactylus* species group (Kluge 1967b), but because it lacks the diagnostic reduced phalangeal character for this group (which probably also includes *Rhynchoedura*, this is unjustified. Perhaps, it should be considered as part of the "left-over" group of *Diplodactylus* whose relationships require further work (p. 78).

Crenadactylus is a small, olive grey, terrestrial/ arboreal genus recognized primarily on the absence of claws (Dixon and Kluge 1964). In most lizards the last phalange is straight and bears a horny claw; in *Crenadactylus* this bone is retained but has lost the claw, and its tip, now

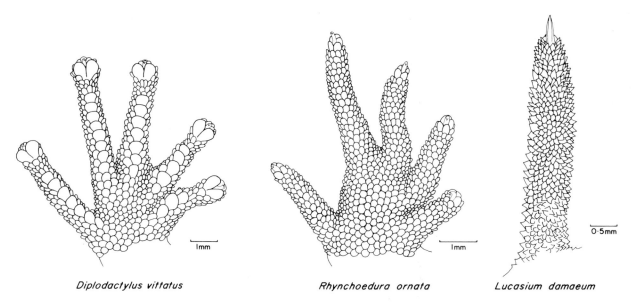

Diplodactylus vittatus   *Rhynchoedura ornata*   *Lucasium damaeum*

Fig. 38. Ventral surface of the rear foot in three closely related geckos of the tribe Diplodactylini to show the probable primitive condition for the tribe *(Diplodactylus vittatus)* and the subsequent reduction *(Rhynchoedura ornata)* and loss *(Lucasium damaeum)* of the terminal pads. Note also the spiny sub-digital scales in *Lucasium*.

completely internal, is bifurcate (Fig. 39). The reason for this modification, especially the terminal bifurcation which is unique in lizards, is unknown.

Crenadactylus is very widespread in western Australia and occurs in a variety of habitats: mallee shrubland (pers. obs.), hummock grass sand plains, hummock grass rocky hills, and *Spinifex*-covered coastal dunes (Storr 1978c; Storr and Hanlon 1980). Where hummock grass occurs, it appears to prefer to shelter amongst the plant's stipes but away from this growth form it shelters on the ground, beneath rock, timber and debris. Despite its broad geographic and habitat distribution, population densities

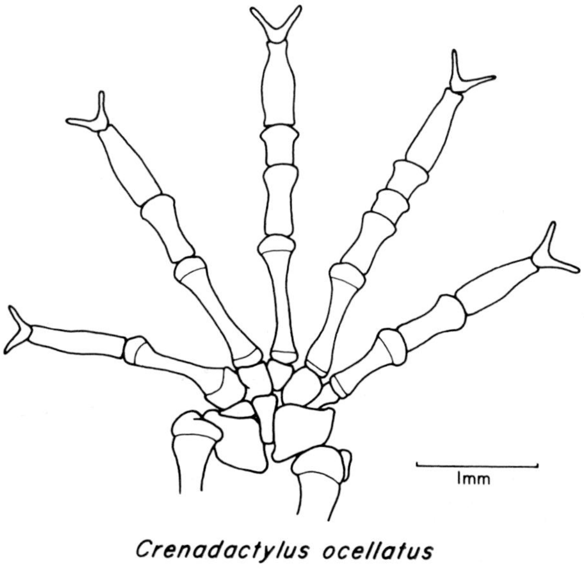

Fig. 39. Front foot of *Crenadactylus ocellatus* (AM R 123338) to show the bifurcate terminal phalanges.

can vary greatly, from abundant to rare, even in the same habitat. The reasons for this are unknown. Field observations suggest "that *C. ocellatus* mainly feeds under logs, litter and rocks, and is rarely encountered in the open at night" (Dell and Harold 1979).

Southwestern populations, which occur in litter and under logs, rocks and debris, tend to be ocellate, whereas more northern and eastern populations, which occur in hummock grass, are striped.

In the *Oedura-Strophurus* lineage of the Diplodactylini the genus *Oedura* comprises 13 species. Most of these are confined to the more equable northern, eastern and southwestern parts of the continent, but one, *O. marmorata* is widespread in the arid interior.

The species are small to medium in size with slightly depressed heads and bodies. The digits are laterally expanded with retractile claws and a pair of terminal scansors are followed by slightly expanded basal lamellae (Fig. 25). The digital pads are very similar to those of *Strophurus*.

Members of the genus are almost exclusively arboreal or saxicolous, but occasionally *Oedura lesueurii* and *O. rhombifer* may be found on the ground amongst debris. Most species appear to occupy one substrate or another, at least in specific areas. For example, *O. coggeri* (Bustard 1966a), *O. filicipoda* (King 1984c), *O. gemmata* (King and Gow 1983), *O. gracilis* (King 1984c), *O. lesueurii* (R. Sadlier and G. Shea, pers. comm.), *O. tryoni* (Broom 1897; Cogger 1957; Bustard 1966a) and *O. marmorata* from central Australia (Bustard 1970a) and *O. obscura* King

Crenadactylus ocellatus

1984c) are usually found on rocks whereas *O. castelnaui* (Bustard 1970a), *O. monilis* (Cogger 1957; Bustard 1971, as *O. ocellata*; Low 1978), *O. reticulata* (Smith and Chapman 1976; How and Kitchener 1983; Chapman and Dell 1985), *O. rhombifer* (Low 1978; R. Sadlier and G. Shea, pers. comm.), *O. robusta* (Czechura and Miles 1983) and *O. marmorata* from the periphery of Australia (Bustard 1970a; Miller 1980) are usually found on trees. *O. lesueurii*, *O. rhombifer* and *O. robusta* also occasionally inhabit buildings. Presumably the depressed head and body of *Oedura* is an adaptation to the narrow crevices and fissures occupied during the day.

What little is known about the food habits of the genus indicates that most species eat a variety of arthropod and other invertebrate prey and may even take an occasional smaller lizard like a gecko or skink (Cogger 1957; Bustard 1971). Although most species are nocturnal, *Oedura marmorata* will apparently make brief forays from shelter during the day to take passing prey (Cogger 1957).

Species of *Oedura* vary greatly in tail shape, and hence presumably in its fat storage capacity. Some species such as *O. gracilis* and *O. obscura* have thin, tapering tails whereas others such as *O. filicipoda* and *O. marmorata* have thick, wide tails. Just how effective these fat stores can be in tiding these species over periods of low food availability, such as drought or winter, is evident from the fact that specimens of *O. lesueurii O. marmorata, and O. monilis* and can live in terraria for between nine and 12 months without food, provided water is available. Without water this period is reduced to two to three months (Cogger 1957).

As far as is known, all *Oedura* species lay two eggs in a clutch (Cogger 1957; Bustard 1967d). The relative clutch mass (RCM1) is about .22-.39, a range fairly typical for geckos (Table 7). At an incubation temperature of 30°C, eggs of five species took 49-88 days to hatch (but only one egg took more than 60 days) (Bustard 1967d). This relatively narrow range in incubation times in *Oedura* is in strong contrast to the relatively wide range in *Gehyra* (Table 9).

The reproductive season in *Oedura* appears to occur during the usual spring-autumn period in the southern and central parts of the continent (Cogger 1957; Bustard 1965c, 1967d, 1971; How and Kitchener 1983), but at various times of year in the north, depending on the species. For example, *O. coggeri* from northeastern Queensland is said to lay at scattered times throughout the year (Cogger 1957, as northern *O. tryoni*); a single *O. castelnaui* is reported to have laid eggs in the wet season (December, but

the specimen was captured in August and probably held in temperate Canberra until laying; Bustard, 1967d), and an *O. coggeri* laid eggs in the late dry season (September; Bustard 1967d). Furthermore, information on *O. gemmata* suggests that this far northern endemic is reproductively active during the mid- to late dry season and quiescent during the mid-wet. Specifically, two females captured on 8 and 10 August were gravid, one with enlarged ovarian eggs and the other with shelled oviducal eggs. An adult male captured 9 August appeared to have testes in the early stages of regression. However, three adult males and six adult females captured 28 February-12 March had extremely regressed gonads (pers. obs.).

Communal egg laying has been reported in *Oedura tryoni*. Two nests were found close together beneath well-embedded pieces of fence paling (Milton 1980b). One nest comprised 20 eggs (requiring 10 females) and the other 12 eggs (six females).

When they hatch, several species of *Oedura* have a distinct and strongly contrasting colour pattern, e.g., *O. marmorata*. However, as they grow, the pattern often becomes more subdued and less contrasting. The adaptive and phylogenetic significance of this pattern difference, which is unique in its magnitude amongst Australian geckos, is unknown.

The only species of *Oedura* for which there is a long-term (two years) population study is *O. monilis*, another species studied by Bustard (1971, as *O. ocellata*) in the Pilliga Scrub (p. 63). This species shows great fidelity to site, rarely leaving once established in a dead tree or stump. Adults usually occur together in pairs or individually and appear to tolerate immatures; however newly mature animals appear not to be tolerated and either must displace an adult or emigrate. Sexual maturity is probably attained in the second or third season after hatching.

There is some anecdotal information that *Oedura rhombifer*, which sometimes inhabitats buildings, has been replaced in such situations by *Gehyra australis*, itself a commensal. Cogger and Lindner (1974) report that in 1966 the former species was found on the buildings at Black Point, on the Coburg Peninsula in the Northern Territory but that by 1970 it "had been virtually replaced" by the latter. Recall that *Gehyra australis*, was itself the early house gecko in Darwin but has now been replaced by *Hemidactylus frenatus*.

Finally, there is an apparently large difference in susceptibility to elapid snake venoms between two species of *Oedura* that is intriguing. Tested with venoms from *Austrelaps superbus*, *Notechis scutatus*, *Pseudechis porphyriacus* and *Pseudonaja textilis*, *Oedura tryoni* was consistently more resistant than *O. lesueurii* (Minton and Minton 1981). The ecological and/or taxonomic significance of this difference remains to be explored.

The genus *Strophurus* comprises 12 species — *S. ciliaris*, *S. elderi*, *S. intermedius*, *S. mcmillani*, *S. michaelseni*, *S. rankini*, *S. spinigerus*, *S. strophurus*, *S. taeniata*, *S. taenicauda*, *S. williamsi* and *S. wilsoni* and ranges widely throughout arid, semi-arid and seasonally dry parts of Australia. All species are climbers and have both basal and distal toe pads. As in many other climbing genera, i.e., *Gehyra*, *Oedura*, *Phyllurus*, *Pseudothecadactylus*, the species of *Strophurus* appear to be specific to either vegetation or rocks: all but one species have been found only in vegetation whereas *S. wilsoni* has been found only on rocks (Storr 1983). The "arboreal" species all seem to be restricted to perches of relatively narrow diameter, twigs and narrow branches for the larger species and grass stems for the smaller. The range in body size is large, extending from a maximum snout-vent length of 48 mm for *S. elderi* up to 88 mm for *S. ciliaris*.

Five species within *Strophurus* appear to form a subgroup. *S. elderi*, *S. mcmillani*, *S. michaelseni*, *S. taeniata* and *S. wilsoni* all lack femoral pores, presumably a derived character within the genus. These species also share small size (maximum SVL = 56 mm), a longitudinally striped colour pattern (remnant in *S. wilsoni*), and in the first three species at least, a strong propensity for inhabiting hummock grass.

As a group *Strophurus* is characterized primarily by a feature that is unique within reptiles: large tail glands which produce a viscous, treacle-like substance varying in colour, perhaps in a species-specific manner, from black to straw-yellow, which in some species can be fired from the dorsal surface of the tail in thin streams for a distance of up to a third of a meter or more (E. Mjöberg, in Lönnberg and Andersson 1913; White, in Zietz 1914, 1915; Morrison 1950, 1951). The streams can be aimed by appropriate curvature of the tail. The extruded substance is sticky, although rapid-drying and is both unpalatable and irritating to eyes. Although direct observations of its function against potential predators are limited (e.g., *Varanus* — Bustard 1979; rats — Richardson and Hinchliffe 1983), the system is almost certainly defensive. Considerable quantities of the material can be extruded from the tail and repeated "firings" can lead to a pronounced shrivelling of the tail (White, in Zietz 1914; Bustard 1964a). The extruded substance does not appear to irritate the gecko itself, because after firing, the residual material may

be calmly licked off the tail surface (Russell and Rosenberg 1981). Preliminary attempts to characterize the material biochemically show that it contains protein and glycoprotein and that electrophoretically revealed components are similar between species (Rosenberg *et al.* 1984).

Dissection of the tail shows that the glands are paired, lying on either side of the tail next to the vertebral column deep to the tail musculature (Fig. 40). They are also segmented, being separated from each other by the transverse septa that define the caudal segments. The contents of the gland are ejected by contraction of the tail musculature and rupture of a slit-like epidermal window overlying an area of thinned muscle and connective tissue. These epidermal windows lie at the dorso-anterior edge of each caudal segment, one per segment, and with a little practice can be seen as transverse slits between the scales on the dorsal midline of the tail. In some species the slits are surrounded by a little rosette of distinctive scales; in others they are found between the enlarged caudal spines, but in a few, there are no markers to serve as a guide to their position.

Regenerated portions of tails have regenerated glands but they are continuous instead of segmented, often unequally developed on the two sides, and have no associated epidermal windows. The glandular substance is extruded either through the skin at the junction between the old and new parts of the tail or through a window in what remains of the original tail (Bourne 1932; Rosenberg and Russell 1980; Richardson and Hinchcliffe 1983).

The evolutionary origin of this tail gland may lie in the modification of the fat bodies that occupy a similar position in many other lizards, including other diplodactyline species (Fig. 40). Most *Strophurus* lack the fat bands in the tail, their place being entirely taken up by the glands, but in one species at least, *S. michaelseni*, the top half of the inside of the tail surrounding the vertebral column is composed of glandular material and the bottom half by fat (Rosenberg and Russell 1980). This could be an example of a intermediate step in the evolution of the complete caudal gland. A reverse evolutionary sequence ending in the tail seen in non-tail squirting diplodactylines was hypothesized by Richardson and Hinchcliffe (1983), but on outgroup comparison the polarity must surely be opposite to what they propose.

There is one observation for *Strophurus elderi* which suggests that the skin itself of this species may be distasteful to predators. Two specimens of this gecko have been observed being grasped

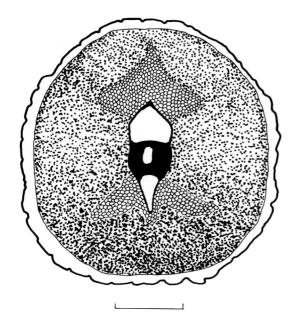

Diplodactylus stenodactylus

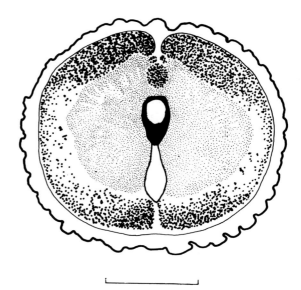

Diplodactylus mcmillani

Fig. 40. Cross-sections of the tails of *Diplodactylus stenodactylus* (AM R 101583) and *D. mcmillani* (AM R 53398) to show the central fat deposit (small cells) in the former and the central exudate material (fine stipple) in the latter. Bar equals 2 mm.

on the head by the flap-footed lizard *Lialis burtonis* and then quickly released without any sign of an exudation from the tail. Whatever the source of its protection may be, *S. elderi* appears to be virtually immune from predation by this lizard prey specialist *par excellence* with which it shares its prefered retreat of hummock grass. In captivity it is never eaten, even by starved *Lialis* known to eat other geckos such as *Gehyra* and *Heteronotia* (Bustard 1979)

Another striking feature of most *Strophurus* species is bright mouth colour, either dark blue or yellow. This colour is shown when the gecko opens its mouth in defensive display (Bustard 1964a). To date, blue mouth colour has been observed in *S. intermedius*, *S. rankini* (pers. obs.), *S. spinigerus* (W. Schevill, in Loveridge 1934), *S. strophurus* (pers. obs.) and *S. williamsi* (Bustard 1964a) and yellow/orange colour has been observed in *S. taeniata* (Schmida 1973a; Swanson 1976). Curiously, some populations of *S. ciliaris* show blue colour (L. Smith, pers. comm.; pers obs.) and some yellow (Cogger 1967, plate 7; L. Smith, pers. comm.; pers. obs.). The blue-mouthed form appears to occur in two areas: across the southern part of the continent and at the head of the Gulf of Carpentaria and the yellow mouthed form elsewhere. One wonders if mouth colour is indicative of a greater degree of differentiation than previously suspected within what is now considered one species, *S. ciliaris*. At least one species of *Strophurus*, *S. elderi*, has the usual pink mouth of most other

geckos[1], and three other species in the genus, *S. michaelseni*, *S. taenicauda* and *S. wilsoni*, remain to be checked.

Little is known about the natural history of most *Strophurus* species. Perhaps the best-known is *S. williamsi* which Bustard (1969c) studied at his Pilliga Scrub site (p. 63). Here the species spent the active season (September-April) high in dead, ring-barked and lightning-struck cypress pines and the dormant season under bark either at the base of these trees or on stumps. Animals occurring on stumps during the winter clearly come from tall dead trees elsewhere and all those animals marked on stumps one winter and recovered the next, were on the same stump. As no suitable summer tree occurred within 25 metres in some of these cases it would appear as if the species has some memory for or means of recognizing specific sites over periods of nearly six months.

Why some *Strophurus williamsi* leave their tree and others do not is unclear, but it may be due to territorial behaviour under crowded conditions. As many as six adults may occur on a single tree during the summer, but this may be too many when they all move to the base for the winter and some may be driven off.

Despite the highly arboreal habits of *Strophurus williamsi*, it appears as if the eggs may be laid on the ground instead of the trees. Bustard

[1]Specimens of *S. ciliaris* from the Edgar Ranges of the south-west Kimberley area are said also to have "pinkish mouths" (Storr and Smith 1981) but the significance of this unique (for the species) exception is unknown.

never found eggs of the species when he was working on his study site, but some years later two pairs of eggs were found in what appeared to be unoccupied *Varanus gouldii* burrows at a site approximately 25 km north of Bustard's site (Shea 1984). Each pair was in a separate burrow approximately 1 m from the entrance and 30 cm below the surface. Possibly the parchment-like shell of this egg is simply too permeable to allow it to be deposited in the relatively dry micro-habitat occupied by the adults i.e., dead trees (Shea 1984).

As in most geckos, the usual clutch size in *Strophurus* is two. However, there is an intriguing observation in *S. spinigerus* that at least some females carry more than two eggs (3-4) and that eggs of different sizes can occur together in one oviduct (Dell and Chapman 1977). This suggests clutch sizes routinely larger than two and perhaps staggered deposition of the eggs in one oviduct. Both events would be unusual in geckos, and hence the original observations should be confirmed in *S. spinigerus* itself and perhaps checked in its close relatives.

Work on *Strophurus* has shown that while there is no relationship between female size and total clutch volume within species, there is a positive relationship amongst species (How *et al.* 1986). A similar result has also been attained in four distantly related overseas species (but using total clutch mass; Vitt 1986). Hence the relationships may be general in geckos: female size is a determining factor in egg size (= offspring size?) amongst species but not within species.

Members of *Strophurus* are the only Australian geckos which appear to expose themselves regularly to daylight (Ehmann 1980), some even apparently seeming to bask in full sun (e.g., *S. ciliaris* — E. Mjöberg in Lönnberg and Andersson 1913; *S. intermedius* — Cogger 1969; *S. spinigerus* — Froudist 1970; *S. strophurus* — Pianka and Pianka 1976). Perhaps correlated with these basking habits is the fact that at least several of the larger species have a jet black parietal peritoneum (Mitchell 1965b; Mutton 1970; pers. obs.), the probable solar radiation screen often seen in fully diurnal lizards (Porter 1967).

Although it is uncertain whether geckos were originally nocturnal or diurnal in their foraging time, most living species, including all the Australian ones, are nocturnal. Furthermore, when foraging, they are generally exposed on the surface. These two factors appear to be fundamental considerations in gecko biology.

The benefits of nocturnal activity are probably reduced predation in terms of either kind or intensity and exposure to a part of the arthropod fauna not available to most other lizards. The costs are requisite activity at reduced levels of light and temperature, in the latter case often well below what is preferred by the animals and in all likelihood what is optimum for many basic functions such as digestion and locomotion. Although it would seem likely that much of gecko biology may be related to methods of dealing with the costs deriving from activity in the dark and at cool temperatures, very little work has been initiated on this basis. Perhaps it would provide a useful approach in the future.

Table 1. Taxonomic distribution of toe pad type within the two subfamilies of geckos in Australia.

Padless	Padded			
	Basal	Terminal	Dual	Continuous
Gekkoninae				
Cyrtodactylus	*Gehyra*	*Christinus*		
Heteronotia	*Hemidactylus*			
Nactus	*Lepidodactylus*			
Diplodactylinae				
Carphodactylus		*Crenadactylus*		*Pseudothecadactylus*
Lucasium[1]		*Diplodactylus*	*Oedura*	
Nephrurus		*Rhynchoedura*	*Strophurus*	
Phyllurus				
Underwoodisaurus				

[1]Secondarily padless (p. 58).

Table 2. Vertebrae number in Australian geckos. Counts are based on personal observation but Holder (1960) and Cogger (1964) provide additional counts.

Taxon	Presacral			Postsacral		
	Range	Mode	N	Range	Mean	N
Gekkoninae						
Christinus						
guentheri	26-27.5	27	26	29	29.0	1
marmoratus	26-28	27	26	29-32	30.3	6
Cyrtodactylus						
louisiadensis	26	26	5	48	—	1
Gehyra						
australis	26	26	11	—	—	—
borroloola	26	26	13	32	—	1
dubia	26	26	10	—	—	—
nana	26-27	26	25	33-35	34.0	3
pamela	26	26	14	34	34.0	1
variegata	25-26	26	19	28-29	28.5	2
Hemidactylus						
frenatus	25-27	26	16	28	—	1
Heteronotia						
binoei	26	26	18	37	37.0	1
Lepidodactylus						
lugubris	26	26	9	26-29	27.1	7
Nactus						
pelagicus	26	26	23	30-31	30.7	3
Diplodactylinae						
Carphodactylus						
laevis	25-26	26	19	44-45	44.5	2
Crenadactylus						
ocellatus	26	26	8	—	—	
Diplodactylus						
alboguttatus	25-27.5	26	13	28-31	29.3	4
byrnei	25-28	26	10	24-29	27.1	21
conspicillatus	27-30	28/29	19	16-23	18.8	14
galeatus	26	26	7	22-23	22.7	3
granariensis	25-27	26	13	21-25	24.8	5
ornatus	25-26	26	5	27	27.0	2
polyophthalmus	25-26	26	10	23-24	23.8	5
pulcher	27-28	27	11	24-26	24.7	4
squarrosus	26	26	2	32	32.0	1
steindachneri	25-27	26	20	27-31	29.2	6
stenodactylus	25-28	26	10	29-34	31.4	9
tessellatus	25-27	26	29	18-23	20.3	15
vittatus	25-27	26	9	20-23	21.9	8
Lucasium						
damaeum	25.5-26	26	8	31-33	31.6	5
Nephrurus						
asper	24-26	25	4	19.5-24	22.2	4
levis	24-26	25	14	26-30	27.2	9
laevissimus	24.5-25	25	9	23-25	24.2	6
stellatus	25	25	6	—	—	—
vertebralis	25-26	25/26	2	—	—	—
wheeleri	25	25	2	26	26	1
Oedura						
castelnaui	26-27	26	14	25-27	26.0	4
gemmata	26	26	12	22-24	22.3	7
robusta	26	26	13	24-27	26.0	3
Phyllurus						
caudiannulatus	25.5-27	26	13	30-37	33.5	6
cornutus	24-25.5	25	14	29-30	29.7	4
platurus	24-26	25	11	29-34	31.8	5
Pseudothecadactylus						
australis	27	27	7	28-31	29.5	4
lindneri	26	26	21	28-31	29.7	8
Rhynchoedura						
ornata	27-29	27	32			
Strophurus						
elderi	26-27	26	10	16-21	17.9	12
mcmillani	26-27	26	24	24	24.0	1
spinigerus	25-27	26	13	23-28	24.8	9
williamsi	26-27	26	8	20-24	22.0	12
Underwoodisaurus						
milii	25-27	26	6	33-42	37.8	12
sphyrurus	26	26	5	25	—	1

Table 3. Distribution of tail breaks in terms of postsacral vertebra number in various Australian gecko species.

Postsacral Vertebra Number

Taxon	1	2	3	4	5	6	7	8	9	10	11	12	13	14	15	16	17	18	19	20	21	22	23	24	25	26	27	28	29	30
Gekkoninae																														
Christinus																														
guentheri						14	4	1			1	1	1				1													
marmoratus						6	1	6	1	2		1	1				1													
Cyrtodactylus																														
louisiadensis					1	1	2																							
Gehyra																														
australis						2	2	1	1	1			1			1			1											
borroloola						1	1	1	1	1	1		1						1	1	1	1							1	
dubia						2	1	1	1	1		2	1	1	1		1						1							
nana						5	3		2	2	3	2		1						1	1				1					
pamela						4	1	1	1	3	1			1	1					1										
variegata						4	1		1			1																		
Hemidactylus																														
frenatus							3	2	2	1			1		1				1		1		1			1				
Heteronotia																														
binoei					2	6			2				2	2	2		2			2										
Lepidodactylus																														
lugubris						4		1		1	2	1	1	1	1	1	2		1											
Nactus																														
pelagicus						1		1			1	1		3	1	2	1	2			1	1						1		
Diplodactylinae																														
Carphodactylus																														
laevis				1	16	1																								
Crenadactylus																														
ocellatus						7	1					1	2																	
Diplodactylus																														
alboguttatus						6	2							1																
byrnei					2	5	1																							
conspicillatus					2	3													1											
galeatus					1	4																								
granariensis						3																								
ornatus						4																								
polyophthalmus						6																								
pulcher						1																								
squarrosus					1	4																								
steindachneri					1	8	2																							
stenodactylus					3	8	2																							
tessellatus				1	1	4																								
vittatus							1																							
Lucasium																														
damaeum					1	11	1																							
Nephrurus																														
levis					1																									
vertebralis						2																								

Table 3 (continued)

Taxon	Postcaudal Vertebral Number																													
	1	2	3	4	5	6	7	8	9	10	11	12	13	14	15	16	17	18	19	20	21	22	23	24	25	26	27	28	29	30
Oedura																														
castelnaui						4	1	1	1																					
gemmata							1	1	1																					
robusta						1	2	2																						
Phyllurus																														
caudiannulatus						1																								
platurus					6	5				1			1																	
Pseudothecadactylus																														
australis						1	1																							
lindneri						6	2	2	1	2	1	2	1			2	2	1											1	
Rhynchoedura																														
ornata					2	9																								
Strophurus																														
mcmillani						6																								
spinigerus						2																								
williamsi				1	1	1																								
Underwoodisaurus																														
sphyrurus				1	4																									
milii					9																									

Table 4. Some general attributes of the karyotype of gecko genera represented in Australia.

Taxon	Chromosome Number		Partheno-genesis	Sex Chromosomes	Reference
	2N	3N			
Gekkoninae					
Christinus	32-36	—	—	+,−	King and Rofe 1976; King and King 1977
Gehyra	38-44	—	—	+,−	King 1979, 1981a, 1982a-b, 1983a-b, 1984a-b; Moritz 1984b, 1986
Hemidactylus	40-46	60, 70	+,−	—	Moritz and King 1985 (for references)
Heteronotia	42	63	+,−	—	Moritz 1983, 1984a; Moritz and King, 1985
Lepidodactylus	44	66	+,−	—	Moritz and King 1985 (for references)
Nactus	28-40	—	+,−	—	Moritz and King 1985
Diplodactylinae					
Crenadactylus	38	—	—	—	King 1977b
Diplodactylus	28, 34, 36, 38	—	—	—	King 1973b, 1977b, 1981a; King *et al.* 1982
Nephrurus	38	—	—	—	King 1977b
Phyllurus	38	—	—	—	King 1977b
Oedura	36, 38	—	—	—	Werner 1956; King 1977b
Rhynchoedura	38	—	—	—	King 1977b
Underwoodisaurus	38	—	—	—	Matthey 1949

Table 5. References to feeding habits of Australian geckos both in the wild (W) and in captivity (C).

Taxon	Reference
Gekkoninae	
Christinus	
guentheri	Cogger *et al.* 1983 (W)
marmoratus	Chapman and Dell 1978a (W); T. F. Houston in Dell 1985 (?)
Cyrtodactylus	
louisiadensis	McCoy 1980 (W)
Gehyra	
nana	James *et al.* 1984 (W)
variegata	Bustard 1968a, 1969a (W); Pianka and Pianka 1976 (W); Dell and Chapman 1979a (W); Chapman and Dell 1980a (W); Dell 1985 (W); Pianka 1986 (W)
Hemidactylus	
frenatus	Cagle 1946 (W, as *H. garnoti* but almost certainly this species, *fide* Kluge and Eckhardt 1969); Tyler 1961 (W); Chou 1974 (W); Sahi 1980 (W)
Heteronotia	
binoei	Bustard 1968e, 1968g (W); Pianka and Pianka 1976 (W); Pianka 1986 (W)
Lepidodactylus	
lugubris	La Rivers 1948 (W, as cited by Oliver and Shaw 1953); Oliver and Shaw 1953 (W); Sabath 1981 (W)
Diplodactylinae	
Crenadactylus	
ocellatus	Dell and Harold 1979 (W)
Diplodactylus	
alboguttatus	Dell and Chapman 1979a
conspicillatus	Pianka and Pianka 1976 (W); Pianka 1986 (W)
granariensis	Dell and Harold 1979 (W); Dell and Chapman 1979a (W)
maini	Chapman and Dell 1979 (W), 1980a (W)
pulcher	Pianka and Pianka 1976 (W); Dell and Chapman 1978 (W), 1979a (W); Pianka 1986 (W)
stenodactylus	Pianka and Pianka 1976 (W); James *et al.* 1984 (W); Pianka 1986 (W)
vittatus	Webb 1983 (W)
Lucasium	
damaeum	Bustard 1965c (C); Peters 1970c (C); Pianka 1986 (W)
Nephrurus	
asper	Longman 1918 (C)
deleani	Harvey 1983 (C)
laevissimus	Pianka and Pianka 1976 (W); Pianka 1986 (W)
levis	Pianka and Pianka 1976 (W); Pianka 1986 (W); Schmida 1985 (C)
stellatus	Galliford 1978 (C)
vertebralis	Pianka and Pianka 1976 (W); Pianka 1986 (W)
Oedura	
marmorata	Cogger 1957 (?)
monilis	Bustard 1968e, 1971 (as *O. ocellata*; W); Rössler 1980
reticulata	How and Kitchener 1983 (w)
tryoni	Firth 1979 (C)
Phyllurus	
cornutum	Bustard 1965c (C); Rösler 1981 (C)
platurus	Green 1973b (W); Mebs 1973 (C); Rose 1974 (W)
Rhynchoedura	
ornata	Pianka and Pianka 1976 (W); Pianka 1986 (W)
Strophurus	
ciliaris	Pianka and Pianka 1976 (W); How *et al.* 1986 (W); Pianka 1986 (W)
elderi	Pianka and Pianka 1976 (W); How *et al.* 1986 (W); Pianka 1986 (W)
intermedius	How *et al.* 1986 (W)
michaelseni	How *et al.* 1986 (W)
rankini	How *et al.* 1986 (W)
spinigerus	How *et al.* 1986 (W)
strophurus	Pianka and Pianka 1976 (W); How *et al.* 1986 (W); Pianka 1986 (W)
Underwoodisaurus	
milii	Davey 1945 (C); Peters 1970b (C)

Table 6. Age at first reproduction for Australian geckos. Age is determined as the number of reproductive seasons since the one in which the individual began life as an egg.

Taxon	Age to Reproduction		Reference
	Males	Females	
Gekkoninae			
Gehyra			
variegata	3	3	Bustard 1968c
Heteronotia			
binoei	2-3	2-3	Bustard 1968g
Lepidodactylus			
lugubris	—	<1	Mau 1978
Diplodactylinae			
Oedura			
monilis	2-3	2-3	Bustard 1971
reticulata	3	4	How and Kitchener 1983
Strophurus			
ciliaris	3	3	How *et al.* 1986
spinigerus	2	2	How *et al.* 1986
williamsi	2	2	Bustard 1969c

Table 7. Relative clutch mass for Australian geckos.

Taxon	RCM 1			RCM 2			RCM 3			Reference
	x̄	R	SD	x̄	R	SD	x̄	R	SD	
Gekkonidae										
Christinus										
marmoratus	.33	.31-.34	.02	.25	.24-.25	.01	—			Bustard 1965 (N=2)
Gehyra										
dubia	—	.20-.30		—			—			Bustard 1965 (as *G. australis;* N=?)
variegata	—						.05		.02	Pianka 1986 (N=92)
	—	.20-.30		—			—			Bustard 1968 (N=?)
Heteronotia										
binoei	—			—			.10		.02	Pianka 1986 (N=3)
	.30			—			—			Bustard 1968 (N=1)
Diplodactylinae										
Diplodactylus										
conspicillatus	—			—			.20		.01	Pianka 1986 (N=30)
pulcher	—			—			.15		.05	Pianka 1986 (N=5)
stenodactylus	—			—			.10			Pianka and Pianka 1976 (N=3)
Nephrurus										
laevissimus	—			—			.16		.04	Pianka 1986 (N=29)
levis	—			—			.16		.00	Pianka 1986 (N=2)
Lucasium										
damaeum	—			—			.13		.03	Pianka 1986 (N=4)
Oedura										
castelnaui	.22			.18			—			Bustard 1967 (N=1)
lesueurii	.31			—			—			Bustard 1967 (N=1)
marmorata	.27	.24-.30	.03	.21	.19-.23	.02	—			Bustard 1967 (N=3)
monilis	.34	.32-.36	.03	—			—			Bustard 1967 (N=2)
	.39			—			—			Bustard 1967 (N=3)
tryoni	.26			—			—			Bustard 1967 (N=1)
Phyllurus										
platurus	.29			—			—			pers. obs. (N=1)
Rhynchoedura										
ornata	—			—			.17		.05	Pianka 1986 (N=35)
Strophurus										
ciliaris	—			—			.11		.03	Pianka 1986 (N=20)
elderi	—			—			.13		.02	Pianka 1986 (N=5)
	.38			.28			—			Bustard 1965 (N=1)
strophurus	—			—			.13		.03	Pianka 1986 (N=18)

Table 8. Dimensions of Australian gecko eggs. Diplodactyline eggs freshly laid unless indicated otherwise. Values of length and width are ranges and means; ratio is of means.

Taxon	Length	Width	L/W	N	Reference and Comments
Gekkoninae					
Christinus					
guentheri	13.4-15.9	11.9-13.6	1.1	20	Cogger *et al.* 1983
	14.4	12.8			
marmoratus	12.5-14.0	9.5-10.0	1.4	4	Bustard 1965c
	13.5	9.7			
	11.0-13.7	9.0-10.0	—	?	King 1977a
	12	9	1.3	1	Armstrong 1979b
Gehyra					
variegata	9.5-12	8.5-11	1.1	20	Bustard 1968a
	11	10			Bustard 1968a
Hemidactylus					
frenatus	9	8	1.2	?	Yashiro 1931 (as quoted by Fukada 1965)
	12	7	1.7	63	Cagle 1946 (as *H. garnoti* but almost certainly this species, *fide* Kluge and Eckhardt 1969; means only)
	8.8-9	6.8-7	?	4	Petzold 1965
	9	8	1.1	?	Fukada 1965
	8.2-10.0	7.2-8.4	1.2	41	Chou 1979
	9.2	7.9			
	8.1-8.9	7.1-7.6	1.2	?	Cogger *et al.* 1983
	8.4	7.2			
	9.8	8.3	1.2	?	Minton 1966 (means only)
Heteronotia					
binoei	8.6-9.0	7.0	1.3	3	Bustard 1968g
	8.8	7.0			
Lepidodactylus					
lugubris	8.8-9.2	6.2-6.8	?	?	Snyder 1919
	10	4	2.5	48	Cagle 1946 (averages; width in error ?)
	9	6-7	1.4	4	Mau 1978
	9.0	6.5			
	8	5	1.6	1	Whitaker 1970
	8.0-9.0	6.8-7.5		3	Schwaner 1980
	8.3	7.1			
	9.7	8.6	1.1	2	Gibbons and Zug 1987 (averages)
Nactus					
pelagicus	12.2	—	—	1	Schwaner 1980
Diplodactylinae					
Diplodactylus					
elderi	12.5-14.0	7.5-9.5	1.5	4	Bustard 1965c (some eggs after 27d)
	12.7	8.5			
Nephrurus					
asper	29-30	15.5-16.5	1.8	2	Gow 1979
	29.5	16			
deleani	25	13	1.9	1	Delean and Harvey 1984
	24.5	13.5	1.8	1	
Oedura					
monilis	21-24.3	10.5-11.9	2.0	6	Rössler 1980
	22.3	10.9			
Strophurus					
williamsi	15.2-15.9	9.5-9.7	1.6	2	Shea 1984 (eggs near term)
	15.5	9.6			

Table 9. Incubation times (days) for Australian gecko eggs. Means in parentheses.

Taxon	Incubation Period	Temperature (°C)	N	Reference
Gekkoninae				
Christinus				
guentheri	210-273	19-29	5	Cogger 1971
	91	25	1	Cogger *et al.* 1983
marmoratus	207±2	?	4	Davey 1924
	85-92	25	3	Bustard 1965c
	70-87	?	?	King 1977a
Gehyra				
dubia				
southern	75-101 (87)	25	8	Bustard 1969c
northern	139-148 (143)	25	12	Bustard 1969
variegata	61-79	25	?	Bustard 1968a
Hemidactylus				
frenatus	46-58 (52.9)	?	?	Chou 1979
	45	?	1	Husband 1980b
	77-88	24-26	4	Schwaner 1980
Heteronotia				
binoei	47-48	25	?	Bustard 1968g
Lepidodactylus				
lugubris	92+	?	1	Oliver and Shaw 1953
	?-73	?	?	Schwaner 1980
	93	22		Mau 1978
	110+	?	2	Schnee 1901
Diplodactylinae				
Diplodactylus				
elderi	43	30	2	Bustard 1965c
Nephrurus				
deleani	55-56	29-30.5		Delean and Harvey 1984
Oedura				
castelnaui	60	30	2	Bustard 1967d
lesueurii	58	30	2	Bustard 1967d
marmorata	88	30	1	Bustard 1967d
monilis	54-55	30	3	Bustard 1967d (as *O. ocellata*)
	50-60	23-30	7	Rössler 1980
tryoni	49-51	30	6	Bustard 1967d

Table 10. Body temperatures (°C; FBT) of Australian geckos active in the field.

Taxon	Mean	Range	SD or SE	N	Reference
Gekkoninae					
Christinus					
guentheri	21.5	19.0-23.5	0.99	?	Cogger *et al.* 1983[1]
	19.0	18.4-21.4	0.64	?	Cogger *et al.* 1983[1]
marmoratus	15.9	—	±0.54	15	Froudist 1970[1]
	14.1	—	±0.27	7	Froudist 1970[2]
	19.6	—	±0.72	5	Froudist 1970[4]
Gehyra					
punctata	—	34.2-40.3	—	7	Licht *et al.* 1966a[4]
"variegata"	20.8	—	3.7	6	Williams 1965[1]
	20.1	—	±2.32	5	Williams 1965[2]
	32.9	—	±0.42	7	Froudist 1970[4]
	20.6	—	±2.82	5	Froudist 1970[2]
	21.2	—	—	1	Froudist 1970[1]
	26.5	—	4.0,±0.24	262	Pianka and Pianka 1976[1]
	34.1	33.0-35.0	1.02	3	Licht *et al.* 1966a[4]
	—	24.0-28.0	—	4	Licht *et al.* 1966a[4]
	26.6	19.5-34.7	—	15	Werner and Werner (in Werner and Whitaker 1978)[3]
	20.3	17.8-22.0	1.44	6	Pers. obs.[1]
Hemidactylus					
frenatus	27.1	25.9-29.1	—	—	Cogger *et al.* 1983[1]
	30.7	—	—	—	Cogger *et al.* 1983[4]
	27.2	26.0-27.8	—	—	Cogger *et al.* 1983[2]
Heteronotia					
binoei	27.5	23.2-31.2	—	7	Webb *et al.* 1972[4]
	27.0	—	3.5,±0.69	25	Pianka and Pianka 1976[1]
Diplodactylinae					
Diplodactylus					
conspicillatus	27.7	—	±3.6	50	Pianka and Pianka 1976[1]
granariensis	20.9	16.1-26.0	4.1	5	Williams 1965[1] (as *D. vittatus* but probably this species)
pulcher	27.7	—	±3.7	24	Pianka and Pianka 1976[1]
steindachneri	16.8	—	—	1	Pers. obs.[1]
stenodactylus	26.6	—	±3.5	36	Pianka and Pianka 1976[1]
Lucasium					
damaeum	21.0	—	—	1	Pers. obs.[1]
Nephrurus					
laevissimus	22.5	—	±4.1	172	Pianka and Pianka 1976[1]
levis	23.2	—	±3.2	30	Pianka and Pianka 1976[1]
vertebralis	24.1	—	±3.5	14	Pianka and Pianka 1976[1]
Oedura					
coggeri	27.5	22.0-31.8	—	18	Webb *et al.* 1972[4]
marmorata	18.0	16.5-20.4	1.71	4	Pers. obs.[1]
rhombifer	26.1	23.4-28.7	2.25	4	Webb *et al* 1972[4]
Rhynchoedura					
ornata	27.4	—	±2.9	237	Pianka and Pianka 1976[1]
Strophurus					
ciliaris	25.4	—	±5.7	71	Pianka and Pianka 1976[1]
elderi	26.2	—	±4.1	13	Pianka and Pianka 1976[1]
spinigerus	16.0	—	±0.98	3	Froudist 1970[1]
	23.1	—	±1.25	14	Froudist 1970[3]
	25.5	—	±1.64	6	Froudist 1970[4]
strophurus	25.3	—	±4.8	52	Pianka and Pianka 1976[1]
Underwoodisaurus					
milii	17.6	—	2.0	14	Williams 1965[1]
	19.4	—	±2.96	5	Williams 1965[2]
	22.8	—	—	1	Froudist 1970[2]
	22.9	—	±1.62	4	Froudist 1970[4]

[1]Nocturnal, exposed. [2]Nocturnal, sheltering. [3]Diurnal, exposed. [4]Diurnal, sheltering.

Table 11. Body temperatures (°C; PBT) of Australian geckos preferred in a laboratory thermal gradient.

Taxon	Mean	Range	SD or SE	N	Reference
Gekkoninae					
Christinus					
marmoratus	27.6	24.4-32.0	0.42	3	Licht *et al.* 1966a[2]
	27.1	24.2-30.4	0.81	3	Licht *et al.* 1966a[4]
	21.9	—	±1.25	4	Froudist 1970[1]
	23.1	—	±3.54	4	Froudist 1970[2]
	27.5	—	±1.2	?	Froudist 1970[3]
	27.1	—	±1.98	?	Froudist 1970[4]
	27.3	—	±0.83	N	Daniels 1984[4]
Gehyra					
punctata	34.6	29.3-37.0	0.66	6	Licht *et al.* 1966a[2]
"*variegata*"	29.3	—	4.71	4	Williams 1965[6]
	32.8	—	3.09	6	Williams 1965[5]
	35.3	30.0-40.3	—	7	Licht *et al.* 1966a[2]
	32.2	—	±0.31	4	Froudist 1970[1]
	30.8	—	±1.06	4	Froudist 1970[2]
	30.9	—	±0.81	?	Froudist 1970[3]
	30.4	—	±1.08	?	Froudist 1970[4]
Heteronotia					
binoei	30.0	26.9-32.4	1.86	6	Licht *et al.* 1966a[2]
Diplodactylinae					
Diplodactylus					
conspicillatus	34.3	31.8-37.2	4.56	3	Licht *et al.* 1966a[2]
granariensis[7,8]	27.3	—	2.71	4	Williams 1965[5,6]
	24.9	—	±1.94	4	Froudist 1970[1]
	25.8	—	±2.78	4	Froudist 1970[2]
	27.2	—	±3.1	4	Froudist 1970[3]
	27.4	—	±1.78	4	Froudist 1970[4]
Rhynchoedura					
ornata	34.0	31.1-39.4	0.69	8	Licht *et al.* 1966a[2]
Strophurus					
spinigerus	35.9	29.8-38.5	2.62	9	Licht *et al.* 1966a[2]
	34.6	30.7-37.7	1.64	3	Licht *et al.* 1966a[4]
	32.9	—	±0.42	4	Froudist 1970[1]
	33.0	—	±2.43	4	Froudist 1970[2]
	27.9	—	±1.76	8	Froudist 1970[3]
	33.2	—	±0.71	12	Froudist 1970[4]
Underwoodisaurus					
milii	27.3	25.0-33.0	—	5	Licht *et al.* 1966a
	26.3	—	2.32	8	Williams 1965[6]
	22.5	—	±0.7	4	Froudist 1970[1]
	23.0	—	±1.14	4	Froudist 1970[2]
	27.1	—	±0.65	—	Froudist 1970[3]
	26.7	—	±0.61	—	Froudist 1970[4]

[1]Night; photothermal gradient. [2]Day; photothermal gradient. [3]Night; dark substrate gradient. [4]Day; dark substrate gradient. [5]Time not specified; photothermal gradient. [6]Time not specified; substrate gradient. [7]As *Diplodactylus vittatus*, but probably this species. [8]Both authors question whether this species actually chose its position solely with regard to temperature.

Table 12. Critical thermal maximum temperatures (°C; TMax) of Australian geckos.

Species	Mean	Range	SD	N	Reference
Gekkoninae					
Gehyra					
variegata	43.8	—	—	1	Spellerberg 1972a
Heteronotia					
binoei	40.6	—	—	1	Spellerberg 1972a
	40.4	39.4-41.5	0.59	14	Pers. obs.
Diplodactylinae					
Diplodactylus					
conspicillatus	43.8	—	—	1	Pers. obs.
steindachneri	43.5	—	—	1	Pers. obs.
vittatus	43.4	—	—	1	Pers. obs.
Lucasium					
damaeum	42.6	—	—	1	Pers. obs.
Oedura					
marmorata	43.2	42.1-44.0	0.98	3	Pers. obs.
robusta	41.0	—	—	1	Pers. obs.

CHAPTER 4

Pygopodidae — Flap-footed Lizards

IN ordinary life we are accustomed to thinking that the more generally similar two organisms are, the more likely they are to be closely related in terms of descent. That this is not always true is graphically demonstrated by the "legless" lizards of the family Pygopodidae (Fig. 41) and their closest living relatives — the geckos, a group in which there is rarely as much as a toe bone missing.

The evidence for the close relationship between these two families is a long list (Table 1) of mostly detailed anatomical characters, such as postcloacal sacs and bones and details of the inner ear, but also behavioural and reproductive features such as the wiping of the sides of the face with the tongue (Linton 1929; Bustard 1963b; Greer 1985b: Fig. 44), a well-developed "voice" (Annable 1983), and a constant clutch size of two (Table 8). In fact, the evidence for the close relationship between these two families is perhaps more convincing than for any other two families of lizards (Boulenger 1885; McDowell and Bogert 1954; Shute and Bellairs 1953; Underwood 1957; Miller 1966; Wever 1974).

Thirty species and eight genera of pygopodids are currently recognized. The family occurs throughout most of Australia (everywhere except the south-east corner), New Guinea, the Aru Islands and possibly the Bismarck Archipelago, but only the two species of *Lialis* extend the range outside Australia. It is the only group of lizards restricted to the Australian Region which is thought distinctive enough to be given a rank higher than subfamily. The taxonomy and systematics of the pygopodids has been reviewed by Kluge (1974, 1976) and his general scheme of relationships is used in this account (Fig. 45).

As a group, pygopodids are marked by three, interrelated morphological features: limb reduction, body elongation and body attenuation. Consider

the hypothetical ancestor of pygopodids, reconstructed by pooling together all the most primitive characteristics of living species. This animal had lost all external trace of the front limbs, but retained a small remnant of the humerus internally. It had the rear limbs reduced to a small flap-like appendage (Fig. 46) which contained the femur, tibia and fibula, some tarsal bones, four of the five metatarsals, and a single phalange in each of the two middle digits (Fig. 47a). The pectoral girdle contained all the basic elements, but in a reduced state; there were only two ribs attached to the sternum instead of three, and there was no mesosternum (Fig. 48). The pelvic girdle also contained all the basic elements but was so reduced in size that the two halves no longer made contact on the midline, being widely separated. The sacral diapophyses had also separated at their distal ends, now being simple lateral projections like the diapophyses on the caudal vertebrae (Fig. 47b). The body was greatly elongate, at least nine times head length (Table 2) which is 2-3 times greater than the figure for geckos. There was also an increase in the number of presacral vertebrae, to at least 44 (Table 3), which is nearly half again as many as in geckos. The body was also very thin in comparison to its length, that is, it was attenuate. Furthermore, it can be inferred from the habits of the more generally primitive, living pygopodids that all these changes — limb reduction, body elongation and attenuation, and their associated effects evolved while the pygopodid lineage was in a basically surface-dwelling mode.

Other morphological features of pygopodids appear to be related to the basic changes of limb reduction and body elongation and attenuation, because they are changes often seen in other elongate, attenuate squamates. As such, they represent parallel evolutionary developments. For example, certain single structures, such as the liver, have become long and thin, while certain

Fig. 41. Pygopus lepidopodus, the species thought to be the most generally primitive living pygopodid.

paired structures, such as lungs, show a reduction of one member and/or a lengthening of the other. For example, the ratio of left/right lung length ranges from about .68 in some *Lialis* to .09 in some *Aprasia* (Table 4). The neck also seems to have become shorter by one vertebra. Primitively in squamates there are eight cervical vertebrae, but in all those pygopodids in which a count can be made the cervical vertebrae are seven or fewer[1]. The only other similarly short-necked Australian lizards are also limb-reduced and elongate, i.e., the *Lerista bipes* species group and *Anomalopus (Vermiseps) brevicollis*. The three elements of the first presacral vertebra, the atlas, are also fused into a solid ring in pygopodids.

[1]The cervical vertebrae are counted forward from the vertebra just in front of the first vertebra attaching to the sternum via a rib. However, not all pygopodids retain a sternal attachment (e.g., *Aprasia*, *Lialis*, *Ophidiocephalus* and *Pletholax*), making it impossible to determine the number of cervical vertebrae in these forms.

There is both direct and indirect evidence that female pygopodids have, on average, more presacral vertebrae than males. The direct evidence is actual counts for populations of *Aprasia* species (Parker 1956). The indirect evidence is a general sexual dimorphism in the same direction in ventral scales in most pygopodids (Kluge 1974) and the strong likelihood that ventral scales and presacral vertebrae are correlated (Kerfoot 1970), both being intimately related to the segmentation of the trunk. In all other cases of sexual dimorphism in presacral vertebrae in lizards, females also have, on average, more vertebrae, and whereas sexual dimorphism is known in some non-elongate lizards (e.g., lacertids — Arnold 1973), it is much more common in elongate groups (e.g., dibamids — Greer 1985a and scincids — see pp. 118). An increase in vertebral number may be a way of increasing body size or suppleness, factors that could be important in increasing brood size or agility during pregnancy, but why it should occur only in certain taxonomic and ecological groups and not others is unclear.

The pygopodid lineage has also experienced two behavioural changes associated with limb reduction and body elongation and attenuation. The first is a switch from limb locomotion to body locomotion, or lateral undulation, the classic method of locomotion in limb-reduced squamates such as snakes. This latter method apparently had no precursors in geckos and evolved *de novo* in the pygopodid lineage. Lateral undulation, especially when it is the only form of locomotion, usually has profound

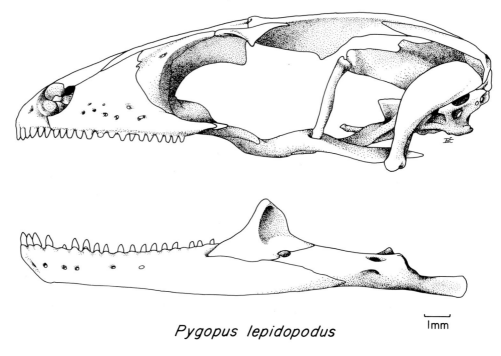

Pygopus lepidopodus |—1mm

Fig. 42. Skull of *Pygopus lepidopodus* (AM R 122134).

consequences for the trunk musculature in squamates. How these muscles have been affected in pygopodids has not yet been studied, but one small manifestation is an anterior process on the dorsal part of the rib just down from the rib's head (Fig. 47b). Presumably this process provides for a discrete muscle attachment which provides fine control of the ribs, the ribs themselves assuming increased importance as the principal means of transferring the force of the trunk muscles to the ventrolateral scales in contact with the substrate.

The second behavioural change involves breathing. Geckos, like all other tetrapods with well-developed limbs, breathe by displacing the axillary region just behind the forelimb, contracting inward on the power stroke to expel air from the lungs and then relaxing to let air passively fill the lungs on the return stroke (opposite to the method of birds and mammals). Pygopodids, like many other squamates lacking front limbs, such as snakes, no longer show any displacement in the axillary area, but rather

Fig. 44. Facial tongue wiping in pygopodids: *Pygopus lepidopodus*. Compare with Fig. 23.

show an overall slight contraction of the entire anterior body — circumferential breathing. This implies a large functional change, the nature of which has yet to be investigated (see also p. 120).

Although all pygopodids retain their hind limbs (Figs 43, 46), it is not clear what function, if any, they have. The flaps are neatly folded back along the sides of the base of the tail during lateral undulation on fairly open ground, but they may be extended on rough terrain (Annable 1983), when the animals are climbing (Annable 1983; pers. obs.) and when they are excited. It has also been suggested that the flaps are used during courtship and mating, but this is apparently only conjecture based on analogy with other limb-reduced squamates, e.g., boid snakes.

A feature which pygopodids may also share as a group but which is unrelated to their form is a distinct sex chromosome arrangement. In the four species karyotyped to date, the males show a group of chromosomes, designated X_1, X_2, Y, which consist of two free chromosomes (X_1 and X_2) and their partners fused to each other ($X_1 + X_2 = Y$), one of which must carry sex determining information (Gorman and Gress 1970; King 1977c).

Pygopodids occur in the arid, semi-arid and seasonally dry parts of Australia and inhabit woodlands, shrublands and grasslands. They do not occur in the cool, wet areas and are rarely, if ever, found in closed forest, closed woodland or swamps. In other words they are closely associated with and have probably evolved exclusively in warm, open and at least seasonally dry habitats.

A careful reading of the literature with an eye to the vertical distribution of pygopodids within habitats reveals greater variation than for almost any other limb-reduced, elongate group of squamates, exclusive of snakes. *Pygopus* is largely terrestrial, but *P. lepidopodus* will sometimes flee

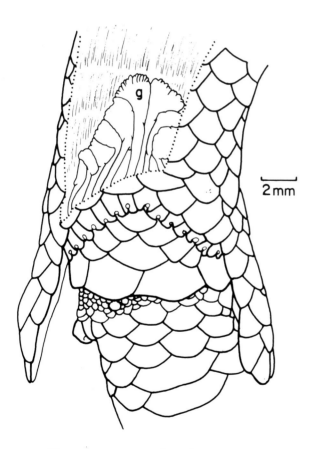

Pygopus nigriceps

Fig. 43. Cut away view of cloacal area of a male *Pygopus nigriceps* (AM R 66543) to show the flap-like rear leg and the bodies and ducts of glands (g) opening through the preanal pores.

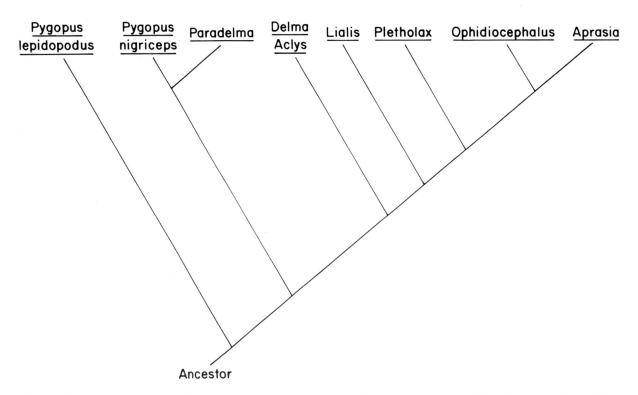

Fig. 45. Diagram of the most popular current hypothesis of pygopodid lizard phylogenetic relationships (after Kluge 1974).

by climbing up into thick vegetation (Bush 1983). There is also an account of a specimen being found a metre and a half up a *Banksia* tree (Anonymous 1944) and another in captivity often climbing half a metre above the ground in vegetation (Mertens 1966). *Lialis* also appears to be largely terrestrial but with some penchant for climbing into hummock grass (Bustard 1970d; pers. obs.) and low shrubs (L. Smith, pers. obs.). *Delma* covers a relatively wide range; it is often seen out on the surface, but it is also frequently uncovered in litter (pers. obs.). Some *Delma* species (e.g., *D. tincta*), but not others (e.g., *D. fraseri*), burrow in loose sand (Kluge 1974) and some (i.e., *D. fraseri, D. impar* and *D. inornata*) have been reported to climb in hummock grass and shrubs and even to sleep in the branches of small plants (Martin 1972; Annable 1983). Similarly, *Aclys* has been collected under litter and has been seen both on the surface and basking and moving in heath approx. 0.3-1.0 m off the ground (Dell and Chapman 1977; Murray 1980). *Pletholax* burrows readily in sand in captivity (Kluge 1974) and in pit traps, but it is also active on the surface and in heath; in fleeing in heath, it has been seen to leap from shrub to shrub (M. Bamford, pers. obs.). *Ophidiocephalus* and *Aprasia* are largely litter/soil inhabitants, and will flee either into deeper earth crevices (*Ophidiocephalus;* Ehmann 1980) or the tunnels of ants and termites (*Aprasia;* Kluge 1974; Shea, pers. comm.). *Aprasia* has also been observed to be active in nature on the surface (Ehmann (1976a) and to

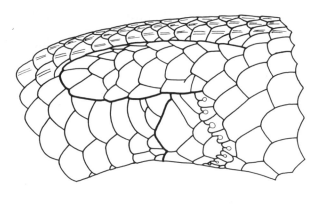

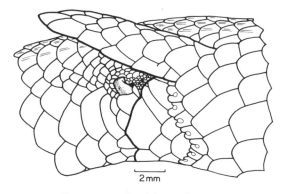

Pygopus lepidopodus

Fig. 46. Right rear leg and vent area of *Pygopus lepidopodus* (AM R 93688; ♂) to show the flap-like rear leg, the preanal pores and the cloacal spur.

bask in captivity on the surface (Martin 1972). *Ophidiocephalus* and *Aprasia* appear to be the only pygopodids that do not climb.

Whether in vegetation or in litter, pygopodids can reverse their direction, i.e., undulate backwards along the identical path just traversed, e.g., *Lialis burtonis* (pers. obs.) and *Ophidiocephalus taeniatus* (Ehmann 1981), respectively. Furthermore, when suspended from a perch (a finger) by its tail, *Lialis burtonis*, at least, can pull itself back up without changing its body shape.

Pygopodids are often characterized as primarily crepuscular to nocturnal (Underwood 1957; Worrell 1963; Wever 1974; Kluge 1976; Peters 1976b; Patchell and Shine 1986b). However, it is increasingly clear that the activity period of most species is quite broad and includes a large diurnal component (Table 5; Littlejohn and Rawlinson 1971). For example, *Lialis* and *Pygopus lepidopodus*, two of the best known species, appear to be encountered abroad about equally frequently during both the day and the night. *Delma* is generally thought to be diurnal and has even been reported sleeping at night (Martin 1972), but there are also a number of observations of nocturnal activity (Table 5). *Delma*'s close relative, *Aclys*, has been trapped both by day and by night (Murray 1980) and has also been reported to bask (Dell and Chapman 1977). *Pletholax* is said to be at least partially, if not exclusively, diurnal (M. Bamford, pers. comm.). In fact, on the basis of present knowledge the only two largely if not exclusively nocturnal pygopodids appear to be *Pygopus nigriceps* and *Paradelma*. The former is almost always found abroad at night and the latter is said to be, at least in captivity, active at night (S. Wilson, in litt., N= 1) but inactive by day (Shea, in press, N= 3). Although *Aprasia* is usually found only under cover, it appears to be clearly active by day (Bustard 1970d; Ehmann 1976a); it is even said to bask (Martin 1972) and to be not shy of daylight (Mertens 1965). *Ophidiocephalus* is surface-active at night and subsurface-active day and night (H. Ehmann, pers. comm.). As in the case of place of activity, perhaps the significant insight regarding the time of activity in pygopodids, is greater scope and variability than heretofore realized.

In view of the obvious diurnal habits of many pygopodids, it is perhaps surprising that there are several features in the pygopodid eye that are thought to be clear adaptations to nocturnal, or at least subdued light, conditions. For example, there is no fovea, the super-sensitive area of the retina present in diurnal forms but generally lacking in nocturnal ones; the visual cells lack the oil droplets which are present in diurnal forms, presumably to filter some of the

incoming light, and the pupil is vertical (Underwood 1970), the hallmark of nocturnal creatures. How can the nocturnal eye morphology be reconciled with the diurnal habits? Could it be that pygopodids, like their gecko relatives, were originally nocturnal and have only become secondarily diurnal? If so, are there any as yet undetected compensatory modifications to readapt the nocturnal eye for at least partial diurnal duty?

Thermoregulation has been studied in only one species of pygopodid, *Lialis burtonis*, and it seemed to control its temperature fairly precisely, at least in the laboratory. Average preferred body temperature was 35.1 ± 0.1°C, while recently fed animals sought heat and averaged about 2.5°C higher and starved animals avoided heat and averaged about 3°C lower (Bradshaw *et al.* 1980). As digestion and general metabolism are temperature dependent in reptiles, one could speculate that recently fed animals raise body temperature in order to digest quickly and starved ones lower it in order to conserve energy. Several pygopodids have been reported as basking in sunlight (e.g., *Aclys* — Dell and Chapman 1977; *Aprasia* — Martin 1972) or under the warm litter directly below a heat source in captivity (*Ophidiocephalus* — Ehmann 1981).

Pygopodids are fairly tolerant of high temperatures. In terms of critical thermal maximum (the body temperature at which an animal initially loses the ability to right itself when turned on its back but from which it recovers when cooled), one *Delma inornata* had a value of 43.3°C, two *Lialis burtonis* 45.9 and 46.4°C and two *Pygopus lepidopodus* 41.4 and 44.3°C (pers. obs.). In terms of "survival time" (i.e., to loss of the righting reflex but with recovery) in a high temperature test chamber, relatively high values were obtained by the three species tested with the two *Aprasia* species proving more durable than the one *Delma* species (pers. obs., Table 6). The greater durability of *Aprasia* was also evident in another survival time experiment (but with slightly different criteria) where an *Aprasia* species attained higher values than a species of *Lialis* and of *Pygopus* (Licht *et al.* 1966b).

The feeding habits of certain pygopodids are well known (Table 7). Some taxa, such as *Aclys*, *Delma*, *Ophidiocephalus* and *Pygopus*, seem to be arthropod generalists, as are most lizards, taking any animal of this group they can find, overcome and swallow. However, others such as *Aprasia* and *Lialis* appear to be specialists: the former is said to eat ants (Kluge 1974; Jenkins and Bartell 1980) and the latter is known to take only lizards, especially skinks (Longley 1945b; Mertens 1966; Webb 1973; Kluge 1974; Robinson

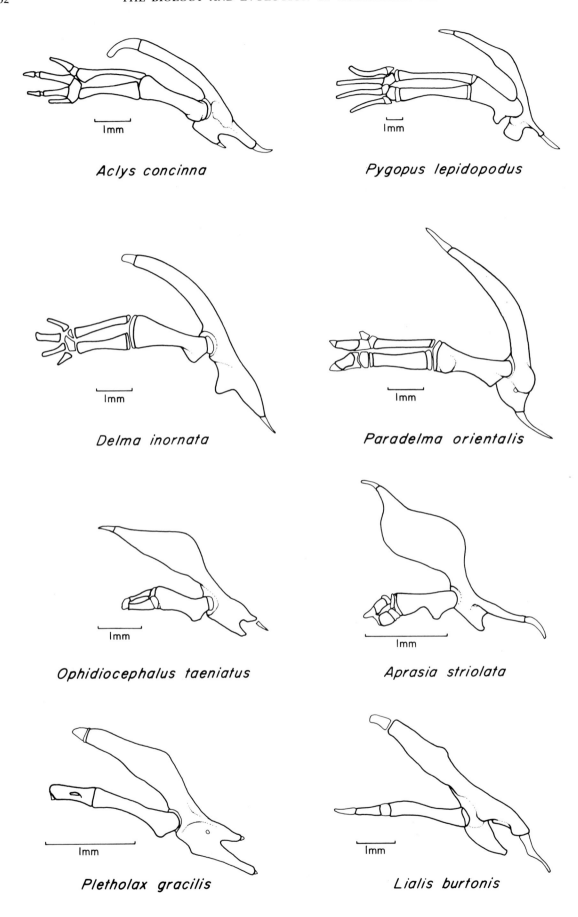

Aclys concinna

Pygopus lepidopodus

Delma inornata

Paradelma orientalis

Ophidiocephalus taeniatus

Aprasia striolata

Pletholax gracilis

Lialis burtonis

Fig. 47(a). The pelvic gridle and rear limb bones in representatives of the eight pygopodid genera.

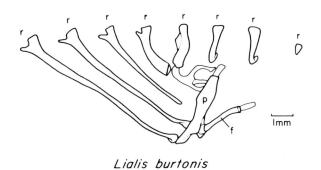

Lialis burtonis

Fig. 47(b). Left side of the pelvic area of *Lialis burtonis* (AM R 39552) to show the separation of the sacral diapophyses and ligament attachments between the diapophyses and the ilium. Note also the anterior process near the head of each rib. Abbreviations: f – femur; p – pelvis; r – rib.

1974; McPhee 1979; Patchell and Shine 1986a) and the occasional snake (Bustard 1970; Kluge 1974; McPhee 1979). *Lialis burtonis* is also cannibalistic in captivity (Peters 1976b). The food habits of *Paradelma* and *Pletholax* are unknown. In nature, the only even partially herbivorous pygopodid is *Aclys*, which had just over 20% (by volume) plant material in its diet (Murray 1980). However, in captivity *Delma*, *Pygopus* and apparently even *Aprasia*, but not *Lialis*, have been observed to eat fruit (Martin 1972; McPhee 1979), and *Delma* and *Pletholax* have been seen to lick sweet fruit (apple and banana) and to lap honey water (H. Ehmann, pers. comm.).

The lizard-eating habits of *Lialis* are worth emphasizing, for it appears to be more than just a strong preference for this kind of prey. *Lialis* may, in fact, not even recognize arthropods as potential food. This is supported by there being no specific first-hand accounts of *Lialis* eating arthropods (e.g., Patchell and Shine 1986a) and by observations of hungry *Lialis* showing no interest in offered arthropods but avidly taking lizards (Mertens 1965). One wonders what sense mediates these specific likes and dislikes — vision or smell/taste. Some simple experiments might answer this question.

Little is known about how pygopodids hunt their prey, other than that *Pygopus* and *Delma* appear to be active searchers (Patchell and Shine 1986a) and *Lialis burtonis* appears to sit in a concealed position (grass clumps or tussocks) and ambush passing prey (Webb 1973; M. Robinson 1974; Patchell and Shine 1986a). *Pygopus* may enter spider burrows in search of prey (Patchell and Shine 1986a).

Delma inornata (Kästle 1969, as *D. fraseri*) and *Pygopus nigriceps* (Philipp 1980) have been observed subduing prey in an unusual fashion — by rotating the body along the long axis two or three times, thereby presumably disorienting and perhaps injuring the prey. As other pygopodids are capable of body rotation in other contexts, i.e., defense when grasped and tunnelling (p. 106), this method of subduing prey may be more widespread than these few observations would indicate.

Courtship in pygopodids has been observed only in *Lialis burtonis* where it is said the male grasps the female almost anywhere on the body or tail with his jaws and then slowly works his way forward, rippling his body over hers, until a nape grip, the primitive one for lizards, is obtained and mating occurs (M. Robinson 1974).

As far as is known all pygopodids are egg-layers usually with a clutch size of two but rarely one (Table 8; Patchell and Shine 1986a). Mean relative clutch mass (1) in the two pygopodids surveyed to date ranged .26-.38 (Table 9), which is similar to the range for geckos. The seasonality of reproduction has been commented upon only incidentally and then only in southern (temperate) species. In these, spring mating and summer egg laying appears to be the norm (references in Table 8). Incubation times, known only for *Delma* and *Pygopus*, are 66-77 days (Table 10). Pygopodid eggs are characteristically much longer than wide (Table 11).

Some pygopodids lay communally. In western Australia, community laying sites of *Pygopus lepidopodus* are usually under well-embedded rocks in sandy soil, with as many as 20 eggs per site (Bush 1981). In the east, a group of six eggs, four hatched and two developing, were found together on one occasion (Wells and Husband 1979), eight developing eggs on another (Hoser 1983), and 76 "hatched/spoiled" eggs on another (Lunney and Barker 1986). In *Lialis burtonis* masses of eight (Robinson 1974) and 20 (McPhee 1979) eggs have been discovered together.

Very little is known about longevity in pygopodids. In captivity, an *Aprasia striolata* lived for two years, a *Pygopus nigriceps* for almost seven years (Mertens 1926 as cited by Mertens 1966) and a *Delma inornata* for over five years (Kästle 1969, as *D. fraseri*).

Almost nothing is known of the population structure or dynamics of any pygopodid. In *Lialis*, but presumably not other species, females are more commonly collected than males, but whether this is a true reflection of the occurrence of the sexes in natural populations remains to be determined. Female pygopodids also generally attain larger size than males, although the cause and significance of this difference is obscure (Kluge 1974; Murray 1980; Patchell and Shine 1986a). Occasionally, certain pygopodids are

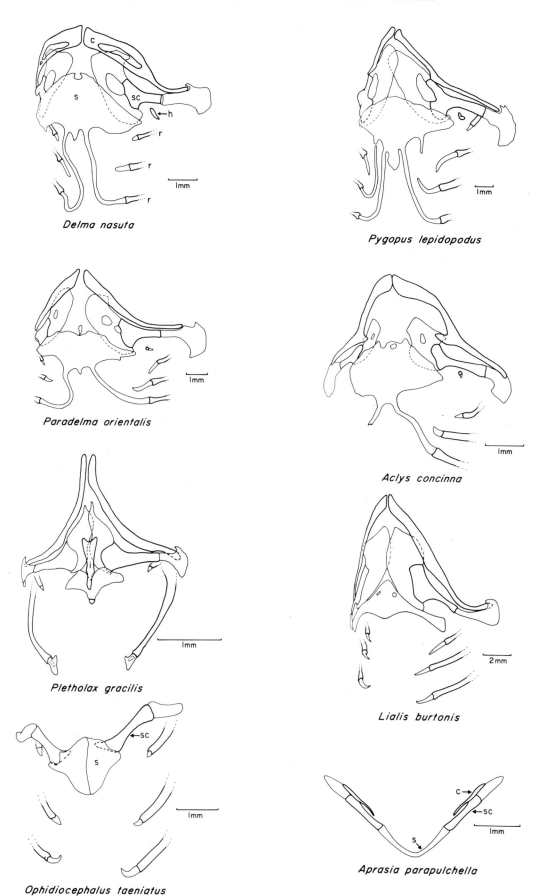

Fig. 48. The pectoral girdle in ventral view in representatives of the eight pygopodid genera. Abbreviations: c — clavicle; h — humerus; i — interclavicle (*Pletholax* only); r — rib; s — sternum; sc — scapula.

found together in large numbers. While these local densities are suggestive of some degree of social behaviour beyond simple mating encounters, they could also simply be the accumulation of individuals in favourable parts of the habitat. Bush (1981, 1986) reports "astounding" local increases of *Delma australis* and *D. fraseri* during December and January and suggests it may be related to breeding. He also notes that it is not uncommon to find several *Aprasia repens* under a single rock.

Next to geckos, pygopodids have the best developed voice of any lizard: a distinct high pitched squeak very similar to two wet balloons being rubbed together. This is usually given only under stress but may also play a role in social behaviour. So far a voice has been recorded in *Aprasia* (Martin 1972), *Delma* (Sonneman 1974; Weber and Werner 1977; Bush 1981; Annable 1983), *Lialis* (Worrell 1963; Bustard 1970d; Weber and Werner 1977) and *Pygopus* (Mertens 1966; J. Coventry, pers. comm.). A voice has not been recorded in *Ophidiocephalus* or *Paradelma*.

When uncovered, surface-active pygopodids usually attempt to flee quickly into the surrounding vegetation while more fossorial forms undulate into the litter. *Delma* often performs a wild tail "dance" when first disturbed (p. 106). When first grasped, pygopodids often bring the tail forward in an attempt to brush-off the attacker, and if firmly grasped, often twist powerfully along the long axis in an attempt to break free. Both behaviours are common in elongate, attenuate squamates. Some species, e.g., *Delma tincta* and *Lialis burtonis*, have been reported to bite when held (Johnson 1975), but I have never experienced this myself.

The species of pygopodids with dark nuchal collars, e.g., some *Pygopus*, some *Delma* and *Paradelma*, have often been thought to be mimics of those surface active, diurnal elapid snakes with similar dark collars e.g., the whipsnakes, *Demansia*, and juvenile eastern brown snakes, *Pseudonaja textilis* (Hall 1905; Kinghorn 1926; Waite 1929; Deckert 1940). This morphological similarity is carried a step further in *Paradelma* and *Pygopus* in that all three species show a defensive behaviour somewhat similar to these snakes. They raise the front part of the body off the ground, simultaneously arch and compress the neck, sometimes extend and flicker the tongue, and often strike (Deckert 1940, as reported in Mertens 1965, 1966; Serventy 1951; Bustard 1968c, 1970d; Bosh 1981; Fig. 49). However, the similarity ends there as they apparently only strike with the mouth closed (Bustard 1968c; Johnson 1975).

Fig. 49. Paradelma orientalis in snake mimicking posture. Photo: S. Wilson.

Even hatchling *P. lepidopodus* are said to show snake mimicry behaviour (Fitzgerald 1983a). Although this "snake mimicry" is very impressive to a human observer, it may be less so to more closely related predators, because a fully displaying *P. nigriceps* was devoured by a brown snake *Pseudonaja textilis* (Johnson 1975), perhaps the one species best suited to detect mimics from models! *Paradelma orientalis* also shows snake-mimicry behaviour in raising the forepart of the body and flicking the tongue (Shea, in press; S. Wilson, *in litt.*), but it has not been seen to compress the neck or to feign-strike.

Ontogenetic colour change occurs in some pygopodid species with light and dark crossbands on the head and nape. These bands are pronounced in the young but become muted or lost as the animals age (e.g., *Delma borea, D. elegans, D. fraseri, D. pax* and *Pygopus nigriceps*. Seasonal colour change has been reported in *Ophidiocephalus* where the bronze dorsal colour is said to become "richer" during the summer (Ehmann 1981).

Short-term colour change has been reported in *Lialis burtonis, Pygopus lepidopodus* and *Delma nasuta*. In the former two species, it is said to consist of an overall lightening upon heating (Mertens 1966), but in the last, it is a little more complex. For example, within seconds of sloughing the yellow belly goes white but returns to yellow a day later. Ventral contact with a moist surface or spraying the whole animal with a fine mist causes a dull yellow venter to turn bright lemon-yellow, but this can be lost within 30 seconds if the animal is disturbed. In contrast to many other lizards with short-term colour change, mild cooling or heating caused no change in ventral colour (Shea 1987c). There appears to be no known case as yet of sexual dichromatism in pygopodids.

THE LINEAGES OF PYGOPODIDS

The genera *Pygopus* (two species) and *Paradelma* (one species) are considered to be the most generally primitive living pygopodids.

Indeed, they are thought to differ relatively little from the hypothetical ancestor of the family. Hence when we look at one of these species we may be looking at an animal not too dissimilar from the ancestral pygopodid.

Pygopus is a familiar, but still poorly known, genus consisting of two of the largest and geographically most widespread species, at least one of which occurs near every mainland capital city. *P. lepidopodus* occurs along the southern temperate edge of the continent and *P. nigriceps* throughout the centre and north. The former has strongly keeled scales but the latter only weakly so, and one wonders if this difference relates to any environmental difference implied by their largely non-overlapping distributions.

Pygopus lepidopodus occurs in two colour forms: plain, and with three longitudinal rows of dark blotches along the back. Both phases are common in the west but the blotched phase is rare in the east. The genetic basis and functional significance of this polymorphism is unknown.

Pygopus lepidopodus has been the subject of one of those observations that is so unusual that it makes one wonder just how much, or little, we know about some very familiar animals. The observation was of a captive specimen apparently excavating a tunnel by rotating its body around the long axis in the hole then removing the loose soil with its mouth (Husband 1980a). Certainly these are heretofore unreported functions for keels (as awls) and jaws (as shovels). But what was the animal doing? Is this a method to obtain tunnel-inhabiting spiders that appear to make up a large part of its diet (Patchell and Shine 1986a)?

Pygopus nigriceps appears to constitute two populations, each morphologically distinct enough to be named a subspecies, *nigriceps* in the west and *schraderi* in the east. The north-south zone of change-over in central Australia is said to be fairly narrow (Kluge 1974), and closer examination of specimens from this area might sharpen our understanding of the nature of the differences between the two populations.

Little is known of the biology of the rare, monotypic genus *Paradelma* from central eastern Queensland. In nature it has been found in woodland under surface cover (Shea, in press; S. Wilson, *in litt.*), and in captivity it shelters under cover rather than burrowing into the litter (Shea, in press). One of the two specimens in the Australian Museum was said to have been killed by a car on the road, which at least suggests that the species is occasionally active on the surface. The specimen was carrying enlarged

ovarian eggs (too scrambled to count) on 14 November which is indicative of reproduction in late spring and early summer.

Delma and *Aclys* are thought to be closely related, with *Aclys* perhaps even being derived within the *Delma* radiation. To date the evidence for the close relationships of these two genera has been based on a suite of morphological characters, none exclusive to the lineage. However, there is one external character that easily diagnoses the group. In all other pygopodids the first pair of infralabials is separated by one or more small chin scales just behind the mental, but in the *Delma-Aclys* group, the first pair of infralabials usually meet behind the mental (pers. obs.).

Delma is the largest pygopodid genus, currently comprising 13 species, just under half the total number in the family. Unfortunately, the diagnosis of *Delma* is not very sound, hence it is uncertain if this is a true lineage or just an assemblage of similar species. However, there is one slightly odd behaviour trait that is known to date among some *Delma* and that may be indicative of a close relationship among these species. It is a wild, repeated thrusting upward of the body and tail on what seems to be the posterior half of the tail when the lizard is uncovered. The action is generally repeated three or four times, often accompanied by high pitched, squeaks before the lizard suddenly comes to ground and heads off into the litter. The whole show takes only a few seconds and is usually over before the startled observer has time to collect his wits and react — which may be the whole point. This behaviour, or an apparent form of it, has been recorded to date in four species: *D. australis*, *D. fraseri*, *D. grayii* and *D. tincta* (Bush 1981; Maryan 1984b, 1985; Bauer 1986).

The genus *Aclys* comprises only one species (*A. concinna*) and occurs only along a very narrow stretch of the south-west coast between the Shark Bay region and Perth. For several years it was only known from two specimens, but as often is the case with rare animals, specimens started to "turn up" more regularly when people learned how and where to look for them. Now several tens of specimens are known. Faunal control authorities take note.

Pletholax is another monotypic genus from the south-west coast. It is unique amongst pygopodids in several regards. For example, it has retained the primitive features of an interclavicle and a low number of presacral vertebrae, 44-49 (Table 3), but it has developed the derived features of keels on the ventral scales (as well as the dorsal) and has the tips of the neural spines expanded dorsally into a kind of winged keel. One wonders if these last two features are not related to the species' arboreal habits.

The genus *Lialis* is the most widespread genus of pygopodids and the only one occurring in New Guinea. There are two species: *L. jicarii* which occurs only in New Guinea and perhaps the Bismarck Archipelago and *L. burtonis* which occurs throughout mainland Australia and southeastern New Guinea and is by far the most widely distributed pygopodid.

Lialis differs most noticeably from the other large, surface-active pygopodids — and probably the common ancestor of pygopodids — in head morphology and function, and these differences are probably related to its most common prey — skinks which are tough and slippery by virtue of their smooth, overlapping scales, each of which contains a tiny osterderm (Fig. 57). For example, whereas all other pygopodids have a short, deep, bluntly rounded head and immovable snout, *Lialis* has a long, flat, slightly pointed head and a snout which bends at the fronto-parietal joint between the eyes. Deflection at this joint may be as much as 40° (Fig. 50). In addition, the middle part of the maxilla is concave ventrally thereby forming a natural curve in the upper jaw (Fig. 51; Patchell and Shine 1986b). Also, whereas other pygopodids have paired parietal bones *Lialis* has them fused. Furthermore, whereas all other pygopodids have a few, relatively bluntly conical straight teeth which are fixed in a vertical position, *Lialis* has many, fine, sharply pointed, recurved teeth which are hinged to bend backwards (Patchell and Shine 1986c).

The longer, lighter skull in *Lialis* (Fig. 51) may allow for a faster, more pincer-like trap for its fast-moving, hard-to-grasp prey. The movement at the fronto-parietal joint may allow the muzzle to curve around the prey anteriorly and prevent

Fig. 50. Lialis burtonis feeding on a skink. Note the strong ventral deflection of the snout at the frontoparietal joint.

escape, while the fused parietals may form a more solid hinge at this joint. The curvature of the maxilla would also act as a passive receptacle for prey. The numerous, fine, backward pointing teeth that bend easily toward the throat but resist movement in the opposite direction may be a passive mechanism for facilitating the one-way movement of prey toward the throat. A similar tooth morphology also occurs in a number of snakes which feed almost exclusively on skinks (Savitsky 1981).

On a per item basis, the lizard prey of *Lialis* appear to be larger than the arthropod prey of other pygopodids (Kinghorn 1924; McPhee 1979; Patchell and Shine 1986a). This has implications for feeding strategies and energy metabolism which have yet to be explored. For example, it may be that in taking relatively large prey, *Lialis* feeds infrequently, or alternatively if it feeds frequently, it may process food more quickly than other pygopodids. Morphologically, the larger prey may "explain" why *Lialis* is the only large pygopodid to have lost the sternal rib attachments — thereby making this part of the body cavity more expandable as large prey passes posteriorly.

Lialis burtonis feeds quite well in captivity, and I have watched entire feeding sequences many times in the laboratory. When a skink is introduced into a terrarium containing a *Lialis*, sitting in hummock grass, the *Lialis* immediately becomes alert and keys on the prey. It then begins moving slowly through the grass toward the skink, the motion being almost imperceptible and seemingly restricted to periods when the skink itself is moving. When *Lialis* reaches the edge of the hummock grass nearest the skink it stops and waits for the skink to move closer. When the skink is within range (anything within a *Lialis* body length away) the *Lialis* throws itself out of the hummock, arrow-straight, to grab the prey in its mouth. The strike is quick and accurate.

Initially the prey struggles violently, but the *Lialis* simply bites down hard with head raised and waits for the prey to exhaust itself. So firm is the grasp that the lizard is squashed flat and unable to rotate. Unlike many other lizards, *Lialis* makes no attempt to bash the prey against the substrate. However, it may carry the prey back into the hummock grass to begin swallowing it. Almost invariably the prey is grasped initially sideways and in this position the *Lialis* begins working its grip forward toward the prey's head. It may stop at the chest as if squeezing the last bit of life from the prey or it may proceed more or less directly to the head. The prey is swallowed head first with a series of vertical neck-kinking motions to push the prey

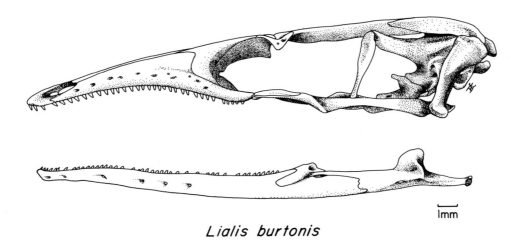

Lialis burtonis

Fig. 51. Skull of *Lialis burtonis* (AM R 105002). Compare with Fig. 42.

into the esophagus and then a series of posteriorly travelling lateral neck waves to propel the prey toward the stomach. Other than the downward deflection of the snout while subduing the prey, no other gross displacement of the head bones seems to occur during feeding. When swallowed the prey may be either dead or alive, and even before the tail has disappeared from the mouth, the *Lialis* will be ready to take another prey if available. After feeding the *Lialis* methodically wipes its face and jaws with its tongue.

Once the feeding sequence has begun, that is, from just after the prey is grasped in the jaws, *Lialis* is almost oblivious to distraction. It can be picked up, handled, photographed, and examined under the microscope, and the feeding will continue, I have noticed the same obliviousness in feeding *Simoselaps*, small semifossorial elapid snakes which feed largely on *Lerista*, and one wonders if it is not a feature of squamate predators that generally feed on large, and, by implication, infrequently caught prey.

Another unusual aspect of *Lialis* biology is that both species occur in a number of different colour patterns, as many as nine having been attributed to *L. burtonis* (Boulenger 1885; Fig. 52). In this species, the patterns can vary from local to geographically widespread and two patterns often occur in one population. The genetic basis for the patterns and their functional significance are unknown. It is not even known whether mothers produce offspring patterned like themselves or if siblings are similarly patterned.

Although pygopodids appear to have evolved their reduced limbs and elongate body while living above ground, they have taken this morphology underground on at least one occasion in their evolution, as evidenced by the litter swimming *Ophidiocephalus-Aprasia* lineage.

This "experiment" shows that although limb-reduction and body elongation are virtual prerequisites for life underground in squamates, they need not evolve only in response to such habits. They may evolve for other reasons and only later be used to advantage in the "new" underground environment.

Ophidiocephalus and *Aprasia* share a number of modifications that are common in burrowers. Externally, they show an overall simplification of the head squamation which has probably resulted from a series of scale fusions, the significance of which may have been to reduce the number of transverse sutures, i.e., the sutures transverse to the primary direction of the net force exerted in locomotion. Internally, the humeral fragment of the fore limb has disappeared, and the pectoral girdle is reduced to remnants of the scapulocoracoids and clavicles. The rear limb lacks all elements distal to the tibia and fibula, and the ilium is expanded in its sacral process. The two taxa also share a slow wriggling of the raised distal end of the tail when disturbed; this is thought to be a predator distraction device (Rankin 1976; Ehmann 1980).

Ophidiocephalus taeniatus, the only species in the genus (Fig. 53) is another good example of how a rare animal can suddenly become common — once one learns how to look for it. The species was described in 1897 from a single specimen received from Charlotte Waters, then a station on the Alice Springs-Adelaide telegraph line, now a pile of rubble. As the actual site is on gibber plain, a very unlikely habitat for a pygopodid, it has always been assumed that the type specimen came from a more suitable nearby habitat and that "Charlotte Waters" was meant only in a general sense. Hence searchers have always concentrated their efforts in more likely looking areas in the general vicinity. Several herpetologists have made trips (pilgrimages?) to

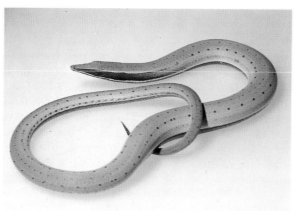

Fig. 52. Three colour forms of *Lialis burtonis*.

the area to look for *Ophidiocephalus* but none was successful until 1978 when two amateurs found the species near Abminga, just over the South Australian border about 26 km south of Charlotte Waters (Ehmann and Metcalfe 1978; Ehmann 1979). The discoverers have since returned to the site and at least 56 specimens have been observed (Ehmann 1981). In addition, another specimen has been found at Coober Pedy extending the range over 300 km to the south (Ehmann 1981).

Ophidiocephalus occurs in the litter mats under trees and shrubs, a microhabitat very similar to that occupied by certain species of *Lerista* . In

captivity it exposes itself on the surface only at night, and in nature its activity is probably restricted to the subsurface. It is a good litter swimmer, and when disturbed undulates into the litter or down an earth crevice. The species can also move backward in its track through the litter. The prey seems to consist of a variety of small arthropods. Conjectural evidence suggests spring mating and summer reproduction, but there is no information as yet on mode of reproduction or brood size. In captivity the species seeks warm spots within the substrate, but without actually exposing itself on the surface (Ehmann 1981).

Aprasia consists of ten species whose present distribution is roughly along the southern quarter of the continent, generally in association with sandy substrates. However, the occurrence of outlying northern populations of two species, *A. rostrata* on Hermite Island in the Monte Bello Group and in the Exmouth region, and *A. striolata* at Mt. Buring in south central Northern Territory suggests that at least these two representatives of the genus once ranged further north, perhaps during a less arid period.

The genus is one of the most highly modified group of lizards known. The internal structure of the eye is simplified in a number of ways; for example, the sulcus, annular pad and scleral cartilages are all lacking (Underwood 1957, 1970). The two halves of the shoulder girdle are fused on the midline at the epicoracoids to form a single transverse structure (Stokley 1947; Kluge 1976). A similar structure occurs only in the North American, sand swimming, legless lizards of the family Anniellidae, but there it is thought to be composed of the fused clavicles. The common carotid artery, the vessel carrying blood from the heart to the head, bifurcates into the interior and exterior carotid arteries, not posteriorly near the heart as in most lizards, but well forward in the neck (Underwood 1957). A similar anterior bifurcation occurs in the south-east

Fig. 53. Ophidiocephalus taeniatus, one of the least well-known pygopodids. Photo: P. R. Rankin.

Asian, North American family Dibamidae, a group which resembles pygopodids in having no external front limbs and only flap-like rear legs (Greer 1985a). It is also said that all *Aprasia* lack maxillary teeth (Parker 1956; Stephenson 1962), a condition otherwise known in squamates only in leptotyphlopid snakes. In fact, just how unexpected the absence of maxillary teeth is amongst lizards is evidenced by one author adding them to his figures using dashed imaginary lines to show what he could not see but felt should be there (Rieppel 1984a: fig. 9 and 1984b: fig. 7)! *Aprasia* is also unique among pygopodids in having a very short tail, 80 percent of SVL or less as opposed to 123 percent or more (data from Kluge 1974: table 12).

The habits of *Aprasia* are only poorly known. It is primarily cryptozoic and is generally found under cover or taken in pit-traps. It is often found in and around termite and ant tunnels under superficial ground cover, suggesting that they may inhabit such tunnel systems. The genus is also said to be "largely, if not exclusively myrmecophagous" (ant eating) (Kluge 1976), but supporting evidence has not yet been advanced for this. Unfortunately, the stomachs of 45 specimens of four *Aprasia* species examined specifically to determine prey type were all empty (Patchell and Shine 1986a). Information on reproduction is limited to data on clutch size (2) for three species (Table 8). In several species the tip of the tail is a vivid orange/red colour (Fig. 54) and is often held upright and slowly writhed when the animal is disturbed; this has led to speculation that the colour and behaviour are a defensive (diversionary) device (Bustard 1970d; Rankin 1976; Ehmann 1981). Sloughed skins have been found tied in knots (Kluge 1974), implying that the animals tie themselves into knots, perhaps to facilitate sloughing. Some sea snakes are known to swim themselves through loose knots when shedding (R. Shine, pers. comm.; pers. obs.) and as a means of scrapping off external parasites.

All *Aprasia* species except *A. aurita* lack an external ear opening and canal. In *A. aurita* and other pygopodids there is an ear canal at the bottom of which is a thin tympanic membrane or eardrum (pers. obs.). Attached to the inner side of this membrane is a thin bone, the stapes, which extends horizontally across the middle ear cavity, to fit by way of a flat circular base, the footplate, into a circular hole in the skull. On the other side of this hole in the skull is the complex inner ear structure for translating the back and forth movement of the stapes into electrical impulses which are transmitted to the brain and perceived as sound. The stapes itself is set in motion when the ear drum vibrates in response to air borne pressure waves — sound.

Fig. 54. Aprasia inaurita (top) and *Aprasia* sp. (bottom) showing the vivid pink to rose tail colour typical of many species in the genus. Photos: P. R. Rankin.

In earless *Aprasia*, the external ear opening and canal seems to have been lost by progressive constriction and then fusion. In most of these species (e.g., *A. parapulchella* and *A. striolata*, pers. obs.) the stapes attaches to the underside of the skin in the same way it would have attached to the tympanum in eared forms. However, in at least one species, *A. repens*, the stapes has lost its horizontal shaft and only retains a nodular remnant of the footplate which has a ligamentous attachment to the posterior side of the quadrate (pers. obs.), the bone that forms the anterior side of the middle ear cavity.

Presumably the one species of eared *Aprasia* can perceive air-borne sound as well as any other pygopodid; the earless *Aprasia* with the stapes attached to the skin can also probably still hear air-transmitted sound but perhaps less acutely because a scaly skin would not respond as sensitively as a delicate membrane. However, the earless *Aprasia* with the remnant of the stapes attached to the quadrate can probably hear air-borne sound only poorly if at all, because the quadrate would appear to be relatively unresponsive to air-borne pressure waves. Instead, this species may be able to hear underground sound, as the larger earth particles which are involved in this sort of vibration could move the quadrate and the attached stapedial remnant.

If this is the way the ear works in different *Aprasia* species we might expect an ecological basis for the difference; for example, the one eared species might be more surface dwelling; the earless species with the stapes attached to the skin, somewhere more subterranean, and the earless species with the stapes attached to the quadrate the most subterranean. Interestingly, the last form has the snout drawn out into a proboscis, as if it is adapted to a more burrowing existence.

Mean presacral vertebral number appears to be directly correlated with mean December (early summer) temperatures in at least two species of *Aprasia: A. repens* and *A. striolata* (Parker 1956; Kluge 1974), but whether the effect is phenotypic or genotypic is unknown. Ventral scale number in snakes, which is highly correlated with vertebral number, often varies clinically and is known to have both a heritable (S. Arnold, pers. comm.) and environmental (Fox 1948; Fox *et al.* 1961) component in development.

It might be fitting to conclude the discussion of pygopodids by asking what they have to tell about the evolution of that large and most difficult to understand group of legless lizards, the snakes. There are perhaps two lessons to be learned. The first has to do with the possible habitat associations of the first snakes. The currently favoured hypothesis about snake origins takes note of the strong correlation in squamates between limb reduction and body elongation on the one hand and burrowing habits on the other. It combines this with certain modifications, especially of the eye, which appear to occur only in burrowers, and proposes that snakes had a burrowing ancestry (Walls 1940; Underwood 1957; Bellairs 1972). However, as we have seen, pygopodids probably did not reduce their limbs and become elongate, i.e.,

become snake-like, in a burrowing mode. Rather they evolved this morphology while still in a surface-active mode and then one lineage (*Aprasia-Ophidiocephalus*) went underground and evolved further modifications typical of burrowers. Perhaps snakes, too, have followed such an evolutionary path and gone through a burrowing phase only after having undergone limb reduction and body elongation.

The second lesson has to do with how quickly and how completely a squamate lineage can go from limbed to virtually limbless. The fact that there are now no "intermediates" between the nearly fully limbed geckos and the nearly limbless pygopodids raises the possibility that such intermediates never existed, i.e., that the transition was sudden and virtually completed in one step. Of course we can still not exclude the possibility that intermediate forms did exist but are now extinct.

Why might pygopodids have undergone this marked transition in body form? One hypothesis suggests that it had to do with entering burrows, especially spider burrows in search of prey (Shine 1986a). However, many geckos use burrows, even spider burrows, and they are stout-bodied and strongly limbed. An alternative hypothesis that focuses on a more general difference between geckos and pygopodids may be the speed with which they are able to move through confined spaces such as litter and vegetation. An elongate, attenuate and limb-reduced form would permit much more rapid passage through these confined spaces than would a short, wide, limbed form. Such increased mobility would enhance both prey capture and predator flight in a very common lizard structural microhabitat — crevices (Gans 1975), and who knows, perhaps a pygopodid-like form is even faster than a gecko-like form in the open. The idea would be worth testing.

Table 1. Derived character states shared by the Gekkonidae and Pygopodidae.

External morphology
  Dorsal crest absent

Behaviour
  Ability to vocalize
  Wiping of face and jaws with tongue

Skull
  Palpebral absent
  Frontals encircle forebrain
  Parietal foramen absent
  Parietals posteromedially convex
  Parietals lack fossa for ascending process of occipital tectum
  Septomaxilla with posterolateral wing
  Lacrimal absent
  Jugal lacks dorsal process
  Postorbital absent
  Supratemporal arch incomplete
  Epipterygoid abuts underside of prootic's alar process
  Stapes lacks dorsal process
  Recessus scalae tympani with two medial apertures
  Pterygoid teeth absent
  Ectopterygoid extends along anterolateral side of infraorbital vacuity
  Parasphenoid process unossified
  Dentary overlaps Meckel's groove

Postcranial skeleton
  Intermedium absent
  Postcloacal bones present

Soft anatomy
  Parietal eye absent
  Extracolumellar muscle present (Wever 1978)
  Cochlear limbus of inner ear dorsally enlarged and curved (Shute and Bellairs 1953)
  Tectorial membrane of inner ear with spindle body (Wever 1974)
  Rectus superficialis muscle absent
  Postcloacal pockets present

Reproduction
  Brood size much reduced (≤ 3), usually a constant 2

Table 2. Snout-vent length/head length in adult pygopodids to show the degree of elongation of the body. Sample size in parentheses.

Taxon	SVL/HL
Delma	
borea	8.8-9.7(5)
inornata	8.9-10.4(8)
plebia	8.8-9.8(6)
Lialis	
burtonis	9.0-10.6 (13)
Pletholax	
gracilis	10.6 (1)
Pygopus	
lepidopodus	10.9-12.7 (12)
nigriceps	11.4-13.0 (14)
Ophidiocephalus	
taeniatus	13.9 (1)
Paradelma	
orientalis	13.3-15.5 (2)
Aprasia	
pseudopulchella	21.1 (1)
repens	21.7 (1)

Table 3. Vertebrae number in pygopodids. Sample sizes in parentheses.

Taxon	Vertebrae	
	Presacral	Postsacral
Pletholax		
gracilis	44-49(19)	119-129(10)
Aclys		
concinna	44-49(12)	124-130(3)
Delma		
australis	52-55(10)	74-77(2)
inornata	52-57(10)	100-114(2)
molleri	53-58(7)	90(2)
plebeia	56-60(10)	91-97(3)
tincta	52-62(13)	109(1)
Ophidiocephalus		
taeniatus	65-66(2)	—
Pygopus		
lepidopodus	74-80(6)	105(1)
nigriceps	76-94(11)	82-84+(3)
Lialis		
burtonis	77-93(15)	69-77(3)
Paradelma		
orientalis	85-92(12)	—
Aprasia		
inaurita	95-110(6)	45-47(2)
pulchella	90(1)	—

Table 4. Lung length in pygopodids.

Taxon	SVL	Lung Length (%SVL)		LL/RL
		Left	Right	
Lialis				
burtonis	192	32 (17)	47(25)	.68
	280	45 (16)	132(47)	.34
Paradelma				
orientalis	88	17 (19)	27(31)	.63
Pygopus				
lepidopodus	207	35 (17)	68(33)	.51
nigriceps	189	37 (20)	62(33)	.60
	160	28 (17)	42(26)	.67
Delma				
borea	86	15 (17)	32(37)	.47
inornata	114	17.5 (15)	38(33)	.48
plebia	95	15 (16)	30(32)	.50
Pletholax				
gracilis	57	6 (11)	15(26)	.40
Ophidiocephalus				
taeniatus	107	11 (10)	28(26)	.39
Aprasia				
inaurita	100	4 (4)	27(27)	.15
striolata	110	5 (5)	57(52)	.09
	101	4 (4)	35(35)	.11

Table 5. Activity times of pygopodids based on observations of animals seen on the surface both in the wild (W) and in captivity (C).

Taxon	Diurnal	Crepuscular	Nocturnal
Aclys			
concinna	Murray 1980(W)	Murray 1980(W)	Murray 1980 (W)
Aprasia	Bustard 1970d(C)		
pulchella	Martin 1972(C)	Martin 1972(C)	Martin 1972(C)
striolata	Martin 1972(C)	Martin 1972(C)	Martin 1972(C)
Delma	Bustard 1970d(C)	Bustard 1970d(C)	Bustard 1970d(C)
australis	Maryan 1984a(C)	Giddings 1983a(W)	
borea			Brunn 1978(W)
fraseri	Martin 1972(C)		Maryan 1984a(W)
grayii	Maryan 1984a(C)		
impar	Martin 1972(C)		
inornata	Sonneman 1974(W; as *D. fraseri*)		Brunn 1980a(W)
nasuta			Brunn 1980a(W)
plebeia	Ludowici 1973(C; as *Delma*, new species), 1975 (C; as *Delma*, new species)		
Lialis			
burtonis	Bustard 1970d (C,W)	Neill 1957(W)	Neill 1957(W)
		Maryan 1984c(W)	Bustard 1970d(C)
			Brunn 1980a(W)
			Hoser 1983(W)
			pers. obs. (W)
Pygopus			
lepidopodus	Martin 1972(C)		Fitzgerald 1983(W)
	Smith and Chapman 1976(W)		Hoser 1983(W), 1984(W)
nigriceps			Brunn 1978(W)
			Fyfe 1979b(W)
			Philipp 1980(W)
			Maryan 1984a(W)

Table 6. Time in minutes to loss of righting reflex (up to 30 minutes) of species of pygopodids at different ambient temperatures (pers. obs.). Sample sizes in parentheses.

Taxon	Ambient Temperature				
	43.0-43.9	44.0-44.9	45.0-45.9	46.0-46.9	47.0-47.9
Aprasia *parapulchella* *striolata*	— 30(7)	— 30(2)	— 30(2)	30(4) 16-17(2)	— 10-15(2)
Delma *australis*	—	—	14(1)	—	—

Table 7. References to food eaten by Australian pygopodids both in the wild (W) and in captivity (C).

Taxon	References
Aclys *concinna*	Murray 1980 (W)
Aprasia *parapulchella*	Kluge 1974 (?) Jenkins and Bartell 1980
Delma *fraseri*	Kästle 1969 (C); Bustard 1970 (?); Dell and Chapman 1978 (W); Chapman and Dell 1980a (W); Patchell and Shine 1986 (W)
inornata	Kästle 1969 (C; as *D. fraseri*); Sonnemann 1974 (C; as *D. fraseri*); Annable 1983 (?); Patchell and Shine 1986 (W)
molleri	Briggs 1973 (C; as *D. fraseri*)
nasuta	Patchell and Shine 1986 (W)
plebeia	Ludowici 1973 (C; as *Delma* new species), 1975 (C; as *Delma* new species)
Lialis *burtonis*	Boulenger 1885 (W); Kinghorn 1924 (?); Longley 1945 (C); Loveridge 1949 (W); Mertens 1965 (C), 1966 (C); Bustard 1968? (C), 1970 (C); Kluge 1974 (W); M. Robinson 1974 (C); Peters 1976b (C); Dell and Chapman 1977a (W); Dell and Chapman 1978 (W); McPhee 1979 (C); Bradshaw *et al.* 1980 (C); Patchell and Shine 1986a, b (C, W)
Ophidiocephalus *taeniatus*	Ehmann 1981 (W)
Pygopus *lepidopodus*	Fitzgerald 1983a (C); Webb 1983 (W); Patchell and Shine 1986a (W)
nigriceps	Philipp 1980 (C); Patchell and Shine 1986a (W)

Table 8. Summary of female size and clutch size in Australian pygopodids. Means in parentheses.

Taxon	Size of ♀♀ (mm)	Clutch Size	N	Reference
Aclys				
concinna	—	2	?	Murray 1980
Aprasia				
parapulchella	—	2	2	Kluge 1974
repens	103	2	1	Pers. obs.
striolata	119-120(119.5)	2	2	Pers. obs.
Delma				
australis	70	2	1	Dell and Chapman 1978
	73-75(74.0)	2	3	Bush 1985
	59	2	1	Pers. obs.
borea	86	1	1	Pers. obs.
fraseri	—	2	?	Davidge 1980
	100	2	1	Dell and Chapman 1977
	120	2	1	Chapman and Dell 1978
	89	2	1	Dell and Chapman 1978
	115	2	1	Bush 1985
grayii	105	2	1	Chapman and Dell 1978
inornata	—	2	1	Sonneman 1974 (as *D. fraseri*)
	114-127(119.3)	2	3	Pers. obs.
molleri	—	2	1	Briggs 1973 (as *D. fraseri*)
nasuta	85-100(?)	2(2.0)	?	Smith 1976
tincta	67	1	1	Smith 1976
	—	2	1	Kluge 1974
Lialis				
burtonis	—	2	1	Loveridge 1949
	187	2	?	Neill 1957
	—	2	2	Webb 1973
	220	1	1	Chapman and Dell 1978a
	—	1-3(2.0)	35	Patchell and Shine 1986a
Pletholax				
gracilis	75	2	1	Mertens 1965
	—	2	?	Davidge 1980
	—	2	?	M. Bramford, pers. comm
Pygopus				
lepidopodus	—	2	2	Fitzgerald 1983a
	—	2	?	Davidge 1980
	175.0-210.0(192.0)	2	8	Wells and Husband 1979
	>160	2	2	Dell and Chapman 1977
	180	2	1	Dell and Chapman 1978
nigriceps	174-189(181.0)	2	3	Pers. obs.

Table 9. Relative clutch mass for pygopodids.

Taxon	RCM 1			Reference
	x̄	R	SD	
Delma				
australis	.38	.30-.44	.07	Bush, 1983 (N=3)
fraseri	.26			

Table 10. Incubation times and temperatures for pygopodid eggs.

Taxon	Incubation Period (days)	Incubation Temperature (°C)	Reference
Delma			
australis	66	28 ± 4	Bush 1985
fraseri	74-77	28 ± 4	Bush 1985
Pygopus			
lepidopodus	71	22-32	Fitzgerald 1983a

Table 11. Dimensions of laid pygopodid eggs.

Taxon	Size (mm)			Reference
	Length	Width	L/W	
Aprasia				
parapulchella	19.2	5.8	3.3	Kluge 1974
	21.3	5.0	4.3	
	17.5	5.6	3.1	
	20.2	5.3	3.8	
Delma				
australis	13	5	2.6	Bush 1985
	14	4	3.5	
	13	5	2.6	
	16	5	3.2	
	16.5	5	3.3	
fraseri	22	8	2.7	
	23	7.5	3.1	
inornata	23	10	2.3	Sonnemann 1974 (as *D. fraseri*)
molleri	21	8	26	Briggs 1973 (as *D. fraseri*)
tincta	13.8	6.0	2.3	Kluge 1974
	13.5	6.6	2.0	
torquata	12.8	5.9	2.2	Kluge 1974
Pletholax				
gracilis	16	5	3.2	Mertens 1965
	17	5	3.4	
Pygopus				
lepidopodus	38	16	2.4	Fitzgerald 1983a
	37	18	2.1	Wells and Husband 1979

CHAPTER 5

Scincidae — Skinks

SKINKS comprise the largest and most diverse family of lizards in the world. There are approximately 1 000 species, which is about 25-30 percent of the world's lizard species. In Australia, the percentage is even higher. The most recent overview (Cogger 1986) recognized 271 skink species and these comprised 57 percent of all lizard species in Australia. Skinks occur virtually everywhere in Australia, and there is no place on the continent which has lizards but lacks skinks. Skinks also cope well with Australian urban environments and therefore are well-known even to city dwellers. For example, in a residential garden less than 4.4 km north of the GPO in Sydney, seven species have been recorded (and eight if one counts another species resident in a nearby park), and even in the inner city suburb of Darlinghurst three species occur commonly and a fourth rarely (pers. obs.).[1]

Skinks are among the most morphologically diverse lizards in Australia and the world (Fig. 55). This is due largely to the great variation in the relative size of the limbs, trunk and tail. Limbs can vary from well-developed to absent, trunks from short and stout to long and thin, and tails from long to short. In addition the scales and hence the whole lizard, can be smooth, keeled, spiny or granular. Skink colours are usually thought to be drab but this view ignores the juvenile hues in many skinks, the seasonally variable hues in others and the ventral, "hidden" hues of still others.

Of the many features that distinguish skinks from other lizards (Table 1), perhaps the most remarkable is a bony secondary palate, a division between the air and food passages, which are confluent in most other lizards (Fig. 56). In primitive skinks the palate is only partial; the

bony shelves which form it (from the palatine bones) project only part way across the top of the mouth, leaving a broad gap between the two passages. In contrast, in advanced skinks — the Lygosominae, of which all the Australian species are members (Greer 1970, 1986a) — the two shelves meet or approach very closely on the midline to form a complete separation between the passages. Furthermore, in some of the advanced skinks the complete secondary palate is carried even further posteriorly by the incorporation of additional bones (the pterygoids) and/or by processes from these bones. The functional significance of the secondary palate in skinks has yet to be understood (Greer 1970).

Skinks, like many lizards, have a bone underlying each scale, called an osteoderm. In most other lizards with osteoderms each one is a single, solid bone, but in skinks it is composed of a mosaic of smaller pieces which form a very characteristic, symmetrical pattern (Fig. 57). This composite osteoderm is characteristic of skinks and can be seen in virtually every scale of the body, tail and limbs. The only other lizard group that shares this osteoderm pattern is the African family Gerrhosauridae, which show it on some of the neck and chest osteoderms. The fact that only skinks and gerrhosaurids show this osteoderm pattern suggests that they may be closely related.

As alluded to above, skinks are remarkable for the evolutionary changes in the relative size of their limbs, trunk and tail (Fig. 55). The most common, or at least, easiest changes to recognize are a shortening of the limbs, an elongation (Table 2) and attenuation of the trunk and either a shortening or elongation of the tail. These changes are in comparison to the relatively unchanging head length.

These changes in shape are also often evident in the skeleton. For example, the primitive number of presacral vertebrae in skinks is 26,

---

[1]The eight species are *Cryptoblepharus virgatus*, *Eulamprus quoyii*, *E. tenuis*, *Lampropholis delicata*, *L. guichenoti*, *Saproscincus mustelina*, *Saiphos equalis* and *Tiliqua scincoides*. The *Cryptoblepharus*, two *Lampropholis* species and *Eulamprus tenuis* occur in Darlinghurst.

Fig. 55. Four skink species to show the variation in basic shape within the family. From left to right: *Egernia napoleonis,*
Glaphryomorphus douglasi, Hermiergis peroni, and *Anomalopus gowi.*

but many Australian lineages show an increase in this number (Table 3), often, but not always, in conjunction with limb reduction, the highest numbers being in two of the limbless species: *Coeranoscincus frontalis* (72-76) and *Lerista apoda* (65-69). In species with an appreciably elevated number of presacral vertebrae, females usually average more vertebrae than males (e.g., Greer, in press; Choquenot and Greer, in press, and Table 4). It is difficult to determine the primitive number of postsacral vertebrae for skinks, but the range is wide, in Australian species from 14 in *Egernia stokesii* to 69 in *Glaphyromorphus gracilipes* (Table 3).

Skinks inherited the primitive squamate number and configuration of bones in the front and rear limbs, but many Australian lineages show loss of limb bones, often, but not always (e.g., *Carlia, Eroticoscincus* and *Lygisaurus*) in conjunction with body elongation and attenuation. Nine species of Australian skinks have lost all external trace of limbs: all four species of *Anomalopus (Vermiseps)*, one species of *Coerano-scincus,* two of *Lerista* and two of *Ophioscincus.*

In general, within any population of Aust-ralian skinks, females either average and/or attain a larger size than males or the two sexes

are equal in size (Pengilley 1972; Murray 1980; Greer 1983b; Brown 1983; Tilley 1984). That this is not true for all skinks is clear from the most generally primitive group of skinks, the northern hemisphere genus *Eumeces,* in which males are larger than females. In some cases, females appear to attain larger size because of greater longevity (e.g., *Eulamprus tympanus* — Tilley 1984) and in others because of faster growth. Also, for any snout-vent length, mature females tend to have smaller heads and larger bodies than males (Bamford 1980; Brown 1983; Tilley 1984; Simbotwe 1985; Bourne *et al.* 1986).

Skinks have two primary modes of propulsion in locomotion: with the limbs and with the body. The former is the mode employed by all limbed lizards and the latter the mode of all limbless lizards and snakes. Skinks with reduced limbs generally use limbed locomotion when on the surface and body locomotion, or lateral undulation as it is also called, when moving under the surface, usually in litter, friable soil or sand. When moving by lateral undulation, the limbs are carried tucked back along the side of the body (front limbs) or tail (rear limbs) in order to streamline the surface. The fact that at least some, perhaps all, skinks with well-developed

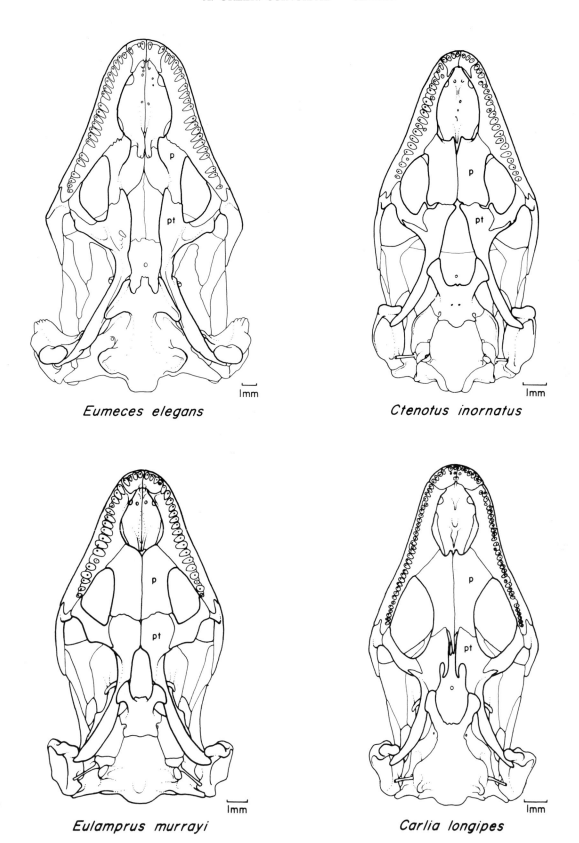

Eumeces elegans

Ctenotus inornatus

Eulamprus murrayi

Carlia longipes

Fig. 56. Ventral view of the skull of one North American and three Australian skinks to show the evolution and variation in the secondary palate: horizontal palatine (p) lamellae only partially developed anteriorly and only partially separating the air passage above and food passage below (*Eumeces elegans* — Museum of Comparative Zoology 28992); palatine lamellae in medial contact, completely separating air and food passages (*Ctenotus inornatus* — AM R 9718); palatal rami of pterygoids (pt) in medial contact behind palatines, extending posteriorly the secondary palate (*Eulamprus murrayi* — AM R 122923); pterygoids with processes also extending the secondary palate posteriorly (*Carlia longipes* — AM R 48357).

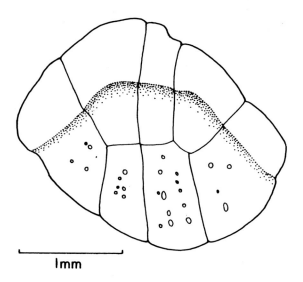

1mm

Coeranoscincus frontalis

Fig. 57. An osteoderm from the pectoral region of *Coeranoscincus frontalis* (AM R 61297) to show the characteristic symmetrical arrangement of the individual components of the skink osteoderm.

limbs can also move by lateral undulation (through vegetation and in litter), shows that the behavioural pattern for this form of locomotion precedes the morphological changes (i.e., body elongation and limb reduction) usually associated with it.

One of the most important taxonomic characters in skinks is the morphology of the lower eyelid (Fig. 58). However, it is only recently that we have begun to understand its adaptive significance (Arnold 1973; Greer 1983c). In skinks, as in other lizards, only the lower eyelid is moveable, being raised to close the eye and lowered to open it. Skinks, also like other lizards, have undergone a number of evolutionary modifications of the eyelid. Phylogenetic and embryological studies (Greer 1983c and references therein) indicate that the modifications have occurred in the following order: 1) the eyelid scaly and freely moveable; 2) the eyelid with a clear disc-like central scale, often called a window, but still freely moveable; 3) the eyelid filled almost completely by an expanded clear area and fused immoveably in the raised position, creating a permanent bubble or spectacle covering the eye, but with the fusion incomplete dorsally, and 4) the same as the preceding condition but with the eyelid fused all around. Among other Australian lizards the first condition, which is the most primitive, is characteristic of all dragons and goannas while the fourth, which is the most highly evolved, is characteristic of all pygopodids and most geckos (some non-Australian species have the first condition). In skinks the third and fourth conditions are sometimes called preablepharine

and ablepharine, respectively, ablepharine meaning "without eyelids", a mistaken interpretation.

In skinks, eyelid type is associated with size and habitat (Greer 1983c). Generally, large skinks and/or ones inhabiting mesic habitats have the primitive scaly eyelid, the small skinks and/or ones inhabiting dry habitats have the spectacle, and intermediate-sized skinks and/or ones inhabiting intermediate type habitats have a clear window in the lower eyelid. The explanation for this association seems to be that as animals become smaller in evolution, their eyes become relatively larger and therefore subject to relatively more evaporative water loss from the ever-moist cornea. Large skinks or ones in mesic habits don't have a big problem and hence retain the eyelid unmodified. Skinks of intermediate size and habitats have some problems, especially when they move out into the sun, but they have the capacity to voluntarily close the eyelid and cap the water loss in high risk situations. Small skinks are constantly at risk and have permanently capped the eye. The relationships between eye type and ecology are especially clear in skinks (and also certain other lizards) and may help explain the origin of the eyelid types in other lizards and snakes where the ecological relationships are not clear, perhaps due to subsequent evolution.

The size and shape of the external ear opening is another important taxonomic character in skinks but the adaptive significance of the several variants have yet to be determined. Primitively the ear opening was moderately wide with a short recess or canal at the bottom of which lay the tympanic membrane or eardrum. Several lineages have reduced or lost the ear opening and this has occurred in one of two ways, either by progressive constriction of the opening and canal or by the encroachment of scales across the tympanum. In the other direction the ear opening may increase in diameter and the canal become quite shallow or virtually non-existent, so that the tympanum is large and almost flush with the side of the head. Almost all the Australian skinks which have lost the external ear opening are cryptozoic to fossorial forms inhabiting humid microhabitats. Similar forms from dry microhabitats usually retain the ear opening.

Skinks have three breathing modes which are related to their morphology and habitat. The primitive mode, not only for skinks but for squamates in general, is lateral displacement of the axilla, the area just behind the forelegs; this area moves in on the power stroke (exhalation), thereby creating a lateral concavity, and out on the passive stroke (inhalation) with a bit of over-shoot before returning to the resting position. This axillary mode occurs in all skinks with well-developed limbs and those limb-reduced but not totally limbless skinks living in humus-rich soils. A second breathing mode is a vertical displacement of the chest just posterior to the level of the forelegs; this involves an upward movement on the power stroke thereby creating a ventral concavity and a downward movement on the passive, return stroke. This vertical mode occurs in skinks inhabiting loose, sandy soils, and may function to prevent sand falling into and filling the air pocket created on exhalation. It is a breathing mode well known in a variety of other sand-swimming squamates (Pough 1969). Both breathing modes occur in certain species in the two Australian sand-swimming lineages — *Eremiascincus* and *Lerista*. A third breathing mode is a slight overall expansion and contraction of the thoracic area. This circumferential mode occurs in the totally limbless, humus-inhabiting skinks and is in contrast to the vertical

chest displacement used exclusively by the totally limbless sand-swimming skinks (two *Lerista* species). It is also the mode used by pypopodid lizards (p. 99) and snakes.

Much is known about the temperature (Tables 13-20) and water relations of skinks and what is known can be succinctly and comprehensively generalized in conjunction with two behavioural/ecological categories and one geographical sub-division. In general, all basking, surface active skinks have high thermal and desiccation resistance parameters, while all shade-loving, crepuscular to nocturnal and cryptozoic to fossorial species can be divided into two groups: those from the arid interior have high parameters, while those from the mesic periphery have low parameters. Alternatively, skinks exposed to high temperatures at sometime during the day are temperature and desiccation resistant and those that experience only continuous low temperatures are temperature and desiccation sensitive. Many

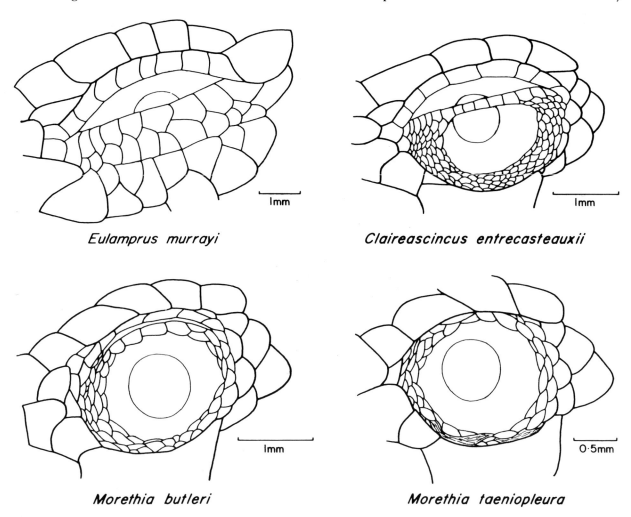

Fig. 58. The left eyelid in four species of skink to show the four eyelid types found in the family: scaly, moveable (*Eulamprus murrayi* — AM R 112287); with a clear central area or window, moveable (*Claireascincus entrecasteauxii* — AM R 67515); preablepharine — with a large clear window, fixed in the raised position but with a papebral slit dorsally (*Morethia butleri* — AM R 70332), and ablepharine — with a large clear window, fixed in the raised position and completely fused dorsally (*Morethia taeniopleura* — AM R 77303).

of the cryptozoic, mesic-adapted skinks along the east and south coasts, e.g., *Calyptotis*, *Hemiergis*, and *Nannoscincus*, are in this second group and are strikingly similar in both habits and physiology to certain terrestrial salamanders of the family Plethodontidae in North America.

Skinks lack a behavioural/physiological attribute that occurs in almost all other lizards: panting (Dawson 1960; Heatwole *et al.* 1973; Greer 1980a). At high temperatures most other lizards open the mouth and pant, thereby possibly cooling the roof of the mouth above which lie large sinuses filled with warm blood (Bruner 1907). Skinks never pant, perhaps because their secondary palate, i.e., the new roof of the mouth, reduces evaporation from the old roof above.

Skinks are generally, drably coloured on the back and sides of the head and body, and the patterns are generally low contrast. As a rule, skinks from mesic, peripheral habitats tend to be black or some shade of grey if diurnal and surface active (Fig. 59) and some shade of brown if nocturnal and/or cryptozoic (Fig. 60). Skinks from xeric and hence generally more open habitats appear to have colour and pattern closely matched to the substrate: brownish to reddish colour on red sand or rock, pale grey or brown on light coloured sand (Fig. 61), and contrastingly striped in tussock grass. No doubt camouflage, rates of heat gain and loss, and protection from short-wave length radiation are all factors in determining these colours and patterns, but no work has yet been done to determine the relative importance of each one.

Colour hues in skinks are usually yellow and red or, occasionally, blue. In juveniles they are usually confined to the tail and may serve either as a general predator deflection device and/or signals to adults of the same species that the little lizard is one of their own and therefore best not eaten. Although these juvenile tail colours are generally lost with age, they are retained in a few species, e.g., *Morethia ruficauda*.

In adults, colour hues are of two kinds, those that are conspicuous at ground level during normal activities and those that are hidden. The conspicuous colours occur on the throat and lower side of the head and on the sides of the body. They are usually sexually dimorphic and seasonal in appearance, and occur in small to medium-sized species with diurnal, ground-dwelling habits living in moderately open habitats, e.g., *Carlia* (some), *Claireascincus*, *Menetia* (some), *Morethia* (most), and *Proablepharus*. All these features suggest that the conspicuous colour probably relates to intraspecific social signalling.

The inconspicuous adult colours of skinks occur on the underside of the body and tail, from the level of the front legs back to the tail base. They are usually sexually dimorphic only in intensity (strongest in males) and are most vivid in semi-fossorial species from dark, humid habitats, i.e., *Coeranoscincus frontalis*, *Hemiergis*, *Nannoscincus*, *Ophioscincus* and *Saiphos*. The fact that these colours occur in both sexes and would be revealed only if the animal were turned on its back suggests that they may serve to startle predators suddenly turning them out of their usual retreats.

There is also a second much less frequent kind of sexual dichromatism in skinks that involves permanent white stripes instead of transient hues. This pattern is also confined to small to medium-sized, diurnal ground-dwelling skinks living in open habitats. It is best developed in *Carlia vivax*, where juveniles and females have white stripes but males lose them, and in northern *Lampropholis delicata*, where only females have them[1]. The permanency of this kind of sexual dichromatism suggest that it serves a more permanent function than the transient type; perhaps it has to do with differences in microhabitat utilization.

Skinks are not usually thought of as possessing the capacity for short-term colour change. However, *Lampropholis guichenoti* are said to lighten somewhat after several days exposure to relatively high temperatures (ca 35°C; Fraser 1980). It is also my impression that *Lampropholis* as well as other small skinks are darker when cool and inactive, and lighter when warmer and more active, parallelling changes in the better known agamids and gekkonids. Short-term colour change in small skinks could easily be studied further with some very basic equipment.

The body scales of Australian skinks are usually smooth, but may be keeled or, rarely, granular. Keeled scales are probably a retained primitive character and amongst Australian skinks are most evident in the genera *Carlia* and *Egernia*. The function of the keels is unknown but could be either structural or thermoregulatory (or both). The only Australian skink with granular scales is *Gnypetoscincus* (Fig. 75); the scales in this cryptozoic, rainforest species appear to help keep the skin evenly moist through capillary spread along the scale edges. The scales in some species of *Carlia* (e.g., *C. schmeltzii*), *Ctenotus*, (e.g., *C. atlas* and *C. brachyonyx*), and *Lerista* (e.g., *L. connivens*), are unusual in that they readily

---

[1]Sexual dichromatism is geographically variable in *Lampropholis delicata*. In northern populations all females have a white lateral stripe, but further south the frequency of striped females declines, to as low as ten percent in some populations.

Fig. 59. Series of diurnal, surface-active skinks from mesic habitats to show the predominately grey colour pattern of the skinks in this habitat. Top row: *Lampropholis guichenoti* and *Morethia obscura;* middle row: *Eulepis trilineata* and *Claireascincus "entrecasteauxii"* form A; bottom row: *Ctenotus labillardieri* and *Egernia formosa.*

tear away when the animal struggles after being grasped. Presumably this acts as a predator escape device.

The scales under the digits, i.e., the subdigital lamellae, are variable in skinks and have long been used as taxonomic characters, primarily at the species or species group level. However, the adaptive significance of the different morphologies remains largely unexplored. The lamellae, which are usually assessed on the fourth toe of the rear foot, may be smooth, grooved longitudinally or raised into keels which vary from obtusely rounded to acutely ridged and pointed.

Most skinks are opportunistic predators with the larger species showing a tendency toward omnivory, or to quote from the most extensive study of skink diet yet undertaken on Australian species: "...scincid diets were found to be basically a function of food availability, microhabitat occupied and scincid size" (Brown 1983:vii). Skinks of the genus *Cryptoblepharus* practice a feeding behaviour unknown elsewhere amongst lizards: piracy. Individuals stand just outside columns of foraging ants and dash in to snatch particularly attractive morsels from the struggling porters (pers. obs.). This behaviour is easily seen in the common species

Fig. 60. Series of nocturnal/cryptozoic skinks from mesic habitats to show the predominately brown colour pattern of the skinks in this habitat. First row: *Eugongylus rufescens* and *Cyclodina lichenigera;* second row: *Saproscincus basiliscus* and *Eroticoscincus graciloides;* third row: *Lygisaurus* cf. *novaeguineae* and *Nannoscincus maccoyi;* fourth row: *Hemiergis quadrilineatum* and *Saiphos equalis.*

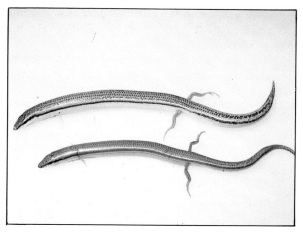

Fig. 61. An example of background matching in a skink from a xeric habitat: *Lerista planiventralis* from a white sand area (top: Carnarvon, W.A.) and a red sand area (bottom: Bullara Station, W.A.).

C. virgatus which is resident on the walls of many suburban and city houses of the east coast. A similar behaviour has been noted in *Morethia boulengeri* (Chisholm 1923 as *Ablepharus lineoocellatus*).

Most skinks have blunt peg-like teeth (Fig. 62). However, *Tiliqua* have a few enlarged, bluntly rounded "molars" (Fig. 64) which presumably serve as a crushing battery for resistant foods. The number of premaxillary teeth, which is more or less constant throughout life (as opposed to the number of maxillary and dentary teeth which usually increases with size) is a very useful taxonomic character in skinks.

Age to maturity in Australian skinks species can vary from the first season after birth to the fifth (Table 8). The determining factor seems to be primarily size, although the third season age for the one mesic, cryptozoic species for which data are available, *Hemiergis peroni*, appears relatively long compared to the other heliothermic species. Perhaps the lower body temperatures of cryptozoic species leads to slower growth and hence later age to maturity. If so, this could have fundamental life history ramifications for such cool temperature forms (this is similar to the explanation given for geckos' relatively long age to maturity, p. 54).

The mating grip in skinks has been little studied, probably due to the small size and cryptozoic habits of most species. A grip in the region just behind the foreleg has been recorded in various *Egernia* species (p.129) and in *Emoia* (Millen 1976), and a grip anywhere from the head back to the shoulder has been noted in various *Tiliqua* species (p.133). Mouth marks found almost exclusively on mature females have been interpreted as resulting from mating grips, and such marks have been recorded in the "pectoral region" of *Claireascincus entrecasteauxii*, *C. triumviratus* and *Pseudemoia spenceri* (Pengilley 1972).

In most skinks, as in most other Australian lizards, copulation occurs shortly before ovulation and hence the sperm fertilize the eggs shortly after entering the reproductive tract. However, in a few southern skinks copulation

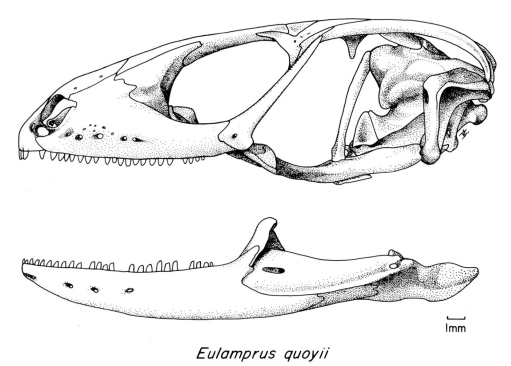

Eulamprus quoyii

Fig. 62. Skull of *Eulamprus quoyii* (AM R 122924).

occurs in the late summer-early autumn post-reproductive period and the sperm are "stored" by the female over winter and then used to fertilize the eggs in spring. This phenomenon occurs in *Claireascincus entrecasteauxii* forms A and B (Pengilley 1972), *Harrisoniascincus coventryi* (Pengilley 1972) and possibly *H. zia* (Ingram and Ehmann 1981), *Hemiergis peroni* (Smyth 1968; Smyth and Smith 1968), *Lampropholis delicata* (Joss and Minard 1985), *L. guichenoti* (some populations, Pengilley 1972; Joss and Minard 1985), and *Pseudemoia spenceri* (Pengilley 1972). In the two forms of *Claireascincus*, the only species studied in this regard, the sperm are stored in the posterior part of the oviduct, although there are apparently no special receptacles for them (Pengilley 1972).

As a group, skinks are both egg-laying and live-bearing (Table 9), and as in most other groups where both modes occur, the live-bearing species tend to be more prevalent in the cooler climates of high altitudes and latitudes (see Shine 1985 for a review). The reason for this seems to be that embryonic development is temperature dependent and in cool climates the developing young will be exposed to generally higher temperatures if they are retained inside the female, which maintains an elevated body temperature through basking, instead of being deposited as eggs in the environment where temperatures are generally lower. Live bearing habits have evolved, on a conservative estimate, at least 11 times amongst Australian skinks (Table 10), and in eight of these the taxa involved are definitely southern in distribution.

This association between reproductive mode and climate occurs not only among species but also within species, although the latter case is very rare. In Australia, both *Lerista bougainvillii* and *Saiphos equalis* show a tendency to retain the young in cooler climates. As a species *Saiphos equalis* retains its developing young until they are almost fully formed, that is at the time of hatching/birth the young are fully shaped and pigmented and clearly recognizable as to species (Bustard 1964b). However, around Sydney *Saiphos equalis* retains its eggs until about 7-9 days before hatching and the eggs are surrounded by a semi-opaque shell (Fig. 79), but in the cooler climate around Point Lookout on the New England Plateau the "eggs" are laid just hours before hatching and are surrounded by only a translucent to transparent membrane (Fig. 79; pers. obs.).

In *Lerista bougainvillii* the situation is even more striking. With one exception the mainland populations are fully egg-laying, i.e., the eggs are laid at the same early stage of embryonic development as are most other lizards and have a normal shell (Smyth and Smith 1974). The island populations, in contrast, are fully live-bearing, the fully formed young being brought forth in transparent membranes which they burst in a matter of hours (Rawlinson 1974a, c). However, in semi-isolated, mainland populations in Western Gippsland, just opposite the live-bearing Bass Strait island populations, the eggs are retained to a stage where only a few days or weeks of additional development are required before hatching (pers. obs.).

In the evolution of ovoviviparity, the retention of the developing young in the female's oviducts has been accompanied by certain changes in the membranes separating mother and young. Initially in evolution, these changes seem to have involved the simple loss of the shell and its membranes, thereby allowing the closer apposition of the maternal and foetal membranes and presumably facilitating diffusion between the two. This type of placenta is seen in a variety of Australian skinks species of the genera *Egernia*, *Eulamprus*, *Hemiergis* and *Tiliqua*. Later, the maternal tissues became more heavily vascularized and the maternal/foetal contact zone interdigitated, both of which changes would increase the area for diffusion. This type of placenta is seen in *Claireascincus* and either *Harrisoniascincus* or *Pseudemoia* (see Rawlinson 1974b for discussion of the confusion of these two taxa in the original paper reporting this placenta type; Weekes 1930).

What exactly passes between mother and developing young is unclear, presumably water, gases and perhaps other small molecules, if better-studied overseas species are any clue. The fact that developing *Eulamprus quoyii* can be raised to term outside the mother's body in tissue culture suggests that at least most if not all the nutrients necessary for development are already available in the yolk (Thompson 1977a-b).

It is interesting to note that much of what we know of the anatomy of the maternal/foetal membranes in ovoviviparous lizards and snakes goes back to the works of H. Claire Weekes working on Australian forms over 50 years ago (Harrison and Weekes 1925; Weekes 1927a-b, 1929, 1930, 1935). Regrettably this interesting work has never been taken up and continued in Australia — despite abundant interesting material.

Brood sizes in skinks vary greatly, ranging 1-53 (Table 9), or possibly as high as 67 (Barnett 1977a), the record holder appearing to be the pink-tongue skink (*Tiliqua gerrardii*). Brood size is generally variable within species, but in some it is a constant 1 or 2. In those species with variable brood sizes there is often, but not always, a correlation between female size and brood size. Similarly in a lineage of species with variable brood sizes, larger species usually have larger broods (e.g., *Ctenotus* — r = .67*, N = 9 and *Lerista* — r = .42*, N = 30). RCM's for Australian skinks with variable brood sizes are higher than for skinks (i.e., *Carlia*) with constant brood sizes (Table 11). This is no doubt a consequence of the fixed, small brood size of the latter.

The number of broods per season is little studied in skinks, and then primarily in temperate species. The only oviparous species that are known to or are very likely to produce more than one clutch per season are temperate populations of *Lampropholis* and *Saproscincus*. Viviparous species, due to their longer gestation times (nearly 2x) are likely to produce only one litter per season (e.g., *Egernia cunninghami* — Barwick 1965), although there is some indirect evidence (a bimodal temporal peak in gravid females) that *E. inornata* may produce two broods per season (Pianka and Giles 1982). The ovoviviparous *Carinascincus greeni* (Rawlinson 1975), and possibly some populations of the oviparous *Nannoscincus maccoyi* (Pengilley 1972) are thought to be biennial in their reproduction.

Communal nesting is known in a number of species, notably, all but one of which, *Carlia*, are southeastern in distribution: *Carlia rhomboidalis* (Wilhoft 1963a), *Eulepis duperreyi* (B. C. Mollison in Green 1965; Ehmann 1976a; Rounsevell 1978); *Lampropholis delicata* (Green 1965; Clarke 1965; Rawlinson 1974a; Wells 1981; M. Wapstra and E. Wapstra 1986), *L. guichenoti* (Mitchell 1959; Clarke 1965; Milton 1980a; Booth 1981; Wells 1981; Shine 1983), *Lampropholis* sp. (Wells 1979b), *Morethia adelaidensis* (pers. obs.), *Nannoscincus maccoyi* (Pengilley 1972) and *Saproscincus mustelina* (German 1986; Fig. 67; pers. obs.). The communal nests of *Lampropholis* and *Saproscincus* are remarkable in several regards. First, they can be large, comprising several tens or, in some cases, well over a hundred eggs, thus representing the efforts of many females. Second, eggs generally appear to be laid more or less simultaneously, often in a matter of hours. This means that females are coming together at the same time, but drawn by what cues is unknown. It also means all the eggs in a nest hatch at about the same time. Third, wherever two or more species or genera occur together the nests always seem to consist entirely of the eggs of only one species, although nests of different species may occur together in the same area. This implies that the cues drawing the females to the nest are probably species-specific.

Gravid females of live-bearing skink species spend more time basking than non-gravid individuals, e.g., *Claireascincus entrecasteauxii* form B, *Eulamprus heatwolei* (as *Sphenomorphus tympanum*; Shine 1980), and *Tiliqua nigrolutea* (Fleay 1931). Such behaviour presumably maintains a more nearly maximally optimal temperature for the developing young. The extent of prolonged basking in other live-bearing skinks and whether or not the phenomenon occurs in oviparous species have yet to be determined. In oviparous skinks there is, not surprisingly, an inverse relationship between the incubation temperature and the time to hatching (Shine 1983).

Very little is known about skink social behaviour, and what is known derives as much from inference as direct observation. Many skinks have been observed to head bob, arch the neck with the head down, and extend the throat, but rarely is the context of the behaviour well enough understood to say what its function may be.

Intraspecific aggression between adults appears to be little-recorded in skinks but whether this reflects the real situation or is an artifact of observation is unclear. Amongst Australian skinks, actual physical fighting has been reported in *Cryptoblepharus virgatus* (Rankin 1973c, as *C. boutoni*), *Eulamprus kosciuskoi* (Done and Heatwole 1977a-b), *E. tympanum* (Rawlinson 1974c), *Tiliqua casuarinae* (Mebs 1974) and *T. gerrardii* (Stephenson 1977). In the latter case adult females and juveniles of both sexes appear to be the most aggressive. The fact that fights occur between individuals of some species (e.g., *Eulamprus kosciuskoi*, *E. tympanum*, and *Tiliqua gerrardii*), but apparently not others (e.g., *Eulamprus quoyi* and other *Tiliqua* species), even though they may be closely related suggests that aggressive behaviour may differ greatly even between close relatives. The reason for this difference is unknown. Other species show nonrandom spacing is suitable habitat, which is also probably achieved through behavioural interactions.

Skinks in the coldest part of the country, i.e., the southeastern highlands, are occasionally found in hibernation aggregations. For example, 16 and 18 *Lampropholis guichenoti* were in two logs of a fire wood pile (Powell *et al.* 1977); approximately 20 and 50 *Pseudemoia spenceri* were under two north facing rock slabs (Rawlinson 1974b), and aggregations of up to eight *Claireascincus entrecasteauxii* (form B) and *Harrisoniascincus coventryi* have been found in logs (Pengilley 1972 and Rawlinson 1975, respectively). Whether the individuals in these aggregations are attracted to each other or to the site itself is unknown, but the latter would appear to be the more likely.

Skinks have two methods of producing sound internally. One is a snorting or hissing resulting from air passing through the nostrils, and the other is a high-pitched squeak resulting, presumably, from air passing through the glottis. Only the latter would seem to qualify as a true voice. Snorting and hissing occur only in the large species of *Egernia* (e.g., *E. major* — Longley 1946a) and in *Tiliqua*. The sound is usually produced when the animal is startled or is facing an intruder. Squeaking has been recorded in *Ctenotus robustus* (Annable 1983), *Egernia cunninghami*, *E. striolata* (Annable 1983), *Eremiascincus richardsoni* (Werner 1973), *Eulamprus amplus* (Covacevich and McDonald 1980), *E. quoyi* (Annable 1983), *E. murrayi* (pers. obs.), *E. tympanum* (Annable 1983), and *Saproscincus mustelina* (Annable 1983). This sound is usually emitted either when the animal is grasped or is in an aggressive encounter.

THE LINEAGES OF AUSTRALIAN SKINKS

The subfamily Lygosominae (p. 4) can be divided into three major groups, each named after a primitive genus in the group: the *Mabuya* group (previously called the *Egernia* group), the *Eugongylus* group and the *Sphenomorphus* group. Each of these major groups is represented in Australia (Greer 1979b; Hutchinson 1981). The *Mabuya* group may be simply an assemblage of relatively primitive species, but the other two groups appear to be lineages. Although the relationships of the three groups are not yet well understood, the available evidence indicates that the *Mabuya* and *Eugongylus* groups are each other's closest relatives (Fig. 63).

The *Mabuya* group has two genera in Australia, *Egernia* and *Tiliqua,* and because of their large size and ease of maintenance they are among the best-known skinks in the world.

The genus *Egernia* (Horton 1968, 1972; Storr 1978a) can only be diagnosed on the basis of a single morphological character: the palatines curl partially around the nasal passage anteriorly above the secondary palate and the vomers send back long posterior processes along the medial edge of the open palatine scroll (Hutchinson 1983). In addition to this morphological feature, *Egernia* also shows one behavioural trait that is unique amongst Australian skinks: a strong attachment to a permanent retreat, be it a rock crevice, hollow log or dead tree, or a burrow dug in earth. All of the skink's activities are centered around these retreats, and they rely on them for both long and short-term shelter (pers. obs.).

There are also two other derived morphological features that are so widespread in *Egernia* that one wonders if the few exceptions may not be reversals that occurred subsequent to the origin of the group. These characters are large size (maximum SVLs for the species range 84-275 mm) and eight or fewer premaxillary teeth (nine in *E. coventryi*, *E. luctuosa* and *E. major*). If these two characters are derived, they, plus the group's ovoviviparous mode of reproduction (Tables 9-10), may serve to link *Egernia* with *Tiliqua,* the only other member of the *Mabuya* group in Australia.

Twenty-six species of *Egernia* are currently recognized and all occur in Australia with one extending into southern New Guinea. The genus occupies a variety of habitats ranging from rainforest (e.g., *E. frerei* and *E. major*) and alpine meadows (e.g., *E. whitii*) to arid sand plains (e.g., *E. kintorei*, *E. inornata*) and saltmarshes (e.g., *E. coventryi* — Schulz 1985). Most terrestrial habitats have at least one species although the population density in any habitat may be low.

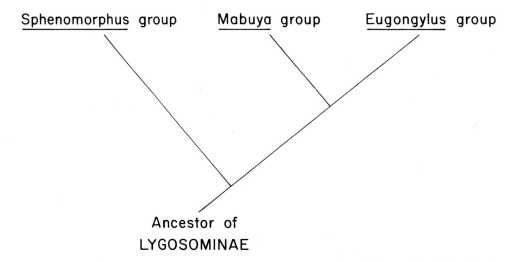

Fig. 63. Schematic diagram of the probable relationships between the three groups of lygosomine skinks. See Greer (1979b) and Hutchinson (1981, 1983) for details and alternatives.

Most species are diurnal but a few are diurnal/nocturnal (e.g., *E. inornata* — Webber 1978, 1979; Pianka and Giles 1982) and two are largely, if not exclusively, nocturnal (*E. kintorei* and *E. striata*). One of these latter species, *E. striata* (Swanson 1976; Pianka and Giles 1982), has the hallmark of nocturnal vertebrates: a vertically elliptic pupil. Depending on the species, *Egernia* are terrestrial, saxicolous, or semi-arboreal (Table 6).

The food habits of a variety of *Egernia* species have been studied both in nature and in captivity, and the broadest generalization one can make is that most species are omnivorous (references in Table 7). There is also some suggestion of both intra- and interspecific dietary shifts to increasing omnivory with increasing size in *Egernia*. In the best-studied species, *Egernia cunninghami* (subsp. *jossae*) from around Canberra, there appears to be a distinct dietary shift from carnivory in juveniles to omnivory in adults. For example, although juveniles in captivity readily ate arthropods, "extreme difficulty was encountered when attempts were made to feed . . . (them) plant material" (Shine 1971) whereas adults both in the laboratory (Shine 1971) and in nature ate both animal and plant material (Barwick 1965). Similarly, the four *Egernia* species in southeast Australia — *E. whitii, E. striolata, E. saxatilis* and *E. cunninghami* show an increase in body size in the order given and they have the percentage volumes of plant material consumed in the diet of 8.4, 39.7, 27.1 and 92.8, respectively (Brown 1983).

The shift from carnivory to herbivory with increasing size is a well-known phenomenon in lizards both within and amongst species. The explanation seems to relate to the difficulty large species have in catching appropriately sized animal prey (Pough 1973; Wilson and Lee 1974). Most animal prey is too difficult for large lizards to catch, other than through the occasional piece of good luck, hence greater reliance is placed on the more accessible plant food.

It has been noted on several occasions that various species of *Egernia* defecate in one spot, thereby creating over time a small pile of scats. This has been noted in *E. cunninghami* (G. Shea, pers. comm.), *E. stokesii* (White 1976), *E. striolata* (Bustard 1970e); and *E. whitii* (Hickman 1960). It is uncertain whether these fecal piles have a purpose in themselves or are simply an artifact of some other aspect of the lizard's activity pattern, such as morning basking.

Copulation, often dificult to observe in small skinks, has been noted on a few occasions in various species of *Egernia,* and in each case it involves an ultimate mating grip in the shoulder

region — just behind the foreleg in those encounters noted most closely. This is a grip a little further posterior than the more usual and primitive neck grip of most other lizards. The *Egernia* species for which we have observations are *E. cunninghami* (Niekisch 1980); *E. depressa* (pers. obs.) and *E. inornata* (Webber 1978).

As far as is known all *Egernia* species are live-bearing, producing 1-8 young per litter (Table 9). This is a relatively small number of young in comparison to female size and may reflect the general low mortality and great longevity of large species. The number of litters per season is known for certain only for *E. cunninghami* where it is one (Barwick 1965); however, it has been suggested that *E. inornata* in the Great Victoria Desert may have two litters (Pianka and Giles 1982).

The time of birth is known only for certain southern species/populations and in all cases appears to occur during the end of summer or the beginning of autumn (e.g., *E. cunninghami* — Barwick 1965; *E. inornata* — Peters 1971c; Pianka and Giles 1982; *E. kingii* — Arena 1986; *E. modesta* — Milton 1987; *E. striata* — Pianka and Giles 1982; *E. striolata* — Bustard 1970e; *E. whitii* — Hickman 1960; Milton 1987. However, one species at a southern locality, *E. stokesii* from west of Wubin, W.A. (30°06′S), gave birth at the end of winter (12 August). This suggests that females in this population, at least, may carry young over winter. This would be a most unusual occurrence and should be checked.

The foetal membranes of young *Egernia cunninghami* may be eaten directly after birth either by the young themselves (Barwick 1965) or by the mother (Niekisch 1975). In *E. stokesii* the young have been recorded to eat the membranes (Zimmerman 1985). Suggestions of a phylogenetic difference between groups either within skinks themselves (Shea 1981) or within lizards as a whole (Rebouças-Spiecker and Vanzolini 1978) have been predicated, at least in part, on a belief that in *Egernia* only the female eats the membranes. The intact placenta in *Egernia* has been described by Weekes (1930).

As might be expected in such large lizards, age to maturity, to the extent that it is known, takes more than one season. In nature, both sexes of *Egernia cunninghami* (subsp: *jossae*) take 5 years to breed (Barwick 1965), *E. whitii* probably about 4 years (Hickman 1960), and *E. striolata* 2 to 3 years (Bustard 1970e). In captivity (outdoors in Florida!), *E. cunninghami* (subsp: *krefftii*) take 1 year 10 months to mature (Bartlett 1981).

In captivity *Egernia* may live many years, which is not surprising given their large size; under these conditions *E. cunninghami* has lived

10 years and 2 months (Conant and Hudson 1949), 15 years and 1 month (Dinardo 1985, but specimen obtained as adult), 17 years (Münsch 1981), and 19 years and 11 months (Flower 1925); *E. stokesii* 12 years (Zimmermann 1985); *E. frerei* 10 years and nearly 10 months (Bowler 1977, as *E. major* but probably this species, *fide* G. Shea, pers. comm.) and *E. major* 8 years and 7 months (Bowler 1977, as *E. bungana*).

A few closely related species of *Egernia* show colour pattern polymorphisms within populations, that is, two or more distinct colour patterns in one population. This phenomenon has been reported in *E. multiscutata* — Hudson *et al.* 1981; *E. pulchra* — Ford 1963b; *E. whitii* — Milton *et al.* 1983; Milton and Hughes 1986; Milton 1987. All these species are members of what is thought to be a single species group — the *E. whitii* group (Storr 1968). Details of the biology of the morphs is best-known in *Egernia whitii* (Milton *et al.* 1983; Milton and Hughes 1986; Milton 1987). In this species there are two morphs (striped-back and plain-back), and they are electrophoretically indistinguishable. However, there is evidence that the morphs associate with their own morph type and occur in slightly different habitats (the plain-backed morph in slightly more open situations). The sex ratio is biased toward females in the stripe-backed morph but not in the plain-backed morph.

Burrowing, the digging and utilization of a true burrow, is probably a primitive trait in lygosomine skinks, because it also occurs among the most primitive species of this subfamily, i.e., the genus *Mabuya*. A number of *Egernia* species retain this habit. The species known to burrow are *E. coventryi* (Smales 1981), *E. inornata* (Davey 1970; Martin 1975; Webber 1979; Chapman and Dell 1979; Coventry and Robertson 1980; Dell and Chapman 1981b; Pianka and Giles 1982; Morley and Morley 1985), *E. kintorei* (Cogger 1986; Davey 1970), *E. multiscutata* (Storr 1960; Ford 1963a-b, 1965a; Dell and Chapman 1977; Dell and Harold 1979; Coventry and Robertson 1980), *E. pulchra* (Ford 1963b, 1965b), *E. slateri* (P. Rankin unpublished; G. Storr, pers. comm. to G. Shea), *E. striata* (Pianka and Giles 1982) and *E. whitii* (Hickman 1960; Ehmann 1976a; Green 1984). In *Egernia* the burrow is used as a retreat from predators and as a refuge during inclement weather.

The burrows that have been investigated closely all show one feature in common — the presence of one or more thinly covered exits which can serve as breakthrough escape hatches when escape from the main exit is blocked. These escape exits, or "pop holes", have been observed to date only in species which burrow in sand (*E. inornata, E. kintorei, E. multiscutata, E.*

striata and *E. whitii*) which is an easy substrate for investigators to excavate. It remains to be seen whether similar exits occur in species which burrow in more firmly bound substrates (e.g., *E. coventryi* and *E. luctuosa*).

The closely investigated burrows show various degrees of complexity. Some simply comprise a single tunnel with two openings (but one thinly closed, e.g., *Egernia whitii* — Hickman 1960) while others comprise several, interconnected tunnels and several openings (e.g., *E. striata* — Pianka and Giles 1982). Some of this variation may be geographic. For example, western Australian *E. inornata* build a two-opening burrow and far eastern Australian specimens a multi-opening system; furthermore distinct but unspecified differences between the burrows of Northern Territory and New South Wales *E. inornata* have been alluded to in the literature (Webber 1979). Different species appear to have different preferences in the orientation of the burrow openings; for example, there is a distinct preference for northerly and north-westerly openings in *E. inornata* (Webber 1979; Pianka and Giles 1982) and for southerly and south-westerly openings in *E. striata* (Pianka and Giles 1982). The reasons for these differences are not yet understood but are thought to relate to the species' thermoregulatory requirements. Eastern Australian *E. inornata* have been noted to use the burrow entrance most directly exposed to sun for their diurnal activity and the entrance last in sun for their nocturnal activity (Webber 1979).

Although body temperatures have been taken on a number of *Egernia* species under a variety of natural and artifical conditions (Tables 13, 15, 17, 19), there is nothing particularly striking in these data, other than perhaps an ecologically understandable trend for species from warmer, drier areas to show higher body temperatures than those from wetter, cooler areas. Similarly, greater endurance at higher experimental temperatures is shown by xeric species compared to mesic forms. For example, when exposed to 43.5°C, *E. pulchra* from the mesic, cool south-west succumbed almost immediately (0 minutes, N = 7); *E. stokesii* from a more northern off-shore island survived longer (0-40 minutes, mean = 16, N = 5), and *E. depressa* from the hot, arid interior survived by far the longest (95-180 minutes, mean = 138, N = 4; Licht *et al.* 1966b).

Although the evidence is almost all indirect and circumstantial, *Egernia* appears to vary greatly in its degree of sociability. For example, *E. striolata* appears to be almost completely solitary, perhaps even actively excluding other individuals from its retreat (Bustard 1970e); *E. cunninghami* (Barwick 1965), *E. depressa* (G. Shea, pers. comm.), *E. saxatilis* (R. Sadlier pers. comm.) and *E. whitii* (Hickmann 1960; D. Milton pers. comm.) appear to occur only in small groups of different-sized individuals, the groups often called "families" in the literature; *E. multiscutata* is described as living in "small discrete colonies" (Coventry and Robertson 1980); in *E. stokesi* "sometimes as many as four or five individuals will be found by turning over a single stone" (Alexander 1922). Direct observations of individual behavioural interactions are rare for the genus, but *E. cunninghami* are said to show no aggressive behaviour to conspecifics either in the wild or in captivity (Barwick 1965).

There is a general relationship in *Egernia* between the degree of keeling on the scales and the place of refuge (Cogger 1960a). The most heavily keeled species tend to shelter in hollow logs or rock crevices; the lightly keeled species are largely arboreal and shelter under exfoliating bark, and the smooth-scaled or only very lightly keeled species retreat to burrows of their own construction.

The heavily keeled species have certain behaviours that bring the keels into play in such a way that it is very difficult for a predator to extract the animal from its crevice or hollow. For example, *Egernia cunninghami* may press the body tightly against the sides of the retreat by arching the back (Barwick 1965). *E. saxatilis* achieves the same effect by inflating the lungs; this species also often turns the body away from the aggressor (Cogger 1960a). This latter position ensures that any force exerted to extract the animal causes the points of the keels to dig deeper into the sides of the retreat (because the posterior edge of the scale is free and can turn out from the base) and also ensures that the force is transmitted along the length of the keel, the orientation in which this structure is strongest. *E. cunninghami*, *E. depressa* and *E. stokesi* press the limbs tightly against the body and *E. depressa* presses its flattened tail, limpet-like, against the substrate (G. Shea, pers. comm.; pers. obs.).

The heavily keeled crevice dwellers also tend to have a short tail. For example, *E. cunninghami* has 42 or fewer postsacral vertebrae, *E. hosmeri* 21-22, *E. depressa* 15-16, and *E. stokesi* 14-15. These last two species, the shortest-tailed skinks in the world, have also lost the ability to drop the tail.

At least one *Egernia* species, *E. cunninghami*, shows a high tolerance to the venom of a potential elapid snake predator. Two adults injected intraperitoneally with *Pseudonaja textilis* venom 10 times the lethal dose for a mammal of equivalent weight survived with no ill effects other than a couple of hours of lethargy (Barwick 1965).

There is one case of hybridization in *Egernia*: a male *E. hosmeri* and a female *E. stokesii* produced three young after being held together in captivity (Peters 1976a). It is not known whether any of these hybrids were fertile — the ultimate test of the potential breakdown for the genetic differences between species.

Tiliqua is probably the best known and most thoroughly studied genus of skinks in Australia, if not the world. The large size, slow gait, docile nature and close association with man have made, at least the larger species, totally familiar. The literature on the genus runs close to 1500 references (G. Shea, in prep.). There are 11 species and all but one are restricted to Australia; the exception, *T. gigas*, occurs in New Guinea and its neighbouring islands (Shea 1982). One of the species, *T. adelaidensis*, has the distinction of being the only probable example, to date, of a native lizard species that has become extinct since European occupation (Ehmann 1982).

Tiliqua has radiated into a wide variety of morphological and ecological types. In size they range from the dwarf *T. adelaidensis* with a maximum SVL of 88 mm up to the giant *T. scincoides intermedia* of 371 mm SVL (G. Shea, pers. comm.). In body form, there are the long and lanky *T. branchialis*, *T. casuarinae*, *T. gerrardii* and *T. maxima* at one extreme and the decidedly stumpy *T. rugosa* at the other. Ecologically, *T. gerrardii* is often found in the east coast rainforests while *T. multifasciata* and *T. occipitalis* occur in the central deserts (Shea and Peterson 1981; Fyfe 1985). All but one of the species are largely diurnal and terrestrial; the exception, *T. gerrardii*, is often crepuscular-nocturnal and semi-arboreal. All species are live-bearing (Table 9; first reported by Haacke 1883, 1885), and litter sizes span the range for the entire family, from one in *T. rugosa* to at least 53 (Barnett, in Anonymous, 1973a; Barnett 1977a) and perhaps even 67 (Barnett 1977a) in *T. gerrardii*.

There appear to be two distinct lineages within *Tiliqua*. One consists of the smaller, elongate species with the head only slightly wider than the neck and with a uniform colour pattern on the body, and the other consists of the larger, stouter species with the head appreciably wider than the neck and a cross-banded pattern on the body. The former group includes *T. branchialis*, *T.*

casuarinae and *T. maxima* and is often regarded as a distinct genus to which the name *Omolepida* (Storr 1976b) or, more appropriately, *Cyclodomorphus* applies. The latter group includes *T. adelaidensis, T. gigas* (non-Australian), *T. multifasciata, T. nigrolutea, T. occipitalis, T. rugosa* and *T. scincoides* to which the generic name *Tiliqua* proper applies. *T. gerrardii* is morphologically intermediate in being elongate but having a wide head; it is sometimes placed in a separate genus, *Hemisphaeriodon*. Here the name *Tiliqua* will be used for the entire group.

Tiliqua as a whole is characterized by a number of morphological features but mention will be made only of those whose adaptive significance seems clear. All but one species have a dark, often blue, tongue and in most species, if not all, the tongue is extruded in some fashion in defensive display (see below). This has led to the common name, for at least the larger members of the group, of "blue tongue" lizards. The one exception, the pink tongue, *T. gerrardii*, is born with a dark tongue like its close relatives (Wilhoft 1960; Miles 1973; Barnett 1977a; Robertson 1980), but generally loses this colour with age (Longley 1938, 1940; Stephenson 1977). However, the age at which tongue colour change begins varies widely, i.e., 2-3 months (Longley 1938) to 2 years (Stephenson 1977). In some cases, adults may retain the dark tongue (Morris *et al.* 1963) and in others, young may be born with a pink tongue (Robertson 1980).

All *Tiliqua* lack one phalange in the fourth toe of the front foot and one in both the fourth and fifth toes of the rear foot giving a phalangeal formula of 2.3.4.4.3/2.3.4.4.3. Possibly in association with this loss of some of the bones involved in locomotion via limbs, they have gained some vertebrae, that is, bones associated with lateral undulation; the lowest recorded number of presacral vertebrae for the genus is 32, six more than the primitive number for skinks. Both the small but attenuate species such as *Tiliqua casuarinae* and the large, stocky species such as *T. scincoides* are adept at lateral undulation.

The genus also has the teeth along the middle of the jaws noticeably enlarged (Fig. 64). This modification has long thought to be related to durophagy, the habit of eating hard prey, which in *Tiliqua* would mean snails. However, more recently, such enlarged teeth in other lizards in general have been related to omnivory (Estes and Williams 1984), which would also apply in *Tiliqua*'s case.

Physiologically, *Tiliqua rugosa* has been characterized as "an unusually slow lizard with

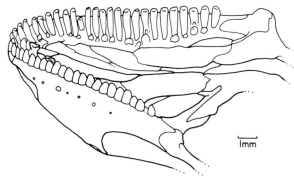

Egernia modesta

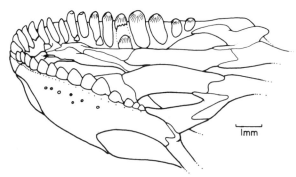

Tiliqua casuarinae

Fig. 64. The left maxilla in lingual view of *Egernia modesta* (AM R 43062) and of *Tiliqua casuarinae* (AM R 37706) to show the enlarged posterior teeth of *Tiliqua* in comparison to the uniformly-sized teeth of most other skinks.

limited stamina" (John-Alder *et al.* 1986). Such a characterization may apply to all the large *Tiliqua* species.

Most *Tiliqua* species are primarily terrestrial, but *T. gerrardii* is semi-arboreal. Morphologically this is manifest in its relatively longer claws and limbs, somewhat prehensile tail (Longman 1915, but see Field 1980), and perhaps also in its more gracile body (at least in comparison to the other cross-banded, large-headed *Tiliqua* which are quite stocky). However, other *Tiliqua* occasionally show surprising climbing abilities which could have provided the basis for *T. gerrardii*'s more confirmed arboreality. For example, *T. branchalis* climbs in tussock grass, *Tiliqua rugosa* can climb wire mesh and low tree stumps, and *T. scincoides* can climb sloping tree trunks, vines and rough vertical brick walls (Ehmann 1982; Lambert 1985; G. Shea, pers. comm.).

Although the large size of *Tiliqua* makes it one of the few lizards groups amenable to the study of individual movements, the only species studied to date in this regard is *T. rugosa* (Warburg 1965b; Bull 1978, in press; Bamford 1980; Bull and Satrawaha 1981; Satrawaha and

Bull 1981; Fergusson and Algar 1986). This work shows that the species has a much smaller home range than predaceous lizards of a similar weight, a result probably due to the greater ease with which it can harvest much of its food, especially plant material and perhaps to a lesser extent, snails. Furthermore within their home range, individuals appeared to spend time in one small area and then suddenly move to another small area but one well-away from the preceding one. Home range size was correlated with body weight in some study years but not others, and gravid females appeared not to differ in home range size from unsexed individuals. Whether these results are typical of the other large, omnivorous species of *Tiliqua* remains to be seen.

There is a large amount of information on the feeding habits of *Tiliqua*, but most of it is based on study of captive specimens of the larger species. In general, it seems as if the smaller species of *Tiliqua* such as *T. branchialis* and *T. casuarinae* prey primarily on arthropods, while the larger species are omnivorous eating a good deal of plant material, carrion, and other slow moving animal material such as snails and slugs, vertebrate eggs and young, and large arthropods. *T. rugosa* may eat larger amounts of plant material than any other species (G. Shea, pers. comm.). It is a very characteristic occurrence in spring, when stopping to examine a browsing *T. rugosa* along the side of the road, to have it suddenly gape revealing a mouth full of green leaves and yellow or blue flowers — which along with the pink mouth lining and dark blue tongue make for a rainbow of colours. This species is said to have a preference for yellow flowers (Anonymous 1954; Meredith 1954), although this has never been substantiated with choice experiments. The omnivory of the larger *Tiliqua* species means they are occasionally attracted to baited mammal traps, with the predictably surprising result (R. Green 1984) for both trapee and trapper.

Although most of the large *Tiliqua* species are fond of snails, they generally attempt to discard the shell before swallowing the fleshy body. Most species crack the shell and then push the pieces out of the mouth with the tongue, but *T. gerrardii* has actually been observed levering the body from the shell using its jaws and the substrate (Field 1980). Another harvesting behaviour has been observed in *T. rugosa*: animals push over and walk along the stalks of long-stemmed flowers to reach the inflorescences (Shugg 1983).

Unfortunately there seems to have been no careful study of the possible change of diet with age in any large species of *Tiliqua* in order to test the theoretical prediction that larger lizards include larger amounts of plant and other easily-harvested material in their diet (p. 129). However, this would make a technically feasible lab/field study.

Courtship and mating are spring-summer phenomena in both temperate and tropical (e.g., *T. multifasciata* — Christian 1977a) populations. There are many literature reports of males fighting during spring (e.g., *Tiliqua* sp. — Hill 1923; *T. gerrardii* — Stephenson 1977; Field 1980; *T. nigrolutea* — Longley 1941; Fleay 1937, 1951; *T. rugosa* — Hill 1923), some bouts lasting an hour or more.

Mating has been observed in several species of *Tiliqua* and usually involves the male grasping the female by the head, neck or shoulder with his mouth (*T. gerrardii* — Stephenson 1977; Field 1980; Robertson 1980; *T. nigrolutea* — Bartlett 1984; *T. rugosa* — Hitz 1983). During the spring in southern Australia, it is common to see pairs of *T. rugosa* crossing the road in tandem (Cogger 1967; Anstis and Peterson 1973), presumably a female in the lead of an amorous male (Fig. 65). The presumed males are more aggressive than the females and will fight if put together. Pairs of *T. rugosa* have been found together after intervals of up 25 days (Bull in press). It has been suggested that males "attend" females until they are ready to mate (Bull in press) or until sufficient time has passed after mating to ensure that the eggs are fertilized by the attending male's sperm (Bamford 1980). *T. rugosa* have been found in the same pairs for as long as three years running (Bull in press); presumably this is more a reflection of their sedentary nature rather than any sort of pair bonding.

The gestation period has been estimated for three *Tiliqua* species in captivity on the basis of the elapsed time between copulation and birth.

Fig. 65. Two *Tiliqua rugosa* in spring from just east of Tibooburra, New South Wales. Note the enlarged, rugose scales.

The results indicate a period of 100-125 days, specifically *T. gerrardii*: 101-110 days (Stephenson 1977; Robertson 1980); *T. rugosa*: 119-125 days (Hitz 1983), and *T. scincoides*: ca 100 days (Longley 1939).

There is a large amount of information on litter size in *Tiliqua*, derived mainly from animals kept in captivity. Regrettably this data often comes without information on female snout-vent length, the most common basis of comparison both within and amongst species. As mentioned above, the range in litter size is 1-53 or perhaps 67. Within species with relatively large litters there appears to be a positive correlation between female size and litter size e.g., *T. casuarinae*, $r = .93$**, $N = 8$), but this relationship probably does not hold amongst species. In *T. rugosa*, which has litter sizes of 1-3, young born singly are larger than young born in pairs (Bamford 1980). The intraspecific relationship between number of young and size of young has yet to be investigated either within or amongst other species. The wide range of litter sizes in this one genus — a range that comes close to spanning the range in brood size for lizards in general — should prove to be an interesting test case for any theory purporting to explain the evolution of brood size in lizards.

Shortly after birth the young of *Tiliqua*, at least the larger species, eat the foetal membranes (e.g., *T. gerrardii* — Wilhoft 1960; Swan 1972; Miles 1973; Mudrack 1974; *T. nigrolutea* — Bartlett 1984; Giddings 1984; *T. scincoides* — Anonymous 1926; Tschambers 1949; Chaumont 1963; *T. rugosa* — Roesch 1956 as cited in Matz 1968; Bamford 1980; Hitz 1983 and *Tiliqua* sp. — Le Souëf 1918). Recall that this phenomenon has been recorded for some *Egernia* species but for no other live-bearing Australian skink as yet. There is no record in *Tiliqua* of the female biting the umbilical cord or eating the foetal membranes as has been recorded for some *Egernia* species.

The mature allantoplacenta is "simple" in that the apposing foetal and maternal membranes are smooth, i.e., non-interdigitating, and the two circulations very close to their respective surfaces (Flynn 1923; Weekes 1930). This is the most widespread allantoplacenta type seen in skinks.

Age at sexual maturity in *Tiliqua* has been assessed in the field for only one species, *T. rugosa*; in three different populations most animals reached sexual maturity in their third season and the rest in their fourth (Bamford 1980; Bull in press). Age to maturity is known or can be inferred for two other species in captivity, although the varying conditions

probably distorts the natural rhythm and hence length of the maturation process: *T. gerrardii* — 1 yr 9 months to 1 yr 10 months (Longley 1940; 1941) and *T. scincoides* — 1 yr 7 months to 1 yr 8 months (Shea 1981).

Certain *Tiliqua* species undergo a pronounced ontogenetic colour change. Amongst the smaller, generally unbanded species, *T. casuarinae* are often (always?) born with a dark nuchal band and often a dark head blotch and smaller posterior neck bands, which are lost with age. *T. branchialis* are profusely white-spotted when born but lose the spots with age (G. Shea, pers. comm.). Amongst the large cross-banded species, embryonic *T. rugosa* are said to have pronounced light crossbands but these become darker, at least dorsally, prior to birth (Gadow 1901). In contrast, *T. gerrardii* are usually solidly crossbanded at birth but the bands become much more diffuse, or are even lost with age, at least in most individuals (Mudrack 1974; Field 1980). Occasionally, *T. gerrardii* are born entirely light (Morris *et al.* 1963), leading one to wonder if the genetic timing of the loss of crossbands may not have had an exceptionally early onset in these individuals. Recall that in *T. gerrardii* tongue colour also lightens with age. Some juvenile *T. gerrardii* have been reported to revert over several months to a darker, more contrasting colour pattern, after having initially lightened in colour (Stephenson 1977).

The large, stocky species of *Tiliqua* such as *T. gigas*, *T. multifasciata*, *T. nigrolutea* (Fleay 1931), *T. occipitalis*, *T. rugosa* and *T. scincoides*, have a very characteristic defensive display (Fig. 66). When intruded upon they simultaneously flatten the body and throw it into an arc facing the intruder, open the mouth, extend the tongue and hiss (Carpenter and Murphy 1978). All this sound and show, along with the short, fat body and often crossbanded colour pattern often results in their being taken for a death adder. This reaction is even developed in near-term but not yet born young (Wakefield 1956).

Fig. 66. Tiliqua occipitalis in threat posture. Photo: P. R. Rankin.

Tiliqua casuarinae also has a snake-like display, but it is more reminiscent of one of the more agile snakes instead of the sedentary death adder. Two Tasmanian specimens were especially good performers in this regard for me. When uncovered and approached, they drew the body and tail into coils, tucked the forelegs back along the body, raised the forepart of the body off the ground, opened the mouth, and protruded the tongue. The tongue was held out and alternately curled slightly up and down, vibrating all the while — just like a snake. Too close an approach would elicit an open-mouthed strike (see also Hewer and Mollison 1974). Similar but less full-blown behaviour has been reported in juvenile mainland specimens, which with their dark neck bands look very much like juvenile brown snakes, *Pseudonaja textilis* (Rankin 1973a). *Tiliqua gerrardii* is also said to extrude the tongue and vibrate it rapidly in display (Longley 1938) but just how similar this behaviour is to that of *T. casuarinae* is unclear.

The larger species of *Tiliqua* are active baskers, and some of the details that have been observed in these large skinks may also apply to smaller species. *T. rugosa* has been observed to begin basking in the morning by exposing only its head from its nocturnal retreat; after maintaining this exposure for a short time and presumably warming the body as a whole with warm blood from the head, it will emerge completely, turn its body at right angles to the sun and even arc its body toward the sun as if forming a parabolic solar collector (Edwards 1978). Experiments on *Tiliqua scincoides* have shown that thermoregulatory behaviour, e.g., shuttling between environments of different temperature, is mediated, at least in part, by brain temperature (Cabanac *et al.* 1967; Hamel *et al.* 1967).

Tiliqua rugosa are occasionally found with a white encrustation of salt around the nostril. This is the exudate of a salt gland located in the nose and capable of excreting both sodium and potassium in higher concentrations than in the blood. The gland is under hormonal control and is most active in the spring (Braysher 1971; Saint-Girons *et al.* 1977; Bradshaw *et al.* 1984). It is unknown whether other *Tiliqua* species also have nasal glands (they seem to never show nasal encrustations of salt), or what special need, if any, this gland meets in *T. rugosa*. Perhaps it relates to the large amount of plant material eaten by this species (G. Shea, pers. comm.).

In addition to excreting salt through the nasal gland *Tiliqua rugosa* is also capable of tolerating high levels of salt in the blood. Presumably this allows them to remain alive during drought or until sufficient drinking water is available

to flush the salt load in the urine. There are numerous reports in the literature of shinglebacks appearing in increased numbers and drinking avidly from puddles after rain (e.g., Simpson 1973). It has been suggested that in order to conserve water, shedding in *T. rugosa* is foregone during dry spells but resumed after rain (Shea 1980).

Although most *Tiliqua*, or at least the larger, diurnal species, will flee to or shelter in burrows of another animal's making, they appear incapable of digging their own burrows. The absence of a burrowing ability is in strong contrast to their close relatives, *Egernia*, in certain species of which the habit is well developed.

The availability and handleability of the larger *Tiliqua* species have made them popular for decades with anatomists and physiologists. However, in most cases the animals were chosen to gain insight into likely general conditions of all lizards or all reptiles, and not of the species *per se*, or even the genus or the family. For an entry into this voluminous literature the bibliography of the genus by G. Shea (in prep.) should be consulted.

Hybridization is relatively common between the larger species of *Tiliqua*. So far the following crosses have been obtained in captivity (C) or inferred from wild-caught specimens (W): *T. nigrolutea* x *T. scincoides* (C; Longley 1939, 1941, 1944b; Worrell 1963; Ehmann 1982), *T. scincoides* x *T. rugosa* (C; Ehmann 1982; pers. obs.) and *T. rugosa* x *T. nigrolutea* (W; G. Shea, pers. comm.). The *T. nigrolutea* x *T. scincoides* crosses are known to have produced fertile offspring (Worrell 1963). On the face of it these hybridizations are surprising, because the species are so morphologically different one would have thought the underlying genetic systems would have been too incompatible to produce a viable organism. However, perhaps the only unusual thing about *Tiliqua* is the large morphological difference between species, which helps us to identify hybrids, and the frequency with which they are kept together in captivity, which increases the opportunities for interspecific matings. It could be that the potential for interspecific hybridization is equally great in other genera but because the species are more similar than in *Tiliqua* and are less often kept together, there is less opportunity for it and less likelihood of its being recognized when it does occur.

Given the apparent high frequency of interspecific hybridization in *Tiliqua* it is perhaps not surprising that a species of this genus figures in the only case of intergeneric hybridization reported to date amongst lizards: a single young produced by a male *T. gigas* and a female *Egernia cunninghami* held in captivity in a private collection in South Africa (Rose 1985). The animal, which is clearly morphologically intermediate between its two parent species, was born 7 February 1979, grew vigorously and was alive at the time of writing (1986). Unfortunately, it is unknown whether the animal is fertile (S. Rose; in litt. 20 April 1985).

There are two peculiarities of locomotion in *Tiliqua* that are worth mentioning. First, in limbed locomotion, *T. scincoides*, shows relatively little side to side swinging or arching of the body (Daan and Belterman 1968); the reason for this is unknown. Second, although all *Tiliqua* may tuck the fore and hind limbs back along the side of the body and tail when undulating through a confined space, *T. casuarinae*, *T. gerrardii* and *T. scincoides* often tuck the hind legs back along the base of the tail while unencumbered and use only the front legs to pull themselves along (Longley 1938; pers. obs.). This may occur either on grass (Wilhoft 1960; pers. obs.) or in climbing on open branches (Field 1980; pers. obs.). The significance of this mode of locomotion is also not clear.

Longevity in *Tiliqua* is commensurate with its large size. In captivity *Tiliqua scincoides* has lived for 12+ years (Longley 1947a) and "over 17 years" (Flower 1925), *T. gerrardii* for 9 years (R. Sadlier and S. Sadlier, pers. comm.), and *T. rugosa* 7 years and 1 month, 7 years and 11 months (Flower 1925) and 14 years and 6 months (Bowler 1977); a field captured *T. rugosa* was estimated to be at least 20 years old (Holmes and Light 1983) and in a field-marked population many animals were thought to live in excess of 9 years (Bull in press).

In all likelihood a *Tiliqua* species figures in what is probably the first detailed account of an Australian lizard species. This is Dampier's often quoted 1699 journal account of what appears to be a *T. rugosa* (Alexander 1914). "And a sort of Guano's [apparently Dampier's term for lizard; it probably derives from "iguana"], of the same Shape and Size with other Guano's, but differing from them in 3 remarkable Particulars: For these had a larger and uglier Head, and had no Tail: And at the Rump, instead of the Tail there, they had a Stump of a Tail, which appeared like another Head; but not really such, being without Mouth or Eyes: Yet this Creature seem'd by this Means to have a Head at each End; and, which may be reckon'd a fourth Difference, the Legs also seem'd all 4 of them to be Fore-legs, being all alike in Shape and Length, and seeming by the Joints and Bending to be made as if they were to go indifferently either Head or Tail foremost. They were speckled black and yellow like Toads, and had Scales or Knobs on their Backs like those of Crocodiles, plated on to the Skin, or stuck into it, as part of the Skin. They are very slow in Motion; and when a Man comes nigh them they will stand and hiss, not endeavouring to get away. Their Livers are also spotted black and yellow: And the Body when opened hath a very unsavoury Smell. I did never see such ugly Creatures any where but here. The Guano's I have observ'd to be very good Meat: And I have often eaten of them with Pleasure; but tho' I have eaten of Snakes, Crocodiles and Allegators, and many Creatures that look frightfully enough, and there are but few I should have been afraid to eat of, if prest by Hunger, yet I think my Stomach would scarce have serv'd to venture upon these N. Holland Guano's, both the Looks and Smell of them being so offensive."[1]

The two other major groups of skinks in Australia, the *Eugongylus* and *Sphenomorphus* groups, differ in a number of ways, and because many of these differences occur widely in each group, it is possible they arose early in the evolution of each group. Morphologically, for example, the *Eugongylus* group increased the number of premaxillary teeth (≥ 11) whereas the *Sphenomorphus* group retained the primitive number (9) or even reduced it (≤ 7). This means in general that the members of the *Eugongylus* group have more and probably slightly smaller teeth in the front part of the mouth than members of the *Sphenomorphus* group. Perhaps this difference relates to a difference in foraging method or prey handling between the two groups.

The members of the *Sphenomorphus* group are generally more secretive than the *Eugongylus* group, that is, they either live under the surface or if surface dwelling, avoid direct sunlight either by being crepuscular or nocturnal or by staying in the deep shade. The genera *Ctenotus* and *Eulamprus* are obvious exceptions to this generalization but whether they retain the primitive day-active, sun-loving habits of skinks in general or have only reacquired them secondarily is impossible to determine at present.

Probably as a result of their general propensity for living in the litter and below the surface where limbs are less useful in locomotion than lateral undulation of the body, the *Sphenomorphus* group has had many more "experiments" with limb reduction than the *Eugongylus* group. In the latter the only essays in this direction have been the loss of the two phalanges in the first toe of the front foot (the "thumb") in *Carlia*, *Eroticoscincus*, *Lygisaurus*, *Menetia*, some *Lampropholis* and some *Saproscincus* and a single phalange in the third and fourth toe of the front foot in *Nannoscincus maccoyi*, but in the former

---

[1] If Dampier's account is not the first recorded description of the stumpy-tail lizard, *Tilqua rugosa*, it is almost certainly the first description of the Australian meat pie.

every genus except *Ctenotus, Eremiascincus, Eulamprus* and *Notoscincus* has experienced phalangeal loss and in several cases the loss of entire limbs.

Perhaps as a further consequence of living under conditions of subdued light, the members of the *Sphenomorphus* group appear to have lost virtually all sexual differences in colouration, whereas such differences are generally retained by the *Eugongylus* group. Presumably sexual differences could not be perceived under reduced-light conditions and hence have been lost as a result of their uselessness.

The two groups in Australia also differ in general in the condition of the lower eyelid (p. 120). In the *Sphenomorphus* group it is generally scaly, except in the three relatively small-bodied taxa inhabiting the arid and semi-arid zone and its southern fringe: *Hemiergis* with a clear window, *Notoscincus* with a spectacle, and *Lerista* with either a window or a spectacle, depending on the species. In the *Eugongylus* group the eyelid generally has a clear window or is ablepharine except in the very largest species, the genus *Eugongylus*, where it is scaly.

As far as is known all members of the *Eugongylus* group retain inquinal fat bodies, whereas most members of the *Sphenomorphus* group have lost them (Smyth 1974; Greer 1986b). Inquinal fat bodies are the large, flat, yellow lobes on either side of the posterior part of the body cavity. They are thought to be energy reserves drawn upon during reproduction, both for courtship activity and yolking of the eggs, and during hibernation/estivation. Why they have been lost in the *Sphenomorphus* group and what compensatory modifications may have occurred, such as increasing the fat stored in the liver and/or tail, are not yet known. One approach to the problem would be to compare those few members of the *Sphenomorphus* group that retain fat bodies with their close relatives that lack them. Within Australia, the species that retain the fat bodies are the *Eulamprus tenuis* species group, three species of *Glaphyromorphus* and *Gnypetoscincus*.

The two groups may also differ in the kind of allantoplacenta formed in ovoviviparous forms. In the *Sphenomorphus* group the tissue contact between the female's oviduct and the embryo's covering membranes is smooth whereas in the *Eugongylus* group the contact is folded due to capillary-carrying ridges in the oviduct. The former condition also occurs in *Egernia* and *Tiliqua*, and in the very generally primitive *Mabuya* (Weekes 1930). As ovoviviparity has evolved independently and repeatedly in both the *Sphenomorphus* and *Eugongylus* groups, one wonders if the difference in the placentae in the two groups may not simply reflect a more widespread difference in the oviduct/shell contact of the more primitive oviparous forms in each group.

The structure of the corpus luteum may also differ between the two groups (Weekes 1934). In the members of the *Eugongylus* group examined (N = 2) there are fibroblastic ingrowths of the theca interna amongst the luteal cells whereas in the *Sphenomorphus* group members (N = 2) there are none. The former condition also occurs in the only other non-skink examined, a lacertid, while the latter also occurs in *Egernia whitii*. Although this difference is intriguing, speculation concerning its phylogenetic and functional significance must wait until the observations have been extended to other species.

There is also another very intriguing possible difference between the two groups that is hinted at in an unpublished Ph.D. thesis on the skinks of the Southern Highlands (Pengilley 1972). In the ecological section of the thesis under the discussion of *Eulamprus heatwolei* (as *Sphenomorphus tympanum* warm temperate form) it was noted that this was the only species studied in which the colour of the testis and pituitary turned yellow, apparently during maximum activity in spring. The telling point is that all the other species studied in similar detail (N = 7) were members of the *Eugongylus* group. Clearly the taxonomic distribution of this feature as well as its functional significance should be pursued.

Karyotypically, both the *Eugongylus* and *Sphenomorphus* groups have, with only a few exceptions, a diploid chromosome number of 30 (King 1973a-b; Donnellan 1985). However, despite the same diploid number, the morphologies of the karyotype are consistently different between the two groups (Donnellan 1985). Unfortunately, it is not yet possible to say which aspects of the two different morphologies are primitive and which derived. It is of further interest to note that the *Mabuya* group seems to be characterized by a diploid number of 32 and to have a karyotype morphology that differs from both the other two groups.

Finally it may be worth noting that as far as is known, communal nesting only occurs in members of the *Eugongylus* group and never in the *Sphenomorphus* group.

Eugongylus is one of the most generally primitive members of the *Eugongylus* group. Among its primitive traits are supranasal scales, a scaly lower eyelid and distinct frontoparietals. The genus consists of three relatively large species centered over the islands of the southwest Pacific

Two species, *E. albofasiolatus* and *E. rufescens*, are very widespread throughout the range, and the latter enters Australia at the tip of Cape York Peninsula.[1] *E. rufescens* is the largest member of the *Eugongylus* group in Australia at a maximum snout-vent length of 169 mm.

Although no studies have been carried out on the Australian population of *Eugongylus rufescens*, it is well known from observations made elsewhere (Schnee 1902; Hediger 1934; McCoy 1980; Cogger 1986). It inhabits closed forests or open areas with abundant ground cover. It tends to be crepuscular to nocturnal, but may also be active during the day in deep shade. When inactive, the lizards shelter under logs and piles of debris, amongst root tangles and occasionally in burrows of other animals, e.g., crabs. The species is primarily terrestrial but with arboreal tendencies. For example, it may be found high off the ground in tree hollows and tall root buttresses.

The species is oviparous, with a clutch size of 2-4. These relatively small clutch sizes (the four females carrying them ranged 121-132 mm, mean = 124.4 mm) are typical of tropical species (Tinkle *et al.* 1970) and are thought to derive from greater competition in the tropics and an increased premium on larger and hence more competitive young. However, whether the eggs and young are relatively larger as predicted has yet to be determined.

One of the most distinctive features of *Eugongylus* other than its large body size, is the colour pattern of the hatchlings. In contrast to the more or less uniformly brown or only indistinctly crossbanded adults, the hatchlings are strongly light and dark crossbanded. The light crossbands become darker with age leading to the more muted pattern of the adult. The significance of this distinctive hatchling pattern is unknown. The light to medium brown colour of adults is a good example of the association of this colour with cryptozoic/nocturnal habits in skinks.

Emoia is a second genus that also just enters Australia at the tip of Cape York. Like *Eugongylus*, it is widespread in the southwest Pacific (and beyond), and its two Australian species, *Emoia atrocostata* and *E. cyanogaster*, include Cape York as part of a much wider distribution (Ingram 1979a). As in the case of the Australian *Eugongylus* species, most of what is known of the Australian *Emoia* is based on observations made elsewhere.

E. atrocostata is a terrestrial/arboreal species of coastal habitats (Brown and Falanruw 1972; Ingram 1979a; Cogger *et al.* 1983), and perhaps for this reason has been able to cross broad

expanses of ocean and achieve a distribution stretching from Taiwan, the Philippines, Singapore and Christmas Island east to the Carolines, Santa Cruz and Vanuatu. The species is strictly diurnal and apparently only comes out to forage when the sun is shining, even though air temperatures on cloudy days often seem high enough to permit activity (Cogger *et al.* 1983).

Along with certain littoral species of *Cryptoblepharus*, is is also one of the most marine species in the family. The species inhabits rocky coasts and mangroves, forages in the intertidal; and will often flee through tidal pools (Ingram 1979a; Cogger *et al.* 1983); they may even forage in these pools, as suggested by small freshly caught fish in the stomachs of some specimens from the Moluccas (Kopstein 1926). Other food items also reflect the littoral habitat (Cogger *et al.* 1983).

The species is oviparous and the eggs take 56-70 (mean=60) days to hatch (conditions unspecified; Alcala and Brown 1967).

The population dynamics of the species has been studied in a mangrove forest in the southern Philippines at a locality only slightly more tropical (9° N latitude) than the Australian population (12° S) (Alcala and Brown 1967). This study found that hatchlings (SLVs = 33-39 mm) and juveniles grew at an average rate of 0.18 mm/day and attained sexual maturity at an age of 9-9.5 months (SVL = 80 mm). As the populations aged, the sex ratio seemed to shift increasingly in favour of males suggesting a higher mortality rate or increasing furtiviness for females. Maximum longevity was between 3 and 4 years. How applicable this study is to the Australian population remains to be seen. It is interesting to note, however, that to date, it is the only study of the population dynamics of a tropical Australian lizard species — but done in the Philippines!

Emoia cyanogaster also occurs in New Guinea, the Bismarck Archipelago and the Solomon Islands. The species is primarily arboreal and one of its remarkable features is the transverse scales on the underside of the digits which have become more numerous (two or three times) and thinner than in close relatives (Fig. 67); by analogy with similar, better-studied skink species (Williams and Peterson 1982), these lamellae probably carry finely branching microscopic filaments to aid in climbing as in geckos.

The species lives mainly in vines and the outer branches and twigs of trees and shrubs (but not on the trunks or main branches). It is completely diurnal and presumably a basker. It comes to the ground to feed and lay its eggs but flees up into vegetation instead of into the litter. It may sometimes occur as high as 6 m above the ground (McCoy 1980).

---

[1]*E. albofasciolatus*, was previously thought to occur possibly in Australia, but this now seems doubtful as no further records have come to light beyond the original literature citation of the last century (Covacevich *et al.* 1982).

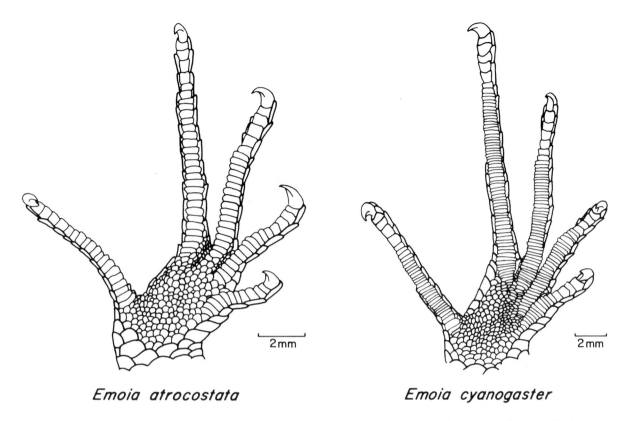

Emoia atrocostata   *Emoia cyanogaster*

Fig. 67. Ventral view of right rear foot in the two species of *Emoia* in Australia to show the two lamellae types in the genus, the primitive, rounded lamellae of *E. atrocostata* (AM R 61587) and the derived, blade-like lamellae of *E. cyanogaster* (AM R 44930).

Both species of *Emoia* have a constant clutch size of two. The relatively small clutch size may be part of the general trend for smaller clutches in the tropics, but the reasons for the constant clutch size are unclear.

Eulepis is a genus of slightly elongate skinks often with bright red throat colour. The group is southern in distribution and like many other southern genera its distribution is disjunct just north of the Bight, where drier country bisects the more mesic southern habitats. There are two species in the south-east, *E. duperreyi* and *E. platynota*, and one in the south-west, *E. trilineata*. *E. duperreyi* and *E. trilineata* are longitudinally striped and very similar to each other whereas *E. platynota* lacks stripes and has a distinctive dark lateral band.

The genus inhabits woodlands and shrublands, and all the species are diurnal, sun-loving and terrestrial. They are also all oviparous. There is a positive intraspecific correlation between female size and clutch size in *Eulepis platynota* and *E. trilineata* (r = .75**, N = 14 and r = .69***, N = 23) but not in *E. duperreyi* (r = .33^NS, N = 12; Greer 1982; pers. obs.). At least one species, *E. duperreyi*, lays communally (B. C. Mollison in R. Green, 1965; Ehmann 1976a; Rounsevell 1978; all as *Leiolopisma trilineatum*).

The slightly elongate appearance of the genus is reflected in an increased number of presacral vertebrae. They have 29-35 vertebrae whereas the primitive number for the *Eugongylus* group as a whole is 27. In addition they show a phenomenon widespread in other elongate skinks with an elevated number of presacral vertebrae: a sexual dimorphism in the number of presacral vertebrae, with females having more (Table 4). The functional significance of this difference may be to increase the suppleness of the female when gravid. However, despite the increase in presacral vertebrae, *Eulepis* shows no reduction in the number of phalanges.

The functional significance of red throat colour in the genus is not understood. In adults it is a seasonal phenomenon, appearing during summer activity and disappearing during the winter quiescence. Both sexes may show it, but males appear to have it more strongly than females. Some individuals show it strongly but others not at all. However, what is most strange about the red throat colour is that hatchlings have it as well (Hewer and Mollison 1974). In some populations all juveniles show it (Rounsevell 1978) whereas in others only some do (pers. obs.). No other skink species, in Australia or elsewhere as far as is known, has bright throat colour in its hatchlings.

Morethia is a group of seven, small, diurnal, largely terrestrial skinks with a spectacled-eye. It occurs throughout Australia except for the mesic south-east (Smyth 1972; Storr 1972b; Rawlinson 1976; Greer 1980b). In addition to the spectacle, the genus is characterized by a strong seasonal sexual dimorphism in which males show bright pink or red colour on the head or body (depending on species but no colour at all in *M. storri*) and by the fusion of the frontoparietal and interparietal scales into a single rhomboid-shaped scale.

There are two subgroups in *Morethia*: the *M. butleri* group which comprises four species and is distributed over the southern half of the country and the *M. taeniopleurus* group which comprises three species and occurs in the northern part. The skinks of the second group tend to be smaller and have completely fused spectacles and distinct light dorsolateral lines.

Most species of *Morethia* have a red or coral pink tail colour as hatchlings. In some species, i.e., *M. butleri*, *M. boulengeri*, *M. taeniopleura* and *M. storri*, the colour fades with age, but in one, *M. ruficauda*, it remains bright or perhaps even becomes more intense. The brilliant tail of the latter has earned it the name of "fire-tail". Indeed, a very characteristic herpetological experience in the arid centre is to be turning rocks or timber looking for reptiles and to suddenly uncover a small dark skink with fiery red tail that moves "like lightening" from one piece of cover to another. This is *M. ruficauda* and when it is "hot", that is with a midday body temperature of around 40°C, it must be one of the most agile skinks in Australia. Three species, *M. adelaidensis*, *M. lineoocellata* and *M. obscura*, lack red tail colour altogether.

The function of the sexually dimorphic red colour in male *Morethia* is not understood, but it is peculiar in that whereas it is confined to the throat in most species, in *M. adelaidensis* it occurs only on the lower sides of the body just below the dark lateral colour. Perhaps the position of the colour is not as important as the colour itself.

Most *Morethia* species (perhaps all) are able to undulate into a loose substrate. They do this in order to flee and also, probably, to avoid inclement surface conditions. This habit may explain a morphological peculiarity of all but one species (*M. butleri*): strongly interdigitating supraciliary and supraocular scales (Fig. 68). This dove-tailing arrangement could act to resist the shearing force on the head engendered during lateral undulation below the surface. The fact that various species show different degrees of interdigitation could further suggest, if this burrowing hypothesis is right, a different

degree of force exerted on the head in different species which could in turn relate to their habits. The only other sand-swimming group widespread in the arid zone, *Lerista*, also has similarly interdigitating supraciliary and supraocular scales. Perhaps the fusion of the frontoparietal and interparietal scales in *Morethia* is also a strengthening device for burrowing.

The stomach contents have been examined in only one species, *Morethia boulengeri* and these indicated that the species is an arthropod generalist (Smyth and Smith 1974).

Age at maturity has been studied in a South Australian population of *Morethia boulengeri* and here it appears that whereas some males reached sexual maturity in their first year and others in their second, females attained maturity only in their second (Smyth and Smith 1974).

All *Morethia* species are oviparous and in all in which sample sizes are adequate there is a positive correlation between female size and clutch size (Smyth and Smith 1974; Greer 1980b).

Although all the southern *Morethia* are known or may be presumed to be summer breeders (Smyth and Smith 1974; Rawlinson 1976; pers. obs.), what little is known of the northern species indicates that reproductive activity occurs at least in part during the winter dry, a time when southern species are still in hibernation. For example, in the Top End, *M. storri* is known to breed from at least the end of the wet season to well into the dry (March-September), with knowledge of its activity during the rest of the year being hampered by the lack of collections (James and Shine 1985). During July, the middle of the dry season, hatchlings, females with yolking ovarian eggs, and females with shelled oviducal eggs all occur (pers. obs.); at this same time of year, southern *Morethia* species would be virtually totally inactive.

In the northern part of the range of *Morethia taeniopleura*, i.e., in the Townsville area, examination of large collections show that females are reproductively quiescent in the middle of the dry season (end of May to end of July) but start showing yolking ovarian eggs thereafter (end of July — first of August) and have oviducal eggs in the late dry (September). Southern *Morethia* species would just be emerging from hibernation as northern *M. taeniopleura* were shedding their yolked ovarian eggs into the oviducts. Unfortunately, events in the rest of the reproductive cycle of these northern *M. taeniopleura* are only hinted at by existing collections. A Cape York female captured in the period 14 April-12 May carried oviducal eggs and eggs

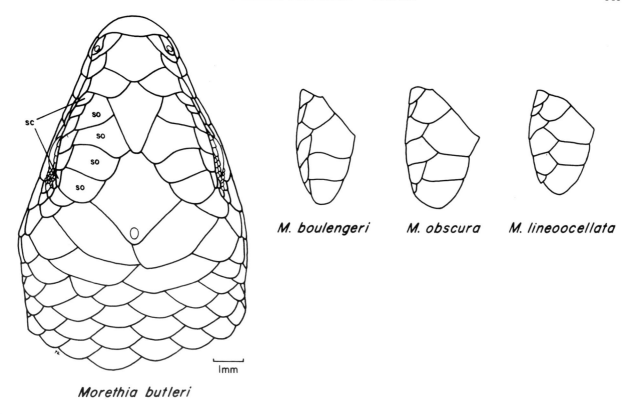

M. boulengeri   *M. obscura*   *M. lineoocellata*

Morethia butleri

Fig. 68. Head of *Morethia butleri* to show the general position of the supraciliary (sc)/supraocular (so) area, and a series of drawings of this area in three other *Morethia* species to show the varying degrees of interdigitation between the two sets of scales.

found in the Townsville area on 17 May hatched between 10-16 July. These observations suggest that reproduction could occur throughout the wet season and up until the dry. Unfortunately, collections of the third northern species, *Morethia ruficauda,* are too meagre to offer any insight into the seasonality of reproduction.

Proablepharus is a poorly diagnosed group of three small species inhabiting the arid, semi-arid and seasonally dry parts of Australia (Copland 1947b, 1952a; Storr 1975c). Like other small skinks in such habitats they have a spectacle covering the eye. *P. reginae* and *P. tenuis* are predominantly plain brown in colour, whereas *P. kinghorni* is light and dark striped and very reminiscent of a small striped *Ctenotus.* Juveniles of all three species have red in the tail but this is lost or much diminished in adults. At least some adults of all species, presumably males, have a red or orange wash on the side of the head (Greer 1980b).

P. kinghorni is an eastern species which inhabits open grassland on black soils (P. Rankin, pers. comm.; Wells 1979a; pers. obs.); *P. reginae* is known only from widely scattered shrubland localities in the centre and west, and *P. tenuis* lives in woodlands in the north. All three species are diurnal and generally terrestrial, although *P. reginae* has been reported 90 cm

above the ground on an *Acacia* trunk (Bradshaw 1976) and one *P. kinghorni* was "collected sunning itself on a dead tree" (S. J. Copland field notes on deposit in the Australian Museum).

Proablepharus reginae and *P. tenuis* are oviparous and lay 1-2 eggs/clutch; one *P. kinghorni* has been found carrying two large, ovarian follicles (Table 9; Smith 1976).

Carinascincus comprises four species restricted to Tasmania and the south-east corner of the mainland (Rawlinson 1974a, 1975; Greer 1982). They are medium-sized, diurnal, basking skinks which are mainly terrestrial but will climb on rocks and logs. All are ovoviviparous and for the species for which there is an adequate sample size, there is a positive intraspecific correlation between female size and litter size (Greer 1982).

All species occur in Tasmania but only one extends to the mainland. This latter, *Carinascincus metallicum,* is the most generally primitive species. It inhabits moist forest and woodlands. The three endemic Tasmanian species, *C. greeni, C. ocellatum* and *C. pretiosum,* are similarly specialized in having slightly squared bodies in cross-section, a relatively large number of small scales and tinted "windows" in the lower eyelid. In contrast to their more generally primitive relative, these three species tend to occur in more open habitats such as forest clearings, heathlands and meadows and to utilize rocks as perches or activity sites; they also occur at higher elevations either attaining or being restricted to the central plateau of the island. One of the species, *C. greeni,* is confined to a few subalpine boulder fields over or around water, and is one of the few Australian reptile species restricted to this habitat.

The exigent climate of the higher elevations in Tasmania seems to have constrained the frequency of reproduction in at least some *Carinascincus.* For example, there is some indication that female *C. greeni* may be biennial in their breeding cycle, i.e., they reproduce only every second year (Rawlinson 1975; pers. obs.). The evidence is that only half the mature females seem to be reproductively active in any one year. Biennial reproduction is known in other high latitude reptiles, notably certain European vipers and a Japanese skink (Hasegawa 1984), and it probably arises because the short growing season does not permit a complete reproductive cycle to occur in one year; instead two years are needed. A second example is the Mt Wellington population of *C. pretiosa* in which the females are said to retain the young over winter in at least some years (Macintosh, in Pengilley 1972).

The cool, open habitats of the Tasmanian species may in part "explain" two of their morphological adaptions: small body scales and tinted lower eyelids. It is a widely demonstrated but poorly understood fact that lizards inhabiting cool climates possess smaller body scales, and hence have higher scale "counts", than their relatives in warmer climates. The tinted windows in the moveable lower eyelid probably act as natural sunglasses to reduce the glare characteristic of open, high altitude habitats.

Although most surface-active members of the *Eugongylus* group are diurnal, there is one surface-active taxon in the group that is distinctly crepuscular and nocturnal, the genus *Cyclodina.* This is a group of seven species centered primarily in New Zealand (Hardy 1977; Gill 1986) but with one species, *C. lichenigera* confined to Philip Island in the Norfolk group and to Lord Howe Island and its surrounding islets and Ball's Pyramid, a volcanic plug rising 548 m straight out of the sea some 18.5 kilometres south of Lord Howe Island (Cogger 1971; Cogger *et al.* 1983). Although most of the islands where *C. lichenigera* occurs have been severely disturbed by man, the effect on the species is difficult to determine due to inadequate understanding of the species' status before man. Today, some populations appear to occur only in low numbers (e.g., Philip I.) or to be geographically restricted

(e.g., Lord Howe I.), whereas others are widespread in the available habitats (e.g., Philip I.) and locally abundant (e.g., Rabbit I.). The wide habitat range (boulder coasts to the interior of some of the islands) implies a broad ecological tolerance.

Like other *Cyclodina, C. lichenigera* is a stocky, robust skink, with the body square in cross section and a basically brown ground colour — this latter colour being typical of *Eugongylus* group taxa that avoid direct sunlight.

The species is unique in the genus in that it is oviparous instead of ovoviviparous, providing yet another example of the north/south-warm/cool dichotomy between egg layers and live-bearers in Australasian skinks. Clutch sizes in *Cyclodina lichenigera* are 2-3, and an incubation period of 68 days has been recorded (Cogger *et al.* 1983).

Animals sheltering by day under boulders on a beach had body temperatures 24.6-26.5°C (mean = 25.9, N = 4) which were higher than the air and substrate temperatures. Body temperatures of skinks active at night have yet to be recorded.

The natural diet has not been studied in any detail, but there is an observation of a specimen on Ball's Pyramid rolling a tern's egg down a rock slope until it broke and then feeding on the contents (D. Rootes, in Cogger 1971). This is interesting both for the implied diurnality of the behaviour and the learning abilities of the species.

Two specimens have been kept alive for seven years in the Australian Museum. They occasionally basked under the light and ate arthropods, fruit and canned pet food (D. Kent, pers. comm.).

Within the *Eugongylus* group there is a subgroup which is recognizable on the basis of a single morphological character that is otherwise found nowhere else in skinks: the ventral fusion of the three elements that make up the first presacral vertebra (the atlas) into a single crescent of bone (Fig. 69). The skinks in this group are all centered over eastern Australia, New Guinea and New Caledonia, that is, basically around the Coral and western part of the Tasman Seas. The unique character plus the geographical proximity of the taxa showing it is strong evidence for the monophyly of the group. There are 12 Australian genera in the group, and we may give it the informal name *Pseudemoia* group after one of its most generally primitive members. All the remaining genera to be discussed here in the *Eugongylus* group are members of the *Pseudemoia* subgroup.

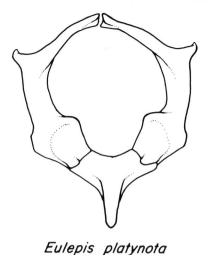

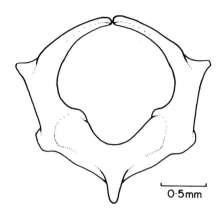

0·5mm

Eulepis platynota   *Pseudemoia spenceri*

Fig. 69. First presacral vertebra (atlas) in *Eulepis platynota* to show the primitive scincid condition and in *Pseudemoia spenceri* to show the characteristic fusion of the intercentrum, with the two arches in the *Pseudemoia* group.

Live bearing members of the *Pseudemoia* group may also share some unusual placental features. In the two species studied, one of the two eastern species of *Claireascincus* (only one species was recognized at the time) and either *Pseudemoia spenceri* or *Harrisoniascincus coventryi* (there is some uncertainty as to the identity of the species actually used, see Rawlinson 1974b), the general dorsal uterine folding which is characteristic of the *Eugongylus* group (see above) becomes concentrated into a distinct elliptical patch directly above the embryo, and the uterine epithelium becomes thicker (Harrison and Weekes 1925; Weekes 1927a-b, 1929, 1930).

Pseudemoia itself occurs only in south-east Australia (Fuhn 1967) and comprises two species, one on the mainland and the other on a tiny island off the south coast of Tasmania (Rawlinson 1974b).

The mainland species, *Pseudemoia spenceri*, ranges from the mountains west of Sydney south to east of Melbourne. It is an arboreal and rock-dwelling species which inhabits forests open enough to permit basking. It has been seen in trees 50-75 m above the ground, the highest level recorded for an Australian skink. The head and body are slightly flattened, a shape that is probably related to its use of narrow crevices such as exfoliating timber strips and rock slabs for retreats. A recent very detailed study of stomach contents showed just what we might expect, i.e., it is an opportunistic forager (Brown 1986). Like many other southern (i.e., cool climate) skinks it is ovoviviparous and litter size increases with female size (Rawlinson 1974b; Greer 1982).

The island species, *Pseudemoia palfreymani*, occurs on Pedra Branca, an uninhabited, near vegetationless rock mass some 26 km south of

the main island of Tasmania. This distribution makes it the most southern lizard in Australia. It occurs primarily in sheltered areas where enough soil has accumulated to permit burrowing. The skinks share the island with colonies of nesting sea birds, and the invertebrates which are inevitably associated with such colonies appear to be a principal source of food. Like its congener, and befitting its extreme southern distribution, it is ovoviviparous. The species apparently likes to bask but due to the local weather conditions has relatively few opportunities. When the sun does shine during the activity season, the lizards quickly emerge to bask, and during one such period a quick count of the population revealed a total of approximately 500 individuals. Thus the species is probably also one of Australia's most numerically restricted species (Rawlinson 1974b; Rounsevell *et al.* 1985).

The other generally primitive genus in the *Pseudemoia* group besides *Pseudemoia* itself is *Claireascincus*. This genus comprises three species, two in the south-east, so-called forms A and B of *C. entrecasteauxii*[1] and one in the south-west, *C. baudini*. All are medium-sized, largely terrestrial lizards which often climb on logs, low shrubs or rocks, usually to bask. All species are live bearing, and the two better-known eastern species show a positive intraspecific correlation between female size and litter size (Pengilley 1972). *Claireascincus "entrecasteauxii"* is said to have "an obvious reduction in yolk content" in the egg (Weekes 1934, a work antedating knowledge that the eastern species was actually composite), an observation that would be worth pursuing.

Males of the two eastern species develop bright breeding colour: in both species, one or two orange/red lateral stripes, and in *Claireascinus*

[1]Work in progress indicates that there are probably four species in the *C. entrecasteauxii* complex (S. Donnellan and M. Hutchinson, pers. comm.)

entrecasteauxii form B, a red belly. Mating follows the birth of the young, the female storing the sperm until ovulation the following spring. In *C. entrecasteauxii* form B, males mate for the first time at about 12 months of age but females not until 19-20 months (Pengilley 1972).

The third species, *Claireascincus baudini*, is known from only four specimens from a remote area along the north-west corner of the Bight and is by far the least known species in the genus and one of the least known members of the family in Australia (Greer 1982).

Claireascincus "entrecasteauxii" has the greatest altitude range of any reptile in Australia, from sea level to the summit of Mt Kosciusko, at 2237 m the highest elevation in Australia (Copland 1947c). Such a distribution would provide a unique opportunity to study altitudinal variation in any number of biological features in a single closely related group.

Nannoscincus is a genus of six species, five of which occur on New Caledonia and one in southeast Australia. Phylogenetic analysis indicates that the basic dichotomy in the lineage is between these two areas (R. Sadlier, in prep.). The genus is unusual within the *Eugongylus* group in being shade-loving and semi-fossorial, features which are much more typical of the *Sphenomorphus* group. The relationships of the genus are unknown. The habits and distribution of the genus recall those of *Cyclodina*: sun-avoiding skinks whose primary distribution is on islands on the east side of the Coral and Tasman Seas but with an outlying species to the west.

The Australian species, *Nannoscincus maccoyi*, is restricted to the cooler, wetter portions of southeastern New South Wales and eastern and southern Victoria. It inhabits forest and is usually only found by turning surface cover; when uncovered it rapidly wriggles into the litter/soil. The species is elongate and has a reduced number of phalanges. The presacral vertebrae range 34-36, the upper figure being the highest known for the *Eugongylus* group, and the phalangeal formula is 2.3.4.4.3/2.3.4.4.3. The dorsum is dark chocolate brown, but the venter of the body is bright lemon or orange yellow (the throat is clear).

Virtually nothing has been published on the diet of *Nannoscincus maccoyi*, but feeding activity has been studied in Victoria populations and these animals appeared to be feeding all year long, albeit at a diminished level during winter (Robertson 1981). However, given the cold climate of this area, it is surprising that any feeding occurred at all during winter.

Reproduction in the species is fairly well known having been studied in some detail in three different areas, one in New South Wales (Pengilley 1972; Shine 1983) and two in Victoria (Robertson 1981). From these studies we know that mating occurs in the spring (October and November), the 1-9 eggs are retained in the oviduct for 8-9 weeks (Pengilley 1972) and laid in midsummer (late December-early January). There is some evidence that in at least some populations females may only reproduce every other year (Pengilley 1972). Egg laying may be solitary or communal; in the latter case as many as 52 eggs have been found together (Pengilley 1972). Clutch size and female size are correlated (r = .39*, N = 37). In the laboratory there was a strong inverse relationship between ambient temperature and both the time between ovulation and deposition and incubation time (Shine 1983; Table 11).

Previously there was apparently some confusion as to the mode of reproduction of *Nannoscincus maccoyi*. Earlier authors reported it to be a live-bearer (Lucas and Frost 1894) or ovoviviparous (Littlejohn and Rawlinson 1971), but all recent studies have clearly established it to be oviparous (Pengilley 1972; Robertson 1981; Shine 1980, 1983; pers. obs.).

There are some very intriguing preliminary data (Shine 1983) which indicate that *Nannoscincus maccoyi* may have a faster rate of embryonic development than two other syntopic, egg-laying skinks, *Lampropholis delicata* and *L. guichenoti*. Compared to these skinks, which have similar-sized embryos and show a similar stage of embryonic development at deposition as *Nannoscincus*, the range of incubation periods at three different ambient temperatures, 15°C, 20°C and 26°C, were all lower in *N. maccoyi*; no sample sizes or dispersion statistics were presented in the original data, so statistical comparison is precluded, but the ranges of days to complete development between *Nannoscincus* and the two *Lampropholis* species either failed to overlap or shared upper and lower values. The significance of this faster developmental rate is obscure, but it raises the question of how widespread in other aspects of *Nannoscincus* physiology this enhanced rate of activity may be.

Like many other cryptozoic species inhabiting the mesic east coast, *Nannoscincus maccoyi* is extremely heat and desiccation sensitive: its preferred body temperature averages only about 21°C; its critical thermal maximum temperature is a relatively low 36.4-38.0 (mean = 37.0), and its rate of desiccation one of the highest known for any Australian skink (Spellerberg 1972a; Robertson 1981).

Eroticoscincus is a monotypic genus endemic to southeastern Queensland. The single species, *E. graciloides*, is a small (maximum SVL 34 mm), plain brown skink with a windowed lower eyelid. It inhabits open and closed forests often near streams. The animals shelter under surface cover such as rocks by day and become active toward dusk and during the evening, foraging through the leaf litter and rarely exposing themselves for any length of time. Their crepuscular/nocturnal habits and association with moist microhabitats are reminiscent of *Nannoscincus* and *Saproscincus*, the only other Australian genera in the *Eugongylus* group with such habits. The species is oviparous (Czechura 1981).

Harrisoniascincus is a poorly diagnosed group of two small species (maximum SVL of the larger species, 59 mm) inhabiting the eastern ranges of New South Wales and Victoria (Rawlinson 1975; Ingram and Ehmann 1981). Both species inhabit woodland forest but usually with open areas for basking. Both are primarily terrestrial, although they will climb on to dead tree trunks lying on the ground. The northern species, *H. zia*, is oviparous and the southern one, *H. coventryi*, ovoviviparous, providing yet another example of the association between ovoviviparity and cool climates. In some Southern Highlands populations of *H. coventryi* it is thought that some males become mature at one year of age but that most males and all females take two years to mature (Pengilley 1972). In contrast to most other southeastern Australian skinks, there is no significant correlation between female size and brood size in *H. coventryi* (Greer 1982).

Menetia is a very distinct group of skinks (Storr 1976c, 1978b; Ingram 1977; Rankin 1979; Sadlier 1984) within the *Pseudemoia* group. They are small in size (maximum SVL of the largest species is only 38 mm) with only four toes on the front foot (they have lost the innermost) and a spectacle covering the eye. They are also unique among Australian skinks in having the frontal greatly reduced in size and the supraciliaries widened. In four species, *M. alanae*, *M. amaura*, *M. concinna* and *M. greyii*, the second supraciliary has extended anteromedially to contact the prefrontal and thereby nearly separate the first supraciliary from the first supraocular — a unique arrangement in skinks (Fig. 70). Two species, *M. amaura* and *M. surda*, have apparently lost the external ear opening (Storr 1976c, 1978b) and a third, *M. maini*, is said to be variable in this regard (Storr *et al.* 1981) — a situation most unusual, if not unique, within skinks. These *Menetia* are the only *Eugongylus* group species in Australia to have lost the ear opening; comparison with the several *Sphenomorphus* group lineages that have lost the ear opening would be worthwhile.

All *Menetia* are terrestrial and diurnal, and oviparous with clutch sizes of 1-3 (Table 9; Bush 1983). Male *M. greyii* develop orange throat colour and yellow ventral body colour during certain parts of the activity season, but the function of these colours is unknown (Smyth and Smith 1974; pers. obs.). As far as is known the only other species to develop ventral colour is *M. alanae* which also has a yellow belly (sex unknown; R. Sadlier, pers. comm.).

Study of a South Australian population of *Menetia greyii* suggested that both males and females become reproductively mature in their first year, that at least some individuals survive to breed again in their second year, and that females may produce two clutches per season (Smyth and Smith 1974).

Although most of the *Menetia* species occur on the more or less mesic periphery of Australia as do most of their near relatives, which makes it seem likely that such habitats are ancestral for the genus, a few species, most notably *M. greyii*, have invaded some of the most arid parts of Australia. How the small, terrestrial, diurnal lizards cope with the high substrate temperatures that prevail in these areas on a summer day is interesting to contemplate. Substrate temperatures easily attain 50-60°C (and 40-43°C is lethal for most lizards), and the lizards' small size would make them very susceptible to almost instant over heating. Thus, in addition to all the normal vagaries of a lizard's life, *Menetia* must also cope with the added hazard of rarely being more than a false step away from being fried.

Cryptoblepharus is a genus of 37 forms (i.e., species and/or subspecies — see Mertens 1931, 1933, 1934, 1958a, 1964; Storr 1961, 1976a; Covacevich and Ingram 1978) whose center of abundance and perhaps early evolution is the islands of the tropical southwest Pacific and Australia; however, from here it also extends west to the small islands of the western Indian Ocean and Madagascar and the east central coast of Africa (Haacke 1977) and east across the Pacific to the west coast of South America (Crombie and Dixon, in press). All the species are small (the largest does not exceed 55 mm SVL), slightly depressed, and with an ablepharine eye. They are usually found off the ground, either on trees and shrubs or rocks, but occasionally, at least some species, occur on the ground.

On first sight it might appear as if the genus *Cryptoblepharus* was one of the most ecologically vagile of skinks. Certainly it has a wide geographic distribution, and within this distribution it occurs in a variety of habitats, both natural, such as mesic woodland, desert shrubland, and rocky intertidal and man-made, such as native huts in Fiji and Paddington (Sydney).

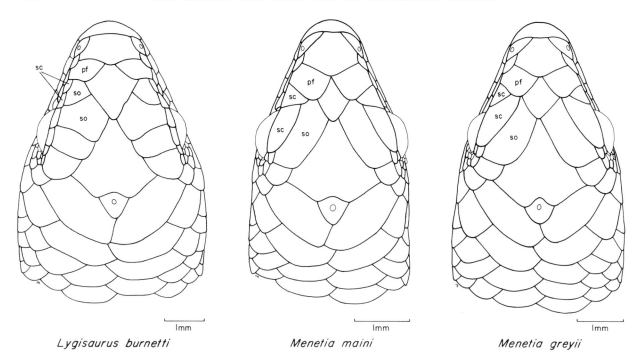

Lygisaurus burnetti   *Menetia maini*   *Menetia greyii*

Fig. 70. Heads of a *Lygisaurus* species and two *Menetia* species to show the differences in the relative size of the supraciliaries in the two similarly sized genera and also the contact between the second supraciliary and the prefrontal in certain *Menetia* species. Abbreviations: pf — prefrontal; sc — supraciliary; so — supraocular.

But are all these habitats and distributions really indicative of a diverse ecological tolerance? Another way of approaching the ecology of these animals is to look for common themes, perhaps less obvious than geographic distribution and gross habitat. For example, what is there in common between a garden wall in Sydney, a dead she-oak tree in Alice Springs and the inter-tidal rocks at Bowen? For one, all are open, sun-drenched habitats; for another they are vertical, basically climbing habitats, and for a third, they are lacking in regular and predictable supplies of fresh drinking water, that is, they are arid.

Thus from the point of view of ecological factors or microhabitat, *Cryptoblepharus* is actually adapted to a rather narrow, and stringent set of ecological parameters: vertical surfaces receiving full sun and lacking or severely restricted in fresh water. Taking these last two features as starting points, it would be interesting to enquire into the behavioural and physiological adaptations the animals may have to solar radiation and dehydration.

The relationships of *Cryptoblepharus* are not clear but may be with *Menetia*. For example, both genera share small size, a spectacled-eye and a small frontal (less so in *Cryptoblepharus*). They also share a seeming ability to cope with habitats low in potable water.

Like certain other "off-the-ground" genera, the species of *Cryptoblepharus* are generally fairly specific with regard to their choice of substrate:

either wood or rock. For example, of the six Australian species, three occur primarily or only on the trunks and main branches of trees and shrubs — *C. carnabyi* (Morley and Morley 1985; pers. obs.), *C. plagiocephalus* (L. Smith, in Storr 1976a; Smith and Chapman 1976; Dell and Chapman 1977, 1979a-b, 1981a-b; Chapman and Dell 1977, 1978, 1979, 1980b; Davidge 1979; Dell and Harold 1979; Smith and Johnstone 1981; Scanlon 1986) and *C. virgatus* (Limpus 1982; Scanlon 1986; pers. obs.) and three occur primarily or only on rocks: *C. fuhni* (Covacevich and Ingram 1978), *C. litoralis* (Mertens 1958a; Limpus 1982; Horner 1984) and *C. megastictus* (L. A. Smith, in Storr 1976a; Smith and Johnstone 1981). *C. virgatus* is a slight exception in that it often occurs on brick walls. It is also the only human commensal species in the genus in Australia (Cook 1973; P. R. Rankin in Anonymous 1973a; Limpus 1982).

Cryptoblepharus litoralis is particularly interesting because, as its scientific name implies, it is an inhabitant of coastal habitats. In fact, its habits are not unlike those of *Emoia atrocostata*, in that it feeds in the upper intertidal zone after the tide has receded (Pearson and Reeves 1978) and has little hesitation about fleeing through tide pools (pers. obs.). It is also dark-coloured like its larger distant relative, but this seems to be a feature common to most, dark-rock inhabiting skinks (Covacevich and Ingram 1978) and may be simply a case of substrate matching. In some areas *C. megastictus* also lives along the coast and enters the splash zone (Storr 1976a).

One of the minor mysteries associated with *Cryptoblepharus* is what it does with its eggs. We know from dissection of museum specimens that most species lay about two eggs per clutch (Table 9; Smith 1976), but the eggs are rarely found in the wild. In fact I know of no record of the eggs of this genus aside from two communal nests, one for the wide-ranging Pacific species *C. poecilopleurus* in Hawaii (McGregor 1904) and the other for a species of uncertain identity on Cosmoledo Atoll in the western Indian Ocean (Honegger 1966). Both nests contained over 70 eggs each.

Seasonality of reproduction has been studied in only one species of *Cryptoblepharus*: *C. plagiocephalus* from the Top End of the Northern Territory. Here the species appears to be reproductively active year around, the only residual uncertainty being for females during the early wet (November-December) for which specimens were lacking (James 1983). Presumably, southern forms of *Cryptoblepharus* are highly seasonal in their reproductive activity.

Within the *Pseudemoia* group there is a subgroup that can be distinguished by a thumb-like projection at the base of the hemipenis which gives the everted organ the shape of a mitten. Four genera share this character, which is found nowhere else in skinks: *Carlia*, *Lampropholis*, *Lygisaurus* and *Saproscincus*, and all occur either exclusively or largely along the eastern seaboard, especially in the north-east.

On the basis of current thinking these four genera can be further subdivided into two groups: *Carlia* and *Lygisaurus* on the one hand and *Lampropholis* and *Saproscincus* on the other. The former is characterized by the loss of the first digit of the front foot (the thumb) and fused frontoparietals, and the latter by smooth body scales. Neither set of characters is very convincing but the groupings are also supported by unpublished biochemical evidence so perhaps they are valid. If they are, some convergent evolution must have occurred between the two groups because *Lampropholis* also has fused frontoparietals and one species each of *Lampropholis* and *Saproscincus* have also lost the first digit on the front foot, while some *Carlia* and *Lygisaurus* have smooth scales. The three losses of this one digit within the group as a whole illustrates a long-appreciated feature of evolution, namely that close relatives often undergo similar morphological changes in evolution. However, whether this is because close relatives have similar mutations or experience similar selective forces is still uncertain.

The genus *Lampropholis* consists of 8-10 species of small, grey to dark brown sun-loving skinks ranging around the coast from the tip of Eyre Peninsula to the base of Cape York Peninsula (Greer and Kluge 1980). They are among the most nondescript of Australian skinks but at the same time among the most widely seen, as they probably inhabit three-quarters of the gardens of Australia and are often very common. In Sydney they are the lizards one sees and hears skittering through the litter in the flower beds of Hyde Park and the Botanic Gardens.

Mark and recapture studies of *Lampropholis guichenoti* indicate that this species is wide ranging and has neither a territory nor a home range (Milton 1980a), an observation certainly in accord with my subjective impression of all the species I have seen.

Stomach contents have been analyzed in the two common southern species of *Lampropholis*, *L. delicata* and *L. guichenoti*, and there was no reason to believe that they are anything other than opportunistic arthropod predators (Belmont 1979; Crome 1981).

The mating season in at least one species of *Lampropholis* may be subject to geographic variation. In *L. guichenoti* from the New England Plateau, copulation has been observed from late winter to mid-summer (August-February) but mostly in mid to late spring (October-November; Simbotwe 1985); in Sydney there appear to be two distinct mating peaks, spring and late summer (Joss and Minard 1985), while in the Southern Highlands there appears to be only one, in mid to early autumn (Pengilley 1972). This suggests that on the New England Plateau the eggs of one season are fertilized by sperm from a mating during that season whereas in the Southern Highlands the eggs are fertilized by sperm stored from a mating of the previous season. In Sydney, both regimes apply, perhaps with females hatched the previous season doing much of the spring mating.

Lampropholis usually lays its eggs in crevices, under logs or rocks, or in burrows, e.g., spider burrows (Humphreys 1976), but it does not bury them or cover them in any way. *L. delicata* in Tasmania are said to lay two clutches per season (Rawlinson 1974a), but *L. delicata* in Sydney only one (Joss and Minard 1985). As this is the opposite of what might be expected from the relative lengths of the growing seasons, it is possibly worth a closer look. *L. guichenoti* on the New England Plateau lays only one clutch per season (Simbotwe 1985) while in Sydney this species lays two clutches during "good" (i.e., wet) seasons and one during bad (i.e., dry; Joss and Minard 1985). In each population there is a positive correlation between female size and clutch size ($r = .66^{***}$, $N = 54$ and $r = .64^{**}$, $N = 22$, respectively; also see Pengilley 1972).

Lampropholis guichenoti eggs are fairly resistant to brief periods of below freezing temperatures, at least in their later developmental stages. For example, near term eggs (with pigmented embryos within 14 days of hatching) at 23°C showed 100% survival and hatching after 5 hours of exposure at -1.0°C, 50% hatching after 15 hours at the same temperature, and 35% hatching after 5 hours at -1.8°C (Shine 1983). Unfortunately, there are no comparative data for other Australian lizards, but at the very least, these results show a capacity for short-term survival at temperatures comparable to those of a night with frost.

At most localities, the two best-known species, *Lampropholis delicata* and *L. guichenoti* are surmised or are known to become sexually mature by the reproductive season following their hatching, i.e., within one year of age (Clarke 1965; Simbotwe 1985; Joss and Minard 1985). However, *L. guichenoti* females in the generally cooler Southern Highlands may take two years (Pengilley 1972).

In the southern half of the generic range, it is not unusual for two species of *Lampropholis* to occur together, and the most frequent case of sympatry is *L. delicata* and *L. guichenoti*. Both species occur in Sydney, often in the same garden, and any astute gardener can verify for himself that one of the ways they divide up the microhabitat is for *L. delicata* to tend to occur more often in the more shaded, moister areas and *L. guichenoti* in the more open, drier areas (Pengilley 1972). As a historical aside, it may be noted that these two *Lampropholis* were the first Australian lizards to be examined electrophoretically. Samples of each collected syntopically in Sydney differed uniquely at two loci (Harris and Johnston 1977) from which the authors concluded that they were indeed two species, a realization that had been widespread in the herpetological community for over a decade (Clarke 1965).

The small size of *Lampropholis* species may make them susceptible to invertebrate predators. For example, the bodies of *L. guichenoti* have been found dead in the webs of red-backed spiders (Copland 1953) and in another case of invertebrate induced mortality, wolf spiders ejected *Lampropholis* eggs laid in their burrows (Humphreys 1976).

A *Lampropholis* species is the only Australian reptile to have been transported overseas by man and gained a foothold in a new country. *L. delicata* has been carried, presumably in European cargo, to both New Zealand and Hawaii and now has resident populations in both places. In New Zealand the species occurs at several localities on North Island (West 1979), and in Hawaii it occurs on a number of islands (Baker 1979) and may even be replacing an earlier transportee, *Lipinia noctua*, a species probably introduced by Polynesians a few hundred years ago (Hunsaker and Breese 1967).

The identity of the Hawaiian population was confused for many years. It was originally described as a new species — *Leiolopisma hawaiiensis* (Loveridge 1939) and then was long thought to be *Carinascincus metallicum*. It was not until 1976, in fact, that the true identity of the lizard was first realized (Baker 1979).

The closest living relative of *Lampropholis* is probably *Saproscincus*, a genus of seven or eight species ranging along the near east coast from the base of Cape York Peninsula to southern Victoria. Morphologically, *Saproscincus* differs from *Lampropholis* in having six supralabials instead of seven, in being medium to light brown instead of grey to dark brown, and in having a white spot on the posterior base of the thigh. Ecologically, *Saproscincus* differs in being shade-loving and very heat sensitive instead of sun-loving and heat resistant (Greer and Kluge 1980; Tables 19-20). They tend to occupy more heavily shaded habitats and microhabitats and to be surface active only during the late afternoon. Thus whereas the skink probing through the sunlit litter or rockery in a Sydney garden at noon is likely to be a species of *Lampropholis*, the one disappearing into the deeper shadows late in the afternoon is likely to be a *Saproscincus*.

Two subgroups are recognizable within *Saproscincus* one of which is clearly a lineage on the basis of derived traits and the other which is the assemblage of species lacking these traits. The lineage comprises the three small, northern species *S. basiliscus*, *S. czechurae* and *S. tetradactyla*. In addition to small size (maximum SVL = 47 mm) these species share a peculiar arrangement of nuchal scales (Greer and Kluge 1980; Ingram and Rawlinson 1981), an elevated number of premaxillary teeth (13 or more vs 11) and a constant clutch size of two; in addition, the last two named species share a very short tail, at least in terms of the number of postsacral vertebrae (32-35; Table 3). The assemblage comprises the large, southern species *S. challengeri*, *S. mustelina* and two or three undescribed species one of which inhabits certain of the more mesic parks of Sydney. The two groups come close or overlap slightly in the mountains west of Mackay. In comparison to two *Lampropholis* species and *S. mustelina*, *S. challengeri* appears to have an unusual form of ovarian steroid synthesis (Joss 1985), but whether it shares this peculiarity with any other species in the genus remains to be learned.

As far as is known all *Saproscincus* species are oviparous. Clutch size is variable in the larger southern species (e.g., *Saproscincus challengeri* and *S. mustelina*) but, as mentioned above, a constant two in the smaller northern species (i.e., *S. basciliscus*, *S. czechurae* and *S. tetradactyla*; Table 9). The most southern species, *S. mustelina*, lays communally (Fig. 71) and shows a positive correlation between female size and clutch size (r = .39*, N = 37).

Within the *Carlia-Lygisaurus* group, only the latter genus can be diagnosed further, leaving *Carlia* as basically "all the other" skinks in this subgroup. In view of the taxonomic problems with *Carlia*, it may simply be best to characterize the group as we have it (Mitchell 1953; Storr 1974d) and to point out some evolutionary trends that are evident within it.

All *Carlia* are diurnal heliotherms and most are inhabitants of woodlands; however, *C. rhomboidalis* occurs in forests, but usually in areas open enough to provide basking sites, while at the other extreme *C. munda* often occurs in open shrubland or woodland and *C. triacantha* in rocky ranges with hummock grass. Most species are terrestrial but some are loosely (e.g., *C. jarnoldae*) or strongly (e.g., *C. coensis*, *C. mundivensis*, *C. rimula*, *C. scirtetis* — Wells 1975, as unidentified *Carlia*) associated with rocks. The strongly saxicolous species usually share a large size; generally dark, non-dichromatic colour patterns and a relatively large number of small body scales.

Fig. 71. *Saproscincus mustelina* and an in situ communal nest of the species.

As far as is known all *Carlia* species are oviparous and lay two eggs at a time (Wilhoft 1963a; Wilhoft and Reiter 1965; James 1983; Table 9). Seasonality of reproduction has been studied in seven *Carlia* species, all populations of which are well within the tropics: *C. amax*, *C. gracilis*, *C. munda* and *C. triacantha* in the Top End of the Northern Territory (James 1983; James and Shine 1985) and *C. longipes* (as *C. fusca*) and *C. rhomboidalis* in northeastern Queensland (Wilhoft 1963a-b, 1964; Wilhoft and Reiter 1965). Very generally, in both areas, both male and female reproductive activity is lowest in the period from the late wet season to the mid- to late dry season; the one exception is male *C. rhomboidalis* which appears to maintain the same level of activity (as measured by gonad weight) throughout the year. In two species, i.e., *C. amax* and *C. rhomboidalis*, female reproductive activity, as measured by the presence of yolking ovarian follicles and/or oviducal eggs, is greatly curtailed during the late wet to late dry but does not cease, as appears to be the case with the other species.

Temperatures taken on free ranging animals have been reported for two species (*Carlia rhomboidalis* and *C. longipes* — Wilhoft 1961, the latter species as *C. fusca*) but as these were not related to the behavioural and physiological states of the animals, they are not very informative. Experiments on thermal tolerance for six species, show that they are moderately tolerant with means for the just below lethal limits ranging 40.3-42.3°C. (Greer 1980a), which is what one would expect given their habitats and heliothermic behaviour.

Many species show some form of adult sexual dichromatism, which in some cases involves striking hues, leading to the common name of rainbow skinks. The sexual dichromatism can be of two kinds: a permanent, year round pattern or a transient, seasonal pattern. The permanent pattern usually involves the retention of light, longitudinal stripes in females and their ontogenetic loss in males, e.g., *Carlia bicarinata* and *C. vivax*. The transient pattern usually involves light colour hues such as red, orange or blue and is carried only by the males. These colours usually occur on the throat (e.g., *C. rhomboidalis*) or side of the body (e.g., *C. gracilis*, *C. rufilatus*, *C. schmeltzii* and *C. tetradactyla*). Some other species develop transient black heads or throats in males (e.g., *C. johnstonei* and *C. munda*, respectively). Sexual dichromatism of both the permanent and transient variety appear to be the primitive condition for lygosomines and hence may be primitive in these *Carlia*. The function of these sexually dimorphic patterns and colours remains to be determined in any skink.

There is an especially interesting case of geographic variation in throat colour in *Carlia rhomboidalis* which would be worth investigating further. The northern population has a red throat and the southern one a blue throat. No morphological or karyotypic differences have been detected between the populations, and the situation now invites biochemical investigation. It would also provide an ideal opportunity to investigate the functional significance of the throat colour both within and between the two different populations.

Some *Carlia* species appear to lack any sexual dichromatism, e.g., *C. coensis, C. mundivensis* and *C. scirtetis*. These species are relatively large and inhabit bare rocks, hence they may have lost the striking patterns and colours of other *Carlia* for reasons of camouflage.

The skinks of the genus *Lygisaurus* are all small (max. SVL of the largest species 42 mm), with smooth scales and a clear window which nearly fills the lower eyelid. Depending on the species, the eyelid is either almost always carried raised, i.e., closed, or is actually fixed in that position. The only species karyotyped to date, *L. burnettii*, had a reduced chromosome number, 28 instead of 30 (Donnellan 1985), which raises the question of how widespread in the genus this derived karyotype may be.

The genus ranges from New Guinea south to the northwestern suburbs of Sydney. It comprises nine species but only three of these are currently recognized, and no doubt others remain to be discovered, especially in New Guinea. Only one species *L. burnettii* has been studied in any detail and then only morphologically (Copland 1949).

In general the species are diurnal, heliothermic and terrestrial, although one of the undescribed species (from Chillagoe, Qld) climbs on limestone outcrops. As far as is known the group is oviparous with a clutch size of two.

We can conclude our review of the *Pseudemoia* subgroup, and also the *Eugongylus* group, with brief comments about the least well-known representative of the subgroup: *Techmarscincus*. The single species in the genus, *T. jigurru*, occurs only near the summit of Bartle Frere in northeastern Queensland. Here it lives on granite boulder fields surrounded by dense rainforest. Almost nothing is known of its biology other than it is a basker and oviparous (Covacevich 1984; Shea 1987b). Also, like many other rock inhabiting members of the *Eugongylus* group, it has a depressed head and body.

The broad relationships within the *Sphenomorphus* group are obscure and hence discussion of the genera along these lines impossible. There are, however, two broad ecological subdivisions that provide a useful basis of discussion: surface dwelling versus semi- to completely fossorial. The surface dwelling forms include five genera, two of which occur along the mesic east coast — *Eulamprus* and *Gnypetoscincus* and three of which are centred over the arid, semi-arid and seasonally dry habitats that cover virtually the rest of Australia — *Ctenotus, Notoscincus* and *Eremiascincus*. In keeping with their surface-dwelling habits these skinks have relatively long hind legs (≥ 28 percent of SVL), show no loss of phalanges and in only one case have more than the primitive number of presacral vertebrae for skinks, 26 (*Notoscincus* has 27-30).

The skinks of the genus *Eulamprus* share only one character — ovoviviparity[1] — and hence their association together as an evolutionary lineage is only tentative. They are all medium to large skinks with well-developed limbs. They are usually associated either with water or very moist habitats and often occur off the ground, either on logs or rocks. In fact, they are the most "supra-terrestrial" skinks of the *Sphenomorphus* group in Australia. The genus, as currently constituted, ranges along the east coast, with the exception of *E. quoyii* which is said to follow the Darling River across the interior of the southeast part of the continent.

Within *Eulamprus* there is an assemblage of six species which share no further derived features with each other or with any other species in the genus. This group, which can be called the *E. tenuis* group after its best known member, includes *E. amplus, E. brachysoma, E. martini, E. tenuis, E. tigrinum* and one undescribed species. These skinks are medium-sized and grey to brown with dark vertical bars on the sides or back. They occur on trees and logs and/ or rocks in woodland. In Sydney, *E. tenuis* is often a commensal of man, inhabiting sandstone foundations and brick walls. In this microhabitat it is even able to live in the inner Sydney suburb of Darlinghurst (pers. obs.). All the species are diurnal (but in patchy or diffuse sunlight) to crepuscular, and perhaps even nocturnal during the warmest part of the year. They are intermittent and furtive baskers but generally only in protected situations (e.g., overhanging vegetation) or close to cover (Rankin 1973b) and never for long periods. *E. amplus* has been found sleeping out on rocks at night (Covacevich and McDonald 1980).

The remaining species of *Eulamprus* can be hypothesized to be a lineage on the basis of the third pair of enlarged chin scales being separated

[1]There is an account of eggs found in a rock crevice in Padstow, New South Wales, hatching out as *Eulamprus tenuis* (De Lissa 1981), but as other local populations are known to be live- bearing and shelled eggs have never been found in preserved specimens, this observation is almost certainly in error.

by five instead of three longitudinal rows of small scales (Fig. 73), an otherwise unique feature amongst the Australian members of the *Sphenomorphus* group (except for *Gnypetoscincus*), and on the absence of inguinal fat bodies. This subgroup includes *E. heatwolei, E. kosciuskoi, E. leuraensis, E. luteilateralis, E. murrayi, E. quoyii* and *E. tympanum.*

These seven *Eulamprus* species can be further divided into two groups: the *E. quoyii* group which contains *E. heatwolei, E. kosciuskoi, E. leuraensis, E. quoyii* and *E. tympanum* (Rawlinson 1969; Shea and Peterson 1985), and the *E. murrayi* group which contains *E. luteilateralis* and *E. murrayi.* The former group is recognizable on the basis of two features of the digits: the terminal dorsal scales consisting of one row only and the subdigital lamellae being longitudinally

grooved (Fig. 72). The latter group is recognizable on the basis of only one infralabial scale in contact with the postmental (Fig. 73). In addition, the *E. quoyii* group differs ecologically from all other *Eulamprus* in being more strictly diurnal and heliothermic and in often being closely associated with water. This latter feature has led to their being called "water skinks".

The members of *Eulamprus quoyii* group constitute one of the best known skink groups outside the *Egernia-Tiliqua* group. There are accounts of its osteology (King 1964), thermal biology (Veron and Heatwole 1970; Brattstrom 1971b; Spellerberg 1972a-d), diet (Veron 1969b), reproduction (Veron 1969a; Pengilley 1972), placentation anatomy (Weekes 1927b, 1935) and physiology (Thompson 1977a,b), maternal and embryonic blood physiologies

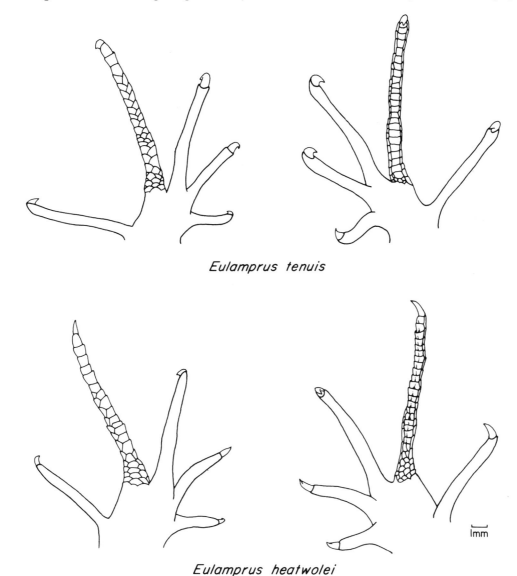

Eulamprus tenuis

Eulamprus heatwolei

Fig. 72. Dorsal and ventral surface of the left rear foot in two species of *Eulamprus* to show the primitive scalation (*E. tenuis* — AM R 107249) and the derived conditions of more than three single scales distally in the dorsal row of the fourth toe, and longitudinally grooved subdigital lamellae (*E. heatwolei* — AM R 94969).

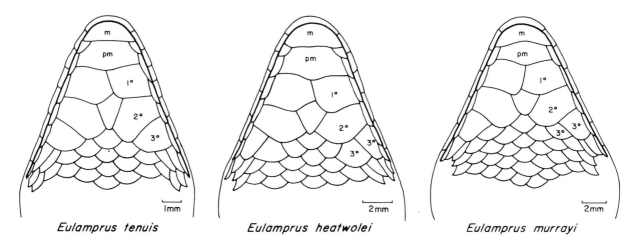

Eulamprus tenuis   *Eulamprus heatwolei*   *Eulamprus murrayi*

Fig. 73. Ventral side of the head in three species of *Eulamprus* to show primitive scalation (*E. tenuis* — AM R 86358) and the derived features of a divided third pair of chin scales (*E. heatwolei* — AM R 94912) and a single infralabial in contact with postmental (*E. murrayi* — AM R 112287). Abbreviations: m — mental; pm — postmental.

Grigg and Harlow 1981), and social behaviour (Done and Heatwole 1977a-b).

As a group, the five species occur in the south-east part of the country and range from sea level to near the top of Mt Kosciusko, the highest point in Australia (Copland 1947c). The species show a certain amount of altitudinal and habitat separation. *Eulamprus kosciuskoi* is restricted to alpine meadows and grassy openings in forest (Copland 1947c — a specimen "was running in wet grass in a small stream carrying melting ice and snow" not far from the summit of Mt Kosciusko; Coventry and Robertson 1980; Mansergh 1982); *E. leuraensis* to moist montane meadows, swamps and woodlands (Wells and Wellington 1984; Shea and Peterson 1985); *E. heatwolei* (Pengilley 1972) and *E. tympanum* to moist meadows and open woodlands, the former at generally lower or more northerly localities than the latter, and *E. quoyii* to creeks and streams in meadows, shrublands and woodlands (Spellerberg 1972b-d; Green 1973a).

Despite the wide habitat and altitudinal range of the group, the thermal biology of the various species is rather uniform. For example, the mean field active temperature and mean pre-ferred temperature for the four best known species (all except *Eulamprus leuraensis*) are each within 3°C of each other (28.1-30.9°C and 28.8-30.8°C, respectively, Tables 13, 15).

Diet has been studied thoroughly only in *Eulamprus quoyii* (Veron 1969b; Daniels 1987) and there is nothing to suggest that these skinks are anything but opportunistic predators, taking anything that they can overcome and swallow. The exception to this may be avoidance of their own young, as not one of these was found in the stomachs analysed. These skinks take not only terrestrial but also aquatic prey, which they seize

when it approaches the surface near the water's edge (Daniels 1987)[1].

Social interactions have been investigated in *E. kosciuskoi* and *E. quoyii* with some interesting differences resulting (Done and Heatwole 1977a-b). Both species show a fairly standard repetoire of skink displays, but aggression appears to be much more intense in *E. kosciuskoi* than *E. quoyii*. This is also borne out by some general observations of G. Swan who knows four of the five species of water skinks well and rates them for aggression: *E. kosciuskoi* > *E. heatwolei* > *E. tympanum* > *E. quoyii*. Indeed, the general feeling of all those who have had experience with *E. kosciuskoi* is that it is a "real little . . .". One might speculate that *E. kosciuskoi* behaviour derives from the relatively short growing season under which it lives, when time is short, it's hard to be Mr. Nice Guy.

Although there are no published details, it appears as if *E. heatwolei*, *E. kosciuskoi*, *E. quoyii* and *E. tympanum* have well-established, long-term home ranges. Spellerberg (1972d), referring to his unpublished PH.D dissertation says of these four: "The strong day to day tenacity of shuttling and feeding localities and basking sites was shown by the lack of dispersal of marked adults throughout the summer and from year to year (Spellerberg 1971). Winter submergence site may also be specific to adult individuals (while not for young animals: Spellerberg 1971)." Similarly, Copland (1947c) speaking of either *E. kosciuskoi* or *E. tympanum*, noted that when disturbed "they took to the water without

---

[1]In the Jenolan Caves area of New South Wales in early autumn (March), I once watched *Eulamprus quoyii* feeding on ants. A very small fast-moving species was avoided but a larger, slower species was eaten. Individual lizards feeding in proximity appeared to stay largely in one place and wait for their prey to appear; they paid no attention to one another. This was in contrast to some nearby *Ctenotus taeniolatus* which were actively searching for their prey and would chase one another when their foraging brought them close together.

hesitation . . .[but] . . .invariably returned to the place from which they had been disturbed, often within a minute".

As far as it is known, mating in the group takes place in the spring (*E. quoyii* — Veron 1969a and *E. heatwolei* — Pengilley 1972, as *Sphenomorphus tympanum* warm temperate form)[1] and birth occurs in the late summer (all, except *E. leuraensis* for which data are lacking as yet). *E. quoyii* appear to be capable of mating in the first season after their birth (Veron 1969a), but *E. heatwolei* not until their second season or, more usually, third season (Pengilley 1972). Litter sizes for the group range 1-8. In the species that have been assessed for the relationship, i.e., *E. heatwolei* (Pengilley 1972, as *Sphenomorphus tympanum* warm temperate form), *E. kosciuskoi* (r=.73*N =9), and *E. quoyii* (Veron 1969a)[2], there is a positive correlation between female size and litter size.

Eulamprus heatwolei (as *Sphenomorphus tympanum*) and *E. quoyii* have hybridized in captivity (Fowler 1974, 1978, 1985). This is the only

---

[1]It has been asserted that *Eulamprus tympanum* copulates in autumn and females store sperm over winter (Rawlinson 1974a). However, as there are no supporting details and other cool climate populations of *Eulamprus* mate in spring, it may be best to seek confirmation of autumn mating in *E. tympanum* before accepting the observation.

[2]Veron (1969a) reported no correlation but calculations based on the data presented in his figure 8 indicate a strong positive correlation (r = .65**, N = 21).

---

other hybridization known amongst Australian skinks outside the genera *Egernia* (p. 131) and *Tiliqua* (p. 135).

Eulamprus luteilateralis and *E. murrayi*, the two species in the genus with only a single infralabial in contact with the postmental, are similar in occurring only in moist forest or closed woodland and living in and under wood (Covacevich and McDonald 1980 and Bustard 1964b as *Sphenomorphus tryoni*, respectively). In captivity adults of each species are said to live compatibly as male/female pairs and with young, but to be very intolerant of each other in larger groups (Covacevich and McDonald 1980; see also Martin 1973).

Strictly speaking on habits alone, the monotypic *Gnypetoscincus* from the moist forests of northeastern Queensland should be placed in the fossorial group, because, as far as is known, it is always found, at least by day, in or under, rotten, often sodden, logs. However, its limbs are very well-developed, and it may be very closely related to *Eulamprus* (see below). In addition, it could be more surface-active than we think, because no one has spent much time looking for lizards in rainforests at night. Another, albeit indirect, indication of the sun-avoiding habits of *Gnypetoscincus* is its total lack of pigment

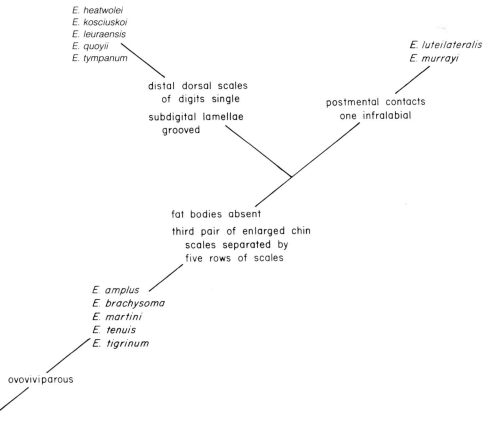

Fig. 74. Diagram of the hypothetical relationships of the species of *Eulamprus*.

in the parietal peritoneum. This is in strong contrast to most other members of the *Sphenomorphus* group, including *Gnypetoscincus'* possible close relatives in *Eulamprus*, which have dark parietal peritonea.

Gnypetoscincus is probably the most unusual looking skink in Australia. Most skinks are covered with smooth or keeled, overlapping scales, but this species is covered with bluntly pointed, non-overlapping, bead-like scales everywhere except the head and venter (Fig. 75). The function of the bead-like scales seems to be to conduct moisture over the surface, because in contrast to other smooth-scaled skinks on which a water droplet will just sit intact, in *G. queenslandiae* any moisture coming in contact with the skin is immediately dispersed over the surrounding skin (Naylor 1980). As these skinks live in virtual constant contact with nearly saturated surfaces, their scale pattern would serve to keep their skin constantly moist.

Gnypetoscincus queenslandiae is ovoviviparous with a litter size of 2-6 (Table 9); there is apparently no relationship between female size and litter size ($r = .19^{NS}$, N = 28).

Fig. 75. Eulamprus brachysoma (top) showing the smooth, flat scales typical of most skinks, and its relative *Gnypetoscincus queenslandiae* (bottom) showing the bead-like scales on the top and sides of the neck, body and tail.

In living in tropical rainforest often in or under rotting logs and probably not emerging until after dark, if at all, *Gnypetoscincus queenslandiae* is one of the most ecologically buffered terrestrial reptiles in Australia. Only the much rarer *Coeranoscincus frontalis* of similar distribution and habits is comparable. Given this presumably relatively aseasonal microhabitat, one could ask what has happened to those biological activities, such as reproduction, that are generally seasonal in less buffered species.

In order to answer this question I surveyed the reproductive state of all the specimens in the Australian Museum for which there were dates of collection. These showed that the smallest gravid female was 61 mm SVL and that of the females this size or larger, the following percentages were gravid in the months for which females of this size were available: June (100%, N = 12), July (73%, N= 15), August (100%, N = 3), September (100%, N = 1), November (0%, N = 2), January (0%, N = 1). Furthermore all the June, July and August specimens carried only yolking ovarian eggs and the September specimen was the only one with oviducal young. This distribution of yolking ovarian eggs and oviducal young along with the reproductively inactive females of November and January is strong circumstantial evidence for ovulation during the early to mid-dry season (June-August), oviducal young during the late dry (September-October?) and birth by the early wet. Mature males are known only from the early to mid-dry, and they are reproductively quiescent. Given this period of quiescence and combining it with what is known of the female cycle, it seems likely that mating occurs during the wet, perhaps the mid to late wet. The general picture therefore is marked seasonality in reproductive activity with females most active in terms of production of young in the dry season and males probably most active in terms of sperm production in the wet.

Gnypetoscincus appears to be related to *Eulamprus* as a whole or at least to one of its subgroups. For example, it shares ovoviviparity and an east coast distribution with all *Eulamprus*. It shares its peculiar chin scales (third pair of enlarged chin scales separated by five instead of three rows of small chin scales) with the *E. quoyii* + *E. murrayi* species groups, and it shares four or more single scales on the dorsal surface of the digits with the *E. quoyii* group. However, in contrast to the *E. quoyii* and *E. murrayi* groups (but like the *E. tenuis* group), it retains inguinal fat bodies, thereby creating a character incompatibility and raising the necessity of hypothesizing either the parallel loss of fat bodies or the parallel acquisition of increased number of single dorsal scales on the digits.

Of the three surface active genera of the *Spheno-morphus* group which inhabit the arid zone, two, *Ctenotus* and *Notoscincus*, are primarily diurnal and one, *Eremiascincus*, is exclusively crepuscular to nocturnal.

Ctenotus has the distinction of being the largest genus of lizards in Australia. There are currently 71 described species, and the discovery and description of new species is still routine (Storr 1964a, 1969, 1970, 1971a, 1973, 1975a, 1978c-d, 1979c, 1980b, 1981b, 1984b; Czechura 1986; Czechura and Wombey 1982; Horner and King 1985; Ingram 1979b; Rankin 1978b; Rankin and Gillam 1979; Sadlier 1985; Sadlier *et al.* 1985). Many species are very similar in appearance. The genus is widespread both geographically and ecologically. It occurs throughout the continent, except in the south-east corner, and in virtually every terrestrial habitat except moist closed forest. At least one species, *C. spaldingi*, also extends into southern New Guinea (Boulenger 1887, as *Lygosoma dorsale*; Copland 1947a).

Several subgroups have been recognized within *Ctenotus*, but these are based on characters whose phylogenetic polarity is unknown. Other characters which vary within the genus and whose polarity is easier to determine, e.g., tongue colour, postorbital bone, Meckel's groove and pre-maxillary teeth number have yet to be exploited in assessing intrageneric relationships. Such an analysis might also lead to additional insight into the well-studied ecological differences between many species and species groups.

Ctenotus are generally diurnal, sun-loving and largely terrestrial, although some species will climb in hummock grass or, occasionally, low shrubs (Smith 1976) after a particularly attractive prey item. They generally run fairly hot, having average active body temperatures of 30-38°C (Table 13) and critical lethal maximum temperatures of 42.5-45.5 (Table 19). When actively foraging, they move "nervously" through the habitat investigating likely hiding places for prey, occasionally pausing to look around or to bask, and dashing for cover at the slightest sign of danger.

Ctenotus appear to be opportunistic predators (e.g., Pianka 1969b, 1981; Smith 197?; Brown 1983; Taylor 1986), but some species are known to take more than 10% plant material by volume. For example, *C. lesuerii* takes approximately 12% (Davidge 1979), *C. fallens* 13-20%, depending on season (Murray 1980), and *C. leae* 33% (Pianka 1969b)[1]. Not surprising, larger individuals and species are able to eat larger prey than smaller forms and hence take a greater variety of prey (Pianka 1969b, 1981; Taylor 1986).

As far as is known all *Ctenotus* are oviparous, with clutch size for the genus ranging 1-9 (Table 9). There is a significant, positive relationship between female size and clutch size in each of the three species in which sample sizes are large enough to make a fair test: the tropical *C. essingtoni* (James 1983), and the temperate *C. labillardieri* (r = .79***, N = 14, pers. obs.) and *C. taeniolatus* (Taylor 1985b). The number of clutches per season is known only for a temperate population of *C. taeniolatus* where it is one (Taylor 1985b).

Information on the seasonality of reproduction is available for several *Ctenotus* species in the extreme northern part of the Northern Territory. Here some species, e.g., *Ctenotus gagudju*, *C. hilli* and *C. robustus*, are reproductively active in the wet to early dry season while others, e.g., *C. arnhemiensis*, *C. essingtoni*, *C. saxatilis*, *C. storri* and *C. vertebralis*, are active in the late dry (Rankin and Gillam 1979; James 1983; James and Shine 1985; Sadlier *et al.* 1985). What phylogenetic and ecological significance this bimodality in reproductive season has for these species remains to be learned.

Several *Ctenotus* species are known to use burrows e.g., *C. allotropis* (Bustard 1968e as *Sphenomorphus strauchii*), *C. borealis* (Horner and King 1985), *C.* cf. *brachyonyx* (R. Sadlier, pers. comm.), *C. essingtonii* (Rankin 1978b), *C. leonhardii* (G. Shea, pers. comm.), *C. regius* (Gillam *et al.* 1978), *C. robustus* (Armstrong 1979), *C. "schomburgkii"*, *C. storri* (Rankin 1978b), *C. taeniolatus* (Taylor 1985a), *C. uber* (Ehmann 1976b), but only in *C. taeniolatus* is it certain that the species digs its own burrow rather than appropriates that of another animal. This species digs burrows under rocks and logs which it uses both in summer and winter (Taylor 1985a). Hibernation posture in the species, the only one studied from this view-point, is characteristic: the tail is curved toward the head and the front and rear legs are bent back and up so that the feet are placed on the back and top of the tail base, respectively (Taylor 1985a).

In many areas, several species of *Ctenotus* occur together in the same habitat, in some cases as many as seven (Pianka 1969b). How similar species, which are presumably in competition for limited resources, actually manage to partition their resources and hence co-exist, is an interesting ecological question, and *Ctenotus* is much studied from this point of view (Pianka 1969b; C. James, in prep.) The results to date indicate that differences in habitat (macro and micro), time of day and/or season, and prey type and size are all important (Pianka *et al.* 1979).

[1]It may be worth noting in passing that this species size vs amount of plant material relationship is the opposite of the trend seen in other lizard lineages i.e., large species eat more plant material; Pough 1973).

Although some *Ctenotus* species retain ventral body colour, e.g., *C. brachyonyx*, some *C. inornata* (Sadlier 1985), *C. labillardieri* and *C. youngsoni*, the majority of species appear to lack it. The function of the colour, and the significance of its absence in the genus is unknown. However, there may be a clue in the fact that in *C. labillardieri* "the abdomen is bright yellow during the breeding season" (Ford 1968).

Although almost nothing is known of intra-specific behaviour in *Ctenotus*, there are some observations on captive *C. robustus* which suggest that aggressive behaviour, either overt or ritualized, does not occur in this species (Done and Heatwole 1977a). This is worth following up in other species because it implies that a major force in social organization in many vertebrates may be lacking in the group. It also sits slightly at odds with the observation that another species, *C. taeniolatus*, never shares its burrow with other individuals of its own or other species (Taylor 1985a) implying a certain level of aggression to maintain this isolation (p. 152). It also conflicts with the fact that *C. brachyonyx* and *C. leonhardii* fight strenuously in collecting bags (G. Shea, pers. comm.).

One species of *Ctenotus* provides an interesting exception to the generally diurnal, high-temperature habits of its congeners. *C. pantherinus* appears to be largely nocturnal and to have relatively low thermal parameters (Table 20). What other evolutionary modifications it may have undergone in becoming nocturnal await discovery.

Ctenotus' uniform basic body plan over a wide range of sizes will make it an excellent group in which to study changes in shape in a single lineage across different species sizes. One measure for which there is already data in the taxonomic literature is relative length of the hind limbs — the more powerful and therefore presumably the more important for locomotion of the two limb pairs. The correlation between maximum species size as expressed as SVL and the low (i.e., adult) end of the range of the hind limb length as a percentage of SVL is negative ($r = -.39**$, $N = 56$).[1] This indicates that as a natural consequence of increasing size, hind limb length becomes relatively smaller in comparison to head + body size. This has consequences for foraging habits in *Ctenotus* specifically (longer-legged, and hence possibly smaller, species forage more in the open — Pianka 1969b, 1981) and perhaps for limb reduction in skinks in general (one way to reduce relative limb size is to become larger).

Notoscincus is a group of three species (two described, one undescribed) widespread in arid and semi-arid Australia (Glauert 1959; Storr 1974b). They are all small, none exceeding 43 mm in snout-vent length, and like other small skinks from arid or semi-arid areas, they have a spectacle.

Little is known of their biology. They have a short, deep head which is unusual among skinks but the significance of this head shape is unclear. Their activity times appear to be very flexible and perhaps primarily dependent on temperature, because individuals have been found abroad both during the day and night (pers. obs.). A difference in foraging habits may exist between the two described species: *Notoscincus wotjulum* forages widely described through the leaf litter (Wells 1979a; pers. obs.), but *N. ornatus* is said to sit and wait in open areas for passing prey — which appears to consist largely, if not exclusively, of ants (Smith 1976).

Some colleagues and I noted some peculiar behaviour in *Notoscincus ornatus* on a trip to northeastern Queensland during the dry season of 1985. We had stopped in a recently burned open woodland just north of Chillagoe and spent about two hours before sunset collecting and observing a variety of small lizards. Up until about a half an hour before sunset we had not seen any *N. ornatus* but then suddenly we started seeing them everywhere. Close observation showed they were coming out of small holes in the bare earth (see also Wells 1979a) and foraging amongst the dead leaves which had accumulated against the downed wood. The activity lasted for about half an hour and then suddenly stopped, presumably because the animals re-entered their holes. We returned to a similar adjacent area the following day at the same time and saw the same thing. The occurrence was peculiar on three counts; first, that the animals should use such small insignificant holes in the ground for retreats; second, that the onset and cessation of the activity period were precise and the duration of activity short, and third, how abundant the animals were.

Mode of reproduction is known for two of the species, *Notoscincus* n. sp. (from the interior) and *N. wotjulum*; both are oviparous. Clutch size for the genus ranges 2-4 (Table 9).

Museum specimens provide a few insights into the seasonality of reproduction in at least one species of *Notoscincus*. In *N. wotjulum* from the Top End of the Northern Territory, females have yolking follicles at the end of the dry (late August) and beginning of the wet (late September). Specifically, mature females from

---

[1]Statistically it would have been better to avoid a measure of limb length that incorporated SVL but in order to profit from the vast amount of published work in gleaning these measurements this "flawed" measure may be taken as an initital indication of the relationship.

the period 7-11 August (N = 2) have only unyolked follicles whereas the mature females from 17-31 August (N = 5) have yolking follicles as do the mature females from 20-26 September (N = 3). This suggests that the onset of reproduction may coincide with the onset of the wet season.

The genus *Eremiascincus* is probably derived from an ancestor like *Glaphyromorphus nigricaudis* or *G. isolepis* (see below), both of which it resembles somewhat in size, shape, and some aspects of colour pattern (Greer 1979a). However, there is a significant difference in the range of the two taxa for unlike *Glaphyromorphus* which is restricted to the periphery of the continent, *Eremiascincus* is widely distributed in the arid interior (Storr 1967a, 1974a; Greer 1979a). The question arises as to what modifications, if any, have enabled this group to enter the arid zone while its relatives have been excluded. Nocturnality is not the answer because this is common to both groups. Perhaps the taxon's heat and water relations would be an interesting line of enquiry.

There are two species of *Eremiascincus* but both show a unusually large amount of morphological variation; this is especially true of *E. fasciolatus* the species that seems to occur in more arid habitats (Storr 1974a). For example, this species, like *E. richardsonii*, is usually deep-headed and has a series of dark brown crossbands on the light brown background of the back and tail, but in certain areas, notably open habitats with fine sand, the head may become markedly flattened, the ground colour pale and the cross-bands much reduced or lost; indeed the colour and pattern can be so reduced that the animals become uniformly pale and are known locally as "ghost skinks".

Only the habits of *Eremiascincus fasciolatus* are well-known. They lie buried in loose sand by day (the mouths of rabbit burrows seem to be favourite resting-places) and begin to stir about sundown. They often begin by just pushing the head through the sand and waiting for passing prey. When an animal such as a smaller lizard comes within range, they burst forth, grab the prey, often by the head, and begin twirling along the long axis of their body, presumably as a means of disorienting the prey or killing it by bashing it on the ground. Later in the evening the lizards will emerge completely and forage on the surface. However, should they be disturbed, they will immediately "dive" into the loose substrate and disappear. Like its better-known congener, *E. richardsonii* is said to be "confined (at least during the day) to caves and deep-crevices where it burrows in soft, moist, karst soils" (Smith 1976).

In the laboratory, well-fed *Eremiascincus fasciolatus* can develop an exceptionally thick tail, which is presumably due to accumulated fat. Fat is both an energy and water reserve, and it would be interesting to know how long fat-tailed specimens could go without either food or water. An ability to store these resources would, of course, be valuable in an environment where they are often in short and unpredictable supply.

Information on reproduction is available only for *Eremiascincus richardsonii* which is oviparous with clutch sizes of 3-7 (Table 9). Literature reports of viviparity in *E. fasciolatus* (Waite 1929; Worrell 1963) appear to be in error (Greer 1979a), but at the moment first-hand observations for this species are lacking.

Although we are perhaps accustomed to thinking of subsurface lateral undulation — sand or litter swimming — as being a trait of elongate skinks with an elevated number of presacral vertebrae, *Eremiascincus* is an excellent example of how this need not be the case. It is neither especially elongate nor does it have an elevated number of vertebrae, and yet it is extremely adept at diving below the surface of the sand and sand-swimming a short distance away from its entry point. It will also sand-swim short distances if pursued below the surface, but if pursued closely will burst out onto the surface, run a short distance and dive below again.

In contrast to the surface-active members of the *Sphenomorphus* group discussed above, the semi-fossorial to fossorial members of this group generally have relatively short limbs (≤ 35 percent of SVL for hind limbs); often a loss of phalanges, digits or even the entire limb, and generally an elevated number of presacral vertebrae. The fossorial group consists of eight genera, seven of which occur along the more mesic northern, eastern and southern periphery of the continent — *Anomalopus, Calyptotis, Coeranoscincus, Glaphyromorphus, Hemiergis, Ophioscincus* and *Saiphos*, and one which has its centre of abundance in the arid, semi-arid interior — *Lerista*. In the following discussion the lineages are discussed more or less in the order of the degree of limb reduction shown by their most highly modified members.

Within the fossorial group, there is an assemblage of 13 species which lack traits distinctive enough to ally them either with each other or with other genera in the *Sphenomorphus* group. However, like all other species, these species have to have a generic name, and as it happens, *Glaphyromorphus* is the one available. However, despite the fact that the group is not monophyletic there is some biological information in the association of the species. This is because their general lack of significant distinguishing characters is, in a sense, just another way of saying they are fairly generalized, or relatively primitive, in their overall morphology, and in theory, it is possible to view at least some of these species as resembling the hypothetical ancestors of some of the other better defined (i.e., more specialized) genera in the fossorial group. The 13 *Glaphyromorphus* species range around the northern and northeastern near-coastal regions of Australia from near Port Hedland to Gladstone, with one out-lying species, *G. gracilipes*, in the far south-west and another, *G. nigricaudis*, extending into southern New Guinea (Copland 1946b). The remaining 11 species are *G. arnhemicus*, *G. cracens* (Greer 1985c), *G. crassicaudus*, *G. darwiniensis*, *G. douglasi* (Storr 1972a), *G. fuscicaudis* (Greer 1979c), *G. isolepis* (Storr 1967a, 1972a), *G. mjobergi* (Greer and Parker 1974), *G. pardalis* (Copland 1946b; Greer and Parker, 1974; Rankin 1978a; Limpus 1982), *G. pumilus*, and *G. punctulatus*.

Various species of *Glaphyromorphus* show features that are important in the context of either the *Sphenomorphus* group as a whole or only the fossorial section itself. Three of the species — *G. douglasi*, *G. isolepis* and *G. pardalis* — retain inquinal fat bodies, and hence are the only *Sphenomorphus* group species in Australia, in addition to *Gnypetoscincus* and certain *Eulamprus*, to retain these structures. There is also the remote chance that the three species are each others closest relatives, because they share an ectopterygoid process, a small strut of bone in the secondary palate which may strengthen the palate (Fig. 76). The three species range across the north of Australia from Port Hedland in the west to Cape York in the east.

Seven of the 13 *Glaphyromorphus* species show an increased number of presacral vertebrae, 28 or more. These are *G. cracens*, *G. crassicaudus*, *G. darwiniensis*, *G. gracilipes*, *G. mjobergi*, *G. pumilus* and *G. punctulatus*. Interestingly, the six other *Glaphyromorphus* species, along with three species of *Calyptotis* (*C. ruficauda*, *C. temporalis* and *C. thorntonensis*), are the only members of the fossorial group to retain 26 presacral vertebrae.

Four of the seven *Glaphyromorphus* species with elevated presacral vertebrae share a further two derived characters: the loss of a phalange in the fifth toe of the rear foot, giving an overall phalangeal formula of 2.3.4.5.3/2.3.4.5.3 and a single infralabial in contact with the postmental. These species are *G. crassicaudus*, *G. mjobergi*, *G. pumilus* and *G. punctulatus*, all from north-east Queensland. In turn, these four species can be subdivided into two groups on the basis of additional characters: loss of a phalange in the

fourth toe of the front foot, giving a further reduced phalangeal formula of 2.3.4.4.3/2.3.4.5.3 in *G. mjobergi* and *G. punctulatus*, and the loss of both the postorbital bone and dark pigment in the parietal peritoneum in *G. crassicaudus* and *G. pumilus*.

The compatibility of the derived characters and the geographic proximity of the above four species — the *Glaphyromorphus crassicaudus* group for want of a better name — suggests that the group may be a lineage within the *Glaphyromorphus* assemblage. However, this possibility should be tempered with the realization that all the derived features discussed also occur in certain other groups within the semi-fossorial and fossorial section of the *Sphenomorphus* group, and hence other schemes of relationship are possible (Fig. 77).

As far as is known all but two, or possibly three, species of *Glaphyromorphus* are oviparous (Table 9). The two certain exceptions are *G. gracilipes* in far south-west Australia and a population of *G. nigricaudis* in southern Cape York (pers. obs.). The two different reproductive modes in the latter species suggests there may be two sibling species — a problem that requires further work. *G. pardalis* has also been reported both as egg-laying (Greer and Parker 1974) and live-bearing (Rankin 1978a), but on circumstantial evidence in both cases (shelled oviducal eggs in the former and young discovered in a female's terrarium with no trace of eggs shells in the latter). This situation requires further investigation as well.

In the only member of the *Glaphyromorphus* assemblage where the sample size was large enough to make a fair test — the temperate, live-bearer *G. gracilipes* — there was a significant positive correlation between female size and brood size: r = 67*, N = 11.

The seasonality of reproduction has been studied in only a few *Glaphyromorphus*. Female *G. cracens* (base of Cape York Peninsula) and *G. punctulatus* (Bowen) were reproductively quiescent during the mid-dry season (19 June-15 July, Greer 1985c and 5 August, pers. obs.[1], respectively) while *G. darwiniensis* (as *Sphenomorphus crassicaudus*), *G. douglasi* and *G. isolepis* from the Top End of the Northern Territory were reproductively active in the mid-wet (January), early to mid-wet (November-January) and late dry to mid-wet (September-February), respectively (James 1983). These results, disparate as they are, suggest reproduction during the wet and quiescence during the dry.

---

[1]Of 14 *G. punctulatus* collected at Bowen on 5 August 1976 all mature females (N = 4, SVL = 62-69 mm) were reproductively quiescent (ovaries with only small, white follicles).

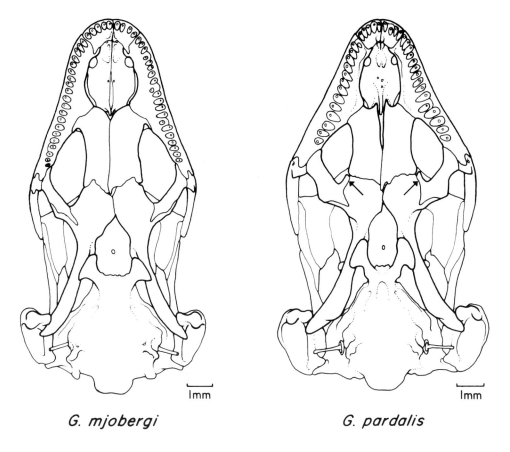

G. mjobergi   G. pardalis

Fig. 76. Ventral view of the skull of two species of *Glaphyromorphus* to show the two conditions of the pterygoid-ectopterygoid relationship in the group: the primitive condition of the pterygoid bordering the infraorbital vacuity (*G. mjobergi* — AM R 81480) and the derived condition of the ectopterygoid sending a finger-like process (arrow) to the palatine to exclude the pterygoid from a position on the infraorbital vacuity (*G. pardalis* — AM R 27042).

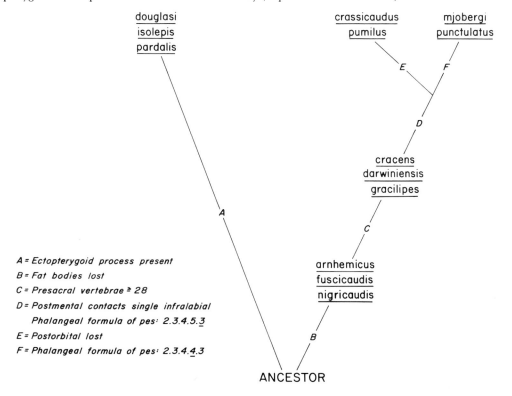

Fig. 77. A diagram of one possible scheme of relationship between the species currently placed in the *Glaphyromorphus* assemblage.

Hemiergis is a genus of five species ranging across the southern part of the continent from the New England Plateau in the east to the Darling Range in the west (Copland 1946a; Storr 1975b; Coventry 1976; Shea and Miller 1986). It is very closely related to *Glaphyromorphus gracilipes* from the mesic south-west corner of Australia (pers. obs.).

Within *Hemiergis* there is an approximate correlation between size and habitat with the smaller species occurring in the drier habitats and the larger in the moister. For example, *H. millewae* and *H. initialis* are the smallest species, maximum SVL 59 mm and 47 mm, and occur in the mallee/hummock grass habitats of the central south and the mallee and shrubland habitats of the south-west, respectively, while *H. peroni* is the largest species, maximum SVL 79 mm, and occurs along the coast and in mesic woodlands of the south and south-west. The two other species, *H. decresiensis* and *H. quadrilineatum* attain 65 mm and 75 mm and occur in woodland habitats. *Hemiergis*' close relative, *Glaphyromorphus gracilipes*, also fits with the trend: it is the largest species of the lineage (max. SVL = 89 mm) and is restricted to the most mesic part of the south-west. The occurrence of *H. decresiensis* under superficial cover varies markedly with surface moisture, during wet periods they can be extremely common but during dry periods very uncommon (Armstrong 1979a).

All *Hemiergis* have a clear window in the lower eyelid and lack an external ear opening. The former modification is typical of small skinks in arid and semi-arid habitats but the ecological basis of the latter is unclear. Despite the loss of the ear opening, the stapes still projects laterally across the middle ear cavity to attach to the underside of the scaly skin in the same general area it would in more primitively-eared forms.

Stomach contents have been analysed in *Hemiergis peroni* (Smyth 1968) and *H. decresiensis* (Robertson 1976, 1981; Crome 1981), and in both it seemed as if the only criterion for "selection" was simply availability. This was even thought to be true of *H. decresiensis* which had taken large numbers of ants.

Mating occurs in the autumn in all species examined to date (*Hemiergis decresiensis* — Pengilley 1972; *H. peroni* — Smyth 1968; Smyth and Smith 1968). All species are live-bearing, and litter size for the group as a whole ranges 1-8 (Table 9). In the only two species for which there were adequate samples, there was a significant positive correlation between female size and litter size for one, *H. peroni* (r = .65***, N = 54) but not for the other, *H. initialis* (r = .15[NS], N = 18).

All species of *Hemiergis* have bright yellow to orange venters from the level of the forelegs back, and in addition, three species — *H. millewae*, *H. peroni* and *H. quadrilineata* — share an orange to red stripe along the upper side of the body. Bright ventral colour is common in many members of the *Sphenomorphus* group but bright dorsal colour is virtually unheard of. Its function in *Hemiergis* remains to be discovered.

Compared to three other small to medium-sized skinks (single species of *Cryptoblepharus*, *Ctenotus* and *Lerista*), two species of *Hemiergis* (*H. decresiensis* and *H. peroni*) showed high rates of water loss (Warburg 1966). *Hemiergis* species are also relatively sensitive to high temperatures, with critical thermal maximum temperatures generally in the range 32.4-39.3°C (Table 20). Assuming these sensitivities extend to all members of the genus, it is remarkable the degree to which certain species have been able to penetrate or persist in seemingly inhospitably hot and dry habitats. For example, *H. millewae* occurs in hummock grasslands on sand plains (Coventry 1976), and *H. initialis* (subsp. *brookeri*) occurs in chenopodeaceous shrublands on well drained limestone (Shea and Miller 1986). In the former case, the animals appear to inhabit the base of the hummock grass and in the latter the shaded portions of the litter halo beneath shrubs. Thus both species exist in a generally unfavourable habitat only by occupying a very specific and well-defined microhabitat.

Among the more mesic adapted fossorial genera, *Hemiergis* provides the best case of limb reduction in terms of the width of the range of variation and the number of intermediate stages within the range. The following digital formulas, arranged in descending and possibly phylogenetic order, have been recorded for various populations or species: 5/5 → 4/4 → 3/3 → 2/2 (Fig. 78). The phalangeal formulas run from the primitive skink condition 2.3.4.5.3/2.3.4.5.4 (in the hypothetical common ancestor of the genus) to the moderately derived 0.0.3.4.0/0.0.3.4.0. As a lineage, *Hemiergis* shows an increased number of presacral vertebrae (≥ 34), but within the lineage there is no significant relationship between the number of presacral vertebrae and the total number of phalanges (Choquenot and Greer, in press).

Calyptotis is a genus of five species which ranges along the east coast from northeastern Queensland to northeastern New South Wales (Greer 1983b). All the species are relatively small (none exceeds 59 mm SVL) and most have a yellowish venter and a reddish underside of the tail. They usually occur in moist woodlands but sometimes extend into closed forest and open paddocks. They are cryptozoic, usually

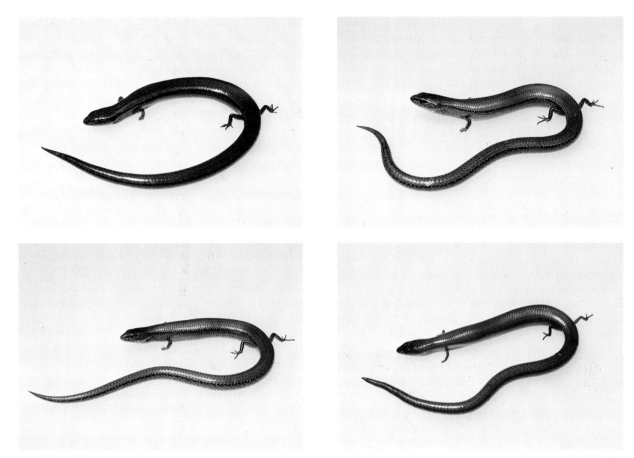

Fig. 78. Representatives of four populations (three species) of *Hemiergis* showing the variation in the number of digits on the front and rear feet. From top to bottom: *H. initialis* with digits 5/5; *H. peroni* with 4/4; *H. peroni* with 3/3, and *H. quadrilineatum* with 2/2.

being found under logs and rocks and when uncovered, disappearing into the surrounding litter or grass roots. They can be very common at some localities, virtually every log or rock harbouring one or more individuals.

The core of the group is made up of three species in the area of north-east New South Wales and south-east Queensland. The three species form an almost linear sequence in their systematically important taxonomic characters; *C. ruficauda* → *C. scutirostrum* → *C. lepidorostrum*. Thus *C. ruficauda* has an external ear opening and well developed prefrontal scales; *C. scutirostrum* has lost the external ear opening, which is now a scaly conical depression, and has greatly reduced the size of the prefrontals, while *C. lepidorostrum* which also lacks the ear opening, has lost the prefrontals (probably by fusion to the frontonasal).

The three species are obviously very closely related and have abutting distributions in a south-north line. In the case of *Calyptotis ruficauda* and *C. scutirostrum* the ranges are separated by an eastern spur of the New England Plateau, which may represent the original geographic barrier that prompted speciation. In the case of

C. scutirostrum and *C. lepidorostrum* there is no obvious geographic barrier and the exclusive but abutting ranges may reflect competitive interactions.

The two other species in the group occur north of the core group, *Calyptotis temporalis* in central eastern Queensland and *C. thorntonensis* restricted to Thornton Peak north of Cairns. Their relationships with the core group are obscure but appear to be fairly basal.

Calyptotis thorntonensis is the smallest species in the group and is also earless. It is an inhabitant of what is virtually a cloud forest and as a result must live under virtually continuously cool conditions. An adaptation to such conditions is evident in the fact that it will perish, presumably from over-heating, if held even gently between the thumb and index finger.

The three southernmost, core-group species of *Calyptotis* are oviparous (Table 9) with females reproductively active in spring-summer (September-January), the typical southern (temperate) pattern. The only northern species for which there is information, *C. temporalis*, showed signs of incipient reproductive activity (a female with yolking ovarian follicles, a male with enlarged testes) in late August, a period slightly ahead of the southern species (Greer 1983b). The mode of reproduction in the two northern species is not known.

In two of the three southern species, *Calyptotis ruficauda* and *C. scutirostrum,* there is a positive correlation between female size and clutch size (r = .78** and .62**, N = 10 and 21, respectively); the correlation in *C. lepidorostrum* was not significant (r = .48^{NS}, N = 10), but, the sample size was small (Greer 1983b).

Calyptotis as a lineage shows one small morphological feature that may be related to its fossorial existence: the loss of a phalange in the fourth toe of the front foot. However, each species or species group within the genus shows its own special further modifications. Each species has lost other, different phalanges, and *C. scutirostrum* and *C. lepidorostrum* have increased the number of presacral vertebrae from 26 to 29.

The loss of the external ear opening in the *Calyptotis scutirostrum-C. lepidorostrum* lineage and in *C. thorntonensis* may also be a modification associated with a fossorial existence. However the ear appears to have been lost in two different ways in the two groups, in the former by gradual constriction of the border, and in the latter by the encroachment of scales onto the tympanic membrane or eardrum.

The closest living relative of *Calyptotis* is often a victim of mistaken identity. It is not unusual for Sydneysiders while working in a shaded, moist part of the garden to uncover a small brown "snake" about 10 cm long with a bright orange belly. Usually the snake slithers away to safety but occasionally it gets chopped into several pieces, scraped into a jar and either brought to the Museum for identification or taken to the house for an excited call to the Museum. It is only when told to look carefully for four tiny limbs each bearing three tiny toes that the realization dawns that what was taken to be a snake is in fact a completely harmless lizard; specifically the three-toed skink *Saiphos equalis.*

Saiphos is a monotypic genus which ranges from southeastern Queensland south to the Wollongong area (Greer 1983a). It is restricted to the highlands north of the Hunter Valley but south of there it occurs also in the coastal lowlands; presumably this far south the lowlands become cool enough for occupation.

In comparison to its relatives in *Calyptotis,* *Saiphos* has become elongate, increased the number of presacral vertebrae and reduced the digits substantially. For example, in adult *Saiphos* the SVL is 8.0-9.2 times the head length but

only 5.2-5.9 times in adult *Calyptotis ruficauda,* the most primitive species in the genus. Furthermore, in *Saiphos* there are 38-40 presacral vertebrae and a phalangeal formula of 0.2.3.3.0/ 0.2.3.3.0 but primitively in *Calyptotis* there are only 26 vertebrae and a phalangeal formula of 2.3.4.4.3/2.3.4.5.4.

In contrast to *Calyptotis,* but in keeping with its occupation of cooler habitats, *Saiphos* is essentially live-bearing (Bustard 1964b; pers. obs.), although as noted in the introduction to this chapter (p. 126), with some geographic variation in the time between when the young are laid/ born and the time when they burst their shells/ membranes (Fig. 79). Litter size ranges 1-7 (Table 9), and there is no significant correlation between female size and litter size (r = .23^{NS}, N = 49).

Anomalopus consists of seven species with limbs ranging from short with digits 3/2 to limbless. The group is elongate and earless. There are two subgenera, *Anomalopus* and *Vermiseps.* The former has minute limbs and often a very pale yellowish green venter while the latter is limbless with a clear venter (Greer and Cogger 1985). For simplicity's sake each group will simply be referred to by its subgeneric name. The three species of *Anomalopus*

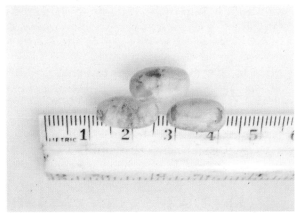

Fig. 79. Saiphos equalis showing the variation in reproductive mode in the species. Top: female with newly born young from the New England Plateau; bottom: freshly laid eggs from Sydney.

occur in central eastern Australia and have contiguous but only slightly overlapping distributions from the Darling Downs east to the coast. Curiously the smallest and strongest-limbed species occurs furthest inland and the weakest-limbed but largest species occurs along the coast. Although the sample size is small (three species), such a geographic pattern to morphological change is intriguing, and at least with regard to the size and aridity gradient (drier inland), is the one also seen in *Hemiergis* (see above). As yet there is no explanation for the trend in either taxon.

Another curious feature of the group is that there are always more toes on the front foot than on the rear. Thus *Anomalopus mackayi* has a digital formula of 3/2, *A. leuckartii* 3/1 and *A. verreauxii* 2/1. This is in contrast to most other limb-reduced lizards in which the front limb is equal to or more reduced than the rear. Again, the significance of this difference is obscure.

Anomalopus leuckartii and *A. verreauxii* occur primarily in woodlands and open paddocks. Unfortunately the habitat of *A. mackayi* is not well known but to judge from its distribution it is probably a black soil savanna species (Shea, Millgate and Peck in press). The first two species are usually found under surface cover such as rocks, logs and debris, and when uncovered, they usually escape by crawling/undulating into the superficial layer of litter and roots. These two species are also known to be oviparous with clutch sizes of 3-11 for *Anomalopus verreauxii* and 3-4 for *A. leuckartii* (Table 9).

The four species of the limbless subgenus *Vermiseps* — *V. brevicollis*, *V. gowi*, *V. pluto* and *V. swansoni* — occur disjunctly along the east coast between Cape York and the area just north of Sydney. They inhabit woodlands, heaths, and vine thickets but appear to be absent from moist, closed forests. They are generally found under surface cover such as rocks, logs and debris and when uncovered, rapidly disappear into the loose substrate or crevices. One species — *V. brevicollis* — is very common on the vine scrub covered limestone reefs around Rockhampton. When the superficial rocks are removed, the animals rapidly disappear down the litter/soil-filled crevices between the fractured blocks.

The mode of reproduction is known for two of the four species: the one temperate species, *Vermiseps swansoni*, is ovoviviparous while one of the three subtropical/tropical species, *V. gowi*, is oviparous. Brood size in the former is three and in the latter 1-3 (Table 9).

When lizards undergo extreme body elongation and limb reduction, they often show unusual internal modifications as well. *Vermiseps* provides some good examples of this. For example, it appears as if the common ancestor of the genus probably had a much reduced right lung. Three of the four species have the left lung about 18-29 per cent of the SVL but the right only about 10-21 per cent. The fourth species, *V. pluto*, has apparently secondarily evolved an elongate right lung; its left lung is 22 per cent of SVL, which is about the same as its close relatives, but its right is much longer, 38 per cent (Table 5). Why *V. pluto* may have evolved an elongate right lung is discussed below. Many elongate squamates have a reduced lung, but in most lizards it is the left lung and not the right as in *Vermiseps*. The only other squamates known to have reduced the right lung instead of the left are amphisbaenians (Butler 1895).

Three species of *Vermiseps* show additional, unusual internal modifications. One of the species, *Vermiseps pluto*, has lost the right oviduct. An oviduct has been lost nine or ten times in reptiles, each time in squamates, and with the exception of the almost legless genus *Dibamus* (family Dibamidae) of the south-east Asia (Greer 1985a), it has always been the left (Greer 1977; Greer and Mys, in press). Only *Dibamus* and *Vermiseps pluto* are known to have lost the right oviduct.

In all other lizards that have lost an oviduct, the brood size is one, and it seems likely that in lizards which produce only one egg at a time and perhaps produce it infrequently enough so that one oviduct can cope, the "superfluous" second oviduct has been lost[1]. Unfortunately, the brood size of *Vermiseps pluto* is unknown, but if it holds true to the previous pattern, it should only produce one egg or young at a time.

Why *Vermiseps pluto* has lost the right oviduct instead of the more usual left is unclear. The usual explanation for the loss of the left oviduct is based on the fact that the stomach normally lies on this side and crowds slightly the left oviduct thereby reducing the number of eggs or young it can carry. In skinks with a brood size of three or more, for example, the left oviduct generally carries one less egg/young than the right. Hence all other things being equal, it would seem most efficent to lose the "crowded" oviduct and retain the "roomy" one. Unfortunately, this seems not to apply to *V. pluto* because its stomach is not "out of place" compared to its relatives. The only other internal rearrangement in *V. pluto* that could be related to the reduction of the right oviduct is the elongation of the right lung, but perhaps this is only a consequence and not a cause of the oviduct's reduction.

---

[1]Lizards that produce several one-egg clutches in a season, usually retain both oviducts, e.g., anolines and certain gekkonids.

Another *Vermiceps* species, *V. brevicollis*, has a reduced number of cervical vertebrae, i.e., a short neck. The neck vertebrae are counted from the first vertebra just behind the skull up to but not including the first vertebra whose ribs actually attach to the sternum. The most common as well as the primitive number of neck vertebrae in skinks is eight; but *V. brevicollis* has seven. Why it has lost one neck vertebra is unknown. The only other skinks with seven neck vertebrae are the *Lerista bipes* species group in Australia (Greer 1986c) and two different groups in Africa.

Vermiseps brevicollis and *V. gowi* share a number of derived character states and are currently thought to be each others closest living relatives. One of the most unusual characters of this two species group is dark pigment in the pulmonary peritoneum. In gross dissection this renders the lung of *V. brevicollis* very dark grey and the lung of *V. gowi* almost black. In all other skinks the lungs are pink in life, due to the circulating blood in otherwise clear tissues, or some very light hue in preserved material — but never dark grey to black, except in these two species of *Vermiseps*.

The genus *Ophioscincus* consists of three limb-reduced earless species: *O. truncatus* which has tiny, styliform front and rear legs (Copland 1952b) and *O. cooloolensis* and *O. ophioscincus* which lack limbs. They occur in the mountains and along the coast in southeastern Queensland and northeastern New South Wales. They usually inhabit moist closed forest but may extend into wallum and woodland. The mode of reproduction is known only for *O. ophioscincus* and *O. truncatus*: both are oviparous with clutches of 2-3 (Table 9; Greer and Cogger 1985).

Unlike other burrowers, for example *Vermiseps*, the internal anatomy and morphology of *Ophioscincus* holds no particular surprises. There is a reduced lung, but it is the usual left (which is 8-18 percent of SVL) instead of the right (Table 5); both oviducts are well developed; the pulmonary peritoneum is clear and the neck has the usual number of vertebrae, eight (cf. *V. brevicollis*).

All three species of *Ophioscincus* have ventral body colour, in *O. truncatus* it is muted yellow and in *O. cooloolensis* and *O. ophioscincus* orange.

Amongst the fossorial skinks, there are two species that share large size (i.e., max. SVL $\geq$ 195 mm), somewhat conical snouts and pointed, slightly recurved teeth (Fig. 88). On this basis they have been grouped together in the genus *Coeranoscincus* (Greer and Cogger 1985). Both species are inhabitants of moist closed forest: *C. reticulatus* which has relatively small limbs,

each with three toes, and occurs in northeastern New South Wales and southeastern Queensland, and *C. frontalis* which is limbless and occurs in northeastern Queensland. Both species lack an external ear opening.

Little is known of their habits other than that they are generally found in leaf litter, under surface cover or in rotten logs. *Coeranoscincus reticulatus* is known to be oviparous and to eat earthworms (McDonald 1977), a rather unusual prey for skinks. Perhaps the pointed, recurved teeth of the genus have evolved to grasp this slippery and powerful burrowing prey. Recall the difficulty in extracting a partially exposed earthworm from its burrow. Curiously, these skinks do not seem inclined to bite in defense, although their teeth could easily cause a serious wound.

Coeranoscincus reticulatus is unique within Australian skinks in the range of adult variation in the degree of cross banding and the intensity of the "ground colour" on the body (McDonald 1977). The bands range from totally banded to plain, with the bands on any one specimen being less well defined posteriorly, and the ground colour ranging from light to dark brown or slaty grey.

Both species of *Coeranoscincus* show colour hues. *Coeranoscincus reticulatus* is pale to light yellow on the throat and the sides of the neck and face (Czechura 1974; McDonald 1977) while *C. frontalis* is pale yellow to dull orange on the ventral part of the body (pers. obs.).

Osteologically, *Coeranoscincus frontalis* may be the most highly modified of all the mesic, fossorial skinks in Australia. For example, in addition to being totally limbless (except for a tiny internal fragment of the humerus), it has the second most elongate body (SVL/HL is 15.3-15.9 in adults; Table 2), the highest number of presacral vertebrae (72-76; Table 3), the highest number of inscriptional chevrons (32-35), no interclavicle, no mesosternum, and only one rib attached to a much shortened sternum; furthermore, the two halves of the pelvic girdle are widely separated and the sacral diapophyses are separated from each other distally.

The genus *Lerista* comprises 54 species of medium to small sized, mostly limb-reduced skinks that occur throughout the arid, semi-arid and seasonally dry parts of Australia in open habitats with sandy substrates (Greer 1967, 1979d, 1983a, 1986c; Greer *et al.* 1983; Storr 1971b, 1976d, 1982a, 1983, 1984a,d, 1985b, 1986a-b). Some species, notably members of the *L. bipes* species group, may even forage along the seaward side of the most seaward terrestrial plants along sandy coasts (Fig. 80).

Fig. 80. Two members of the *Lerista bipes* species group (top), their habitat at Cape Lafeque, W.A. (middle) and tracks typical of the species group (bottom). The larger, darker species, *L. griffini,* occurs on the heavily vegetated back dune area whereas the smaller, lighter species, *L. bipes,* occurs in the dune area itself, seaward as far as the last shrub. The tracks *(L. bipes)* show where the animal undulated beneath the surface (continuous track) and walked over it (faint, intermittent foot prints interrupting the continuous trace).

Lerista is perhaps most interesting because it provides the best case of limb reduction and loss in tetrapods (Mitchell 1958; Figs 81-83). By that, is meant it has the greatest range of limb conditions, with the largest number of intermediate stages, all within a small taxonomic unit. The range goes from the primitive condition for

lizards to a condition more reduced than the most primitive living snakes, i.e., totally limbless; the number of intermediate stages is 12 for the front limb and 13 for the rear, and the taxonomic unit is the genus, with its clear implication of common ancestry and overall basic genetic similarity. Preliminary studies on limb reduction in *Lerista* are just now being published (Greer, in press) and are summarized below in comparison with *Hemiergis*.

A second reason the genus is interesting is that its overall wide-distribution and local species richness (as many as seven in one area near Shark Bay —*fide* P. Kendrick) make it ideal for studying speciation and comparative ecology. Serious study of these topics has only just begun but offer great promise (P. Kendrick, in prep.).

The species of *Lerista* with more well-developed limbs tend to be diurnal and surface-active whereas the more limb-reduced species tend to be crepuscular to nocturnal and active just below the surface as well as on top of it. The change-over seems to come in the stages between three toes on the front limb and a styliform front limb. The more surface-active group forages by walking on the surface or through the top layers of litter, whereas the more subsurface group generally forages either by walking on the surface (using the less-reduced rear limbs) or undulating just below it. The tracks of the latter are very distinctive, being a combination of faint, intermittent little imprints during surface walking and a continuous sinusoidal furrow during subsurface undulation (Fig. 80). The surface-active group is also capable of going subsurface but usually only to escape danger or inclement surface conditions.

The two southernmost species *Lerista bougain-villii* (southernmost populations only) and *L. microtis* are ovoviviparous whereas all other species for which information is available are oviparous. Brood size in the group ranges 1–7. Adequate data on the intraspecific relationship between female size and brood size are available for only four species and the results show no clear trend. In the oviparous *L. bougainvillii* and *L. wilkinsi* there is a significant positive correlation ($r = .48**$, $N = 29$ and $r = .67**$, $N = 19$, respectively) whereas in ovoviviparous *L. bougainvillii* and the oviparous *L. muelleri* there is no significant relationship ($r = .21^{NS}$, $N = 11$ and $r = .07^{NS}$, $N = 18$, respectively). The interspecific relationship between female size and brood size is also positive ($r = .42*$, $N = 27$). As far as can be determined, reproduction occurs during the spring and summer in southern species and during the mid to late dry in most

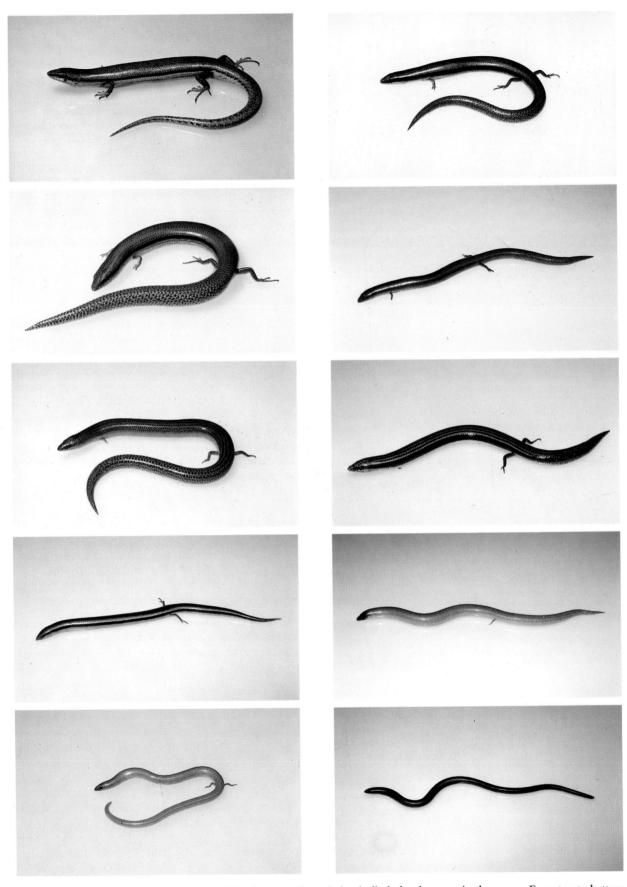

Fig. 81. Ten *Lerista* species showing some of the interspecific variation in limb development in the genus. From top to bottom and left to right: *L. arenicolor, L. orientalis, L. terdigitata, L. borealis, L. neander, L. picturata, L. onsloviana, L. lineopunctulata, L. wilkinsi, L. ameles.*

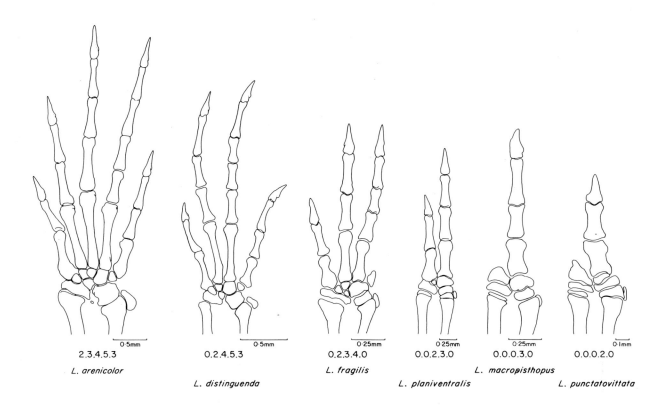

| 2.3.4.5.3 | 0.2.4.5.3 | 0.2.3.4.0 | 0.0.2.3.0 | 0.0.0.3.0 | 0.0.0.2.0 |
| L. arenicolor | L. distinguenda | L. fragilis | L. planiventralis | L. macropisthopus | L. punctatovittata |

Fig. 82. Six *Lerista* species showing some of the interspecific variations in the number and configuration of bones in the front foot.

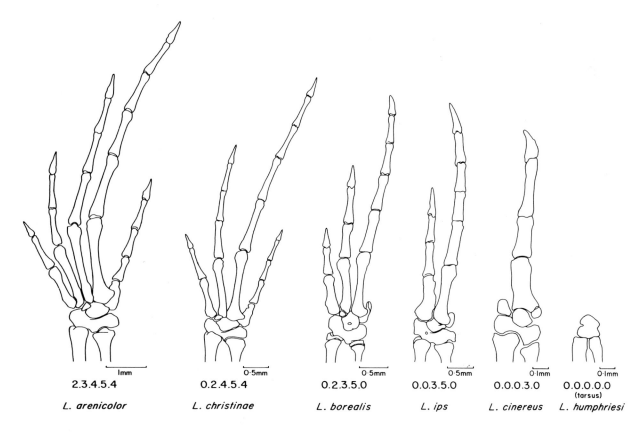

| 2.3.4.5.4 | 0.2.4.5.4 | 0.2.3.5.0 | 0.0.3.5.0 | 0.0.0.3.0 | 0.0.0.0.0 (tarsus) |
| L. arenicolor | L. christinae | L. borealis | L. ips | L. cinereus | L. humphriesi |

Fig. 83. Six *Lerista* species showing some of the interspecific variation in the number and configuration of bones in the rear foot.

far-northern species, the only exception being *L. orientalis* which is quiescent during the dry (Table 21).

Two peculiar morphological features within certain *Lerista* species are worth noting. First, the *L. bipes* group is characterized by the loss of one neck vertebra; it has seven instead of the usual and primitive eight. Recall that *Anomalopus (Vermiseps) brevicollis* is the only other short-necked skink in Australia. Second, *Lerista apoda,* the most highly modified species in terms of the loss of limbs and girdles as well as other features, has greatly reduced the left lung. Generally primitive *Lerista* have both lungs about equally well-developed and comprising about 27-32 percent of the snout-vent length. *L. apoda* has reduced the left lung to 8 percent of this length and the right to 22 percent. Recall also that reduction of the left lung is more common in lizards and snakes than reduction of the right. *L. apoda* is also the most elongate Australian skink, with SVL/HL ranging 15.3-18.1 (Table 2).

Limb reduction and loss has occurred independently in many different skink lineages, and the two best lineages in Australia are *Hemiergis* and *Lerista* (Choquenot and Greer, in press and Greer, in press, respectively). Each represents one of two different ecological categories of limb-reduced skinks in Australia: litter/humus-swimmers (*Hemiergis*) and sand-swimmers (*Lerista*). Reviewing limb reduction in each not only allows us to look for general features in the process but also to identify features specific to these two habitat types. The similarities outweigh the differences.

In both groups, there is variation in the number of phalanges within populations, although this does not exceed 4.8 percent of the sample in any *Hemiergis* population or 9.5 percent in any *Lerista* population. There is also variation in modal phalangeal formulas amongst populations in both groups; one of the five species in *Hemiergis* showed such variation and seven of 54 species showed it in *Lerista*.

When species are arranged in series by decreasing number of phalanges for the front

and rear limbs separately, the order of digit loss on both feet is in the order 1 >5 > 2 > 3 > 4 (only the first three digits are lost in *Hemiergis*). This represents a trimming of the digits from the sides in to the larger central digits.

In both groups there is also a sexual difference in the number of presacral vertebrae within populations, the females having, on average, more vertebrae than males.

In the humus-swimmers there is no relationship between the number of phalanges and the number of presacral vertebrae; however, there is an inverse and curvilinear relationship in the sand-swimmers, that is, as the number of presacral vertebrae increases, the number of phalanges decreases, and as vertebrae number increases, the number of phalanges decreases at an ever slower rate. The humus-swimmers have lost the external ear opening whereas the sand-swimmers retain it. Perhaps humus-swimmers run a high risk of having the ear opening clogged with the moister substrate they inhabit. Humus- and sand-swimming skinks occur elsewhere in the world and should make interesting future comparisons with *Hemiergis* and *Lerista*.

To conclude this chapter, it might be worthwhile highlighting what appears to be a fundamental theme in skink biology, namely, a close association with the litter and uppermost layers of the soil. Skinks forage in this litter for food, they flee into it when startled and they shelter in the litter/soil interface during inclement periods. Such habits have possibly set the stage for the family's repeated evolutionary excursions into an even more intimate association with this microhabitat and its consequent changes of both habits (i.e., more cryptozoic and/or nocturnal behaviour) and morphology (i.e., body elongation and attenuation, and limb reduction). Such habits may also partially account for the flexible bony covering provided by the osteoderms, for also resident in the litter/soil layer is a rich fauna whose chelicerae and teeth are formidable weapons of both offense and defence. What other aspects of skink biology that might be understood in the context of this litter/soil interface remain to be discovered.

Table 1. Derived character states shown by the Scincidae in comparison to the hypothetical common ancestor of all lizards and snakes.

External morphology

    Dorsal crest absent
    Femoral and preanal pores absent
    Osteoderms present, each composed of separate plates arranged in symmetrical
        fashion (Fig. 57)

Skull and hyoid

    Parietals fused
    Vomerine and palatine teeth absent
    Palatine with ventral, horizontal laminae projecting medially to form incomplete
        secondary palate (Fig. 56)
    Stapes imperforate
    Second branchial arch interrupted

Postcranial skeleton

    Vertebrae procoelous, anotochordal
    Intercentra absent on mid and posterior presacral vertebrae
    Sacral diapophyses fused
    Clavicle attaches to suprascapula
    Mesosternum fused in its two parts distally

Soft anatomy

    Tongue with flat, scale-like plicae

Physiology

    Panting response to high temperatures absent

Table 2. Range of snout-vent length/head length (SVL/HL) ratios, a measure of elongation, for Australian limb-reduced and limbless skinks. These may be compared to those for *Eulamprus tenuis* a typical robust-limbed, pentadactyl species in which the SVL/HL range is 3.9-5.3 (N=14).

Taxon	Number of toes	SVL/HL	N
Anomalopus (*Anomalopus*)			
mackayi	3/2	9.7-11.6	6
verreauxii	3/1	12.8-16.3	5
leuckartii	2/1	9.4-12.0	6
Anomalopus (*Vermiseps*)			
brevicollis	—	11.5-12.8	10
swansoni	—	11.1-12.9	7
gowi	—	12.5-14.0	5
pluto	—	12.5-14.3	3
Coeranoscincus			
reticulatus	3/3	7.6-12.5	5
frontalis	—	11.8-15.9	5
Hemiergis			
peroni	4/4	5.4-7.9	50
decresiensis	3/3	5.0-7.9	11
quadrilineatum	2/2	6.0-8.7	10
Lerista			
ameles	—	10.4-12.4	2
apoda	—	15.3-18.1	17
Ophioscincus			
truncatus	1/1	10.4-11.5	9
ophioscincus	—	11.7-12.4	2
Saiphos			
equalis	3/3	8.0-9.2	9

Table 3. Summary of the number of presacral and postsacral vertebrae in Australian skinks. Numbers are range, mode (whole number) or mean (decimal), sample size. For data on *Lerista*, see Greer (in press).

Taxon	Vertebrae		Taxon	Vertebrae	
	Presacral	Postsacral		Presacral	Postsacral
Anomalopus			*Egernia* (continued)		
Anomalopus			slateri	26-27, 26, 7	45-47, 46.0, 2
leuckartii	51-55, 52.8, 16	50-54, 52.0, 4	stokesii	26-28, 27, 7	14-15, 14.1, 2
mackayi	51-58, 53.9, 9	—	striata	26, 26, 5	—
verreauxii	53-57, 54.7, 31	57-62, 57.7, 11	striolata	26-27, 26, 18	31-36, 33.0, 12
Vermiseps			*Emoia*		
brevicollis	46-49, 47.5, 51	48-51, 50.1, 14	atrocostata	27-28, 28, 18	66-70.5, 68.3, 2
gowi	52-57, 54.6, 25	46-52, 49.2, 11	longicauda	28, 28, 18	—
pluto	59-63, 61.0, 6	47-48, 47.7, 3	*Eremiascincus*		
swansoni	53-58, 55.7, 28	40-44, 41.5, 9	fasciolatus	25-27, 26, 11	41-44, 42.8, 5
Calyptotis			richardsonii	26, 26, 11	—
lepidorostrum	29, 29, 9	—	*Eroticoscincus*		
ruficauda	26, 26, 10	40-43, 41.8, 10	graciloides	28-29, 28, 5	—
scutirostrum	29-30, 29/30, 10	—	*Eugongylus*		
temporalis	26, 26, 4	—	rufescens	28, 28, 11	—
thorntonensis	26, 26, 5	—	*Eulamprus*		
Carinascincus			amplus	26, 26, 2	—
greeni	27-28, 28, 9	41-45, 43.3, 6	brachysoma	25.5-26, —, 2	—
metallicum	28-30, 29, 17	43, 43.0, 2	heatwolei[2]	25-27, 26, ?	—
ocellatum	28-30, 29, 9	47, 47.0, 1	kosciuskoi	—	—
pretiosum	27-30, 28, 8	—	leuraensis[3]	26, 26, 5	45-49, 47.0, 2
Carlia			luteilateralis	26-27, 26, 13	—
mundivensis	27, 27, 9	45-48, 46.5, 2	martini	26, 26, 4	—
pectoralis	27, 27, 1	—	murrayi	26, 26, 9	—
tetradactyla	26-27, 27, 9	47, 47.0, 1	quoyii[2]	26-27, 26, ?	—
vivax	27, 27, 11	50-51, 50.5, 2	tenuis	26, 26, 11	46, 46.0, 1
Claireascincus			tigrinum	26, 26, 8	48, 48.0, 1
baudini	28, 28, 2	—	tympanum[2]	26-27, 26, ?	—
entrecasteauxii			*Eulepis*		
form A	28-31, 30, 9	—	duperreyi	31-34, 31.9, 10	44.5-48, 46.3, 2
Coeranoscincus			platynota	31-34, 32.7, 9	50, 50.0, 2
frontalis	72-76, 73.4, 8	62+, 1	trilineata	30-34, 31.9, 15	—
reticulatus	52-55, 54.4, 3	38, 38.0, 6	*Glaphyromorphus*		
Cryptoblepharus			arnhemicus	26, 26, 1	—
plagiocephalus	27, 27, 1	—	cracens	31-33, 32, 13	59, 59.0, 1
virgatus	27-28, 27, 10	44-45, 44.5, 2	crassicaudus	29-31, 31, 12	52-57.5, 54.5, 3
Ctenotus			darwiniensis	28, 28, 3	—
brooksi	26, 26, 11	56, 56.0, 1	douglasi	26, 26, 9	—
essingtoni	26, 26, 7	60, 60.0, 1	fuscicaudis	26, 26, 8	52+, 1
grandis	26, 26, 2	—	gracilipes	34-37, 36, 10	69.5, 69.5, 1
lesueurii ·	26, 26, 1	—	isolepis	26, 26, 4	42, 42.0, 1
pantherinus	25-27, 26, 8	54, 54.0, 1	mjobergi	29, 29, 3	—
pulchellus	26, 26, 9	53-55, 54.0, 3	nigricaudis	26, 26, 8	—
rubicundus	26, 26, 4	—	pardalis	26, 26, 8	41, 41.0, 1
schevilli	26, 26, 5	51, 51.0, 1	pumilus	32-36, 36, 6	50-59, 55.3, 3
taeniolatus	26-27, 26, 13	54, 54.0, 1	punctulatus	33-34, 33, 10	52-53, 52.5, 2
youngsoni	26, 26, 2	—	*Gnypetoscincus*		
Egernia[1]			queenslandiae	26, 26, 10	37-45, 41.2, 6
cunninghami	25-27, 26, 58	24-42, 31.2, 53	*Harrisoniascincus*		
depressa	26, 26, 1	15-16, 15.5, 2	coventryi	28-30, 29, 11	43-45, 44.0, 2
formosa	26, 26, 8	44, 44.0, 2	zia	29-30, 29, 11	—
frerei	26, 26, 9	48-51, 49.5, 8	*Hemiergis*		
hosmeri	26-27, 26.5, 2	21-22, 21.5, 2	decresiensis	35-38, 36.9, 49	43-48, 45.3, 12
inornata	26, 26, 11	36-43, 39.0, 3	initialis	35-39, 37.2, 16	40-42, 41.0, 6
kingii	26, 26, 1	43, 43.0, 1	millewae	35-37, 35.6, 19	51, 51.0, 1
kintorei	26, 26, 3	37, 37.0, 1	peroni	34-38, 35.6, 75	48-61, 51.8, 31
luctuosa	26, 26, 3	57-59, 58.0, 2	quadrilineatum	35-39, 37.2, 20	50-51, 50.5, 2
margaretae	26, 26, 15	50-62, 57.0, 3	*Lampropholis*		
modesta	26, 26, 8	53-56, 54.5, 2	amicula	28, 28, 2	—
multiscutata	26, 26, 3	41, 41.0, 2	caligula	30-31, 31, 7	41-42, 41.5, 2
pilbarensis	25, 25, 1	47, 47.0, 1	delicata	26-27, 26, 26	40-45, 42.6, 5
napoleonis	26, 26, 5	42-50, 44.2, 5	guichenoti	27-28, 27, 16	45-49, 47.0, 3
richardi	26, 26, 5	34-35, 34.5, 2	mirabilis	26, 26, 7	48, 48.0, 1
rugosa	26, 26, 2	—			
saxatilis	25-26, 26, 10	37-40, 38.3, 2			

Table 3 (continued)

| Taxon | Vertebrae | | Taxon | Vertebrae | |
	Presacral	Postsacral		Presacral	Postsacral
Lygisaurus			*Proablepharus*		
burnettii	26.5-28, 27, 10	—	tenuis	28, 28, 1	—
heteropus	27-28, 27, 9	—	*Pseudemoia*		
Menetia			palfreymani	29-30, 29, 7	45, 45.0, 1
alanae	30, 30, 2	—	spenceri	28-30, 29, 10	42-45, 43.7, 3
amaura	29-31, 29, 4	—	*Saiphos*		
greyii	28-33, 29/30, 17	—	equalis	38-40, 38, 17	—
maini	29-31, 30, 9	—	*Saproscincus*		
surda	29, 29, 1	—	basiliscus	27, 27, 8	46, 46.0, 2
Morethia			challengeri	27-29, 27, 7	49, 49.0, 1
butleri	29-31, 30, 10	48-50, 49.0, 3	czechurae	26-28, 27, 11	33-35, 34.0, 3
taeniopleura	29-31, 30,10	—	mustelina	28-31, 29, 9	48, 48.0, 1
Nannoscincus			tetradactyla	26, 26, 24	32-33, 32.3, 3
maccoyi	34-36, 35/36, 11	37-39, 37.7, 3	*Tiliqua*[1]		
Notoscincus			adelaidensis	33-35, 34, 6	34-36, 34.8, 5
ornatus	28-29, 28, 5	40-42, 40.7, 3	branchialis	40-44, 42.1, 67	35-41, 38.6, 24
wotjulum	28-30, 28/30, 2	—	casuarinae	37-44, 40.6, 32	42-57, 51.7, 10
sp. nov.	28-29, 28, 7	41, 41.0, 2	gerrardii	34-36, 34.9, 10	45-48, 46.3, 6
Ophioscincus			maxima	41-42, 42, 5	—
cooloolensis	43-44, 43.6, 5	—	multifasciata	33-34, 33, 4	23, 23.0, 1
ophioscincus	49-51, 50.0, 14	42, 42.0, 1	nigrolutea	36-38, 37/38, 15	26-29, 27.4, 9
truncatus	45-50, 47.2, 32	40-42, 41.1, 11	occipitalis	34-35, 35, 4	28-29, 28.7, 3
			rugosa	35-37, 36, 26	18-21, 19.8, 10
			scincoides	37-39, 39, 12	27-30, 27.9, 10

[1]Data from G. Shea, pers. comm.
[2]Tilley 1984.
[3]Shea and Peterson 1985.

Table 4. Sexual dimorphism in presacral vertebral number in single populations of various skink species with an elevated number of presacral vertebrae. For data on *Hemiergis* and *Lerista*, where the trend is similar to that shown here, see Choquenot and Greer (in press) and Greer (in press), respectively.

Taxon	Sex	Range	Mean	SD	N	t
Anomalopus brevicollis Rockhampton, Qld; 6 September 1984	♂♂ ♀♀	46-48 47-49	47.3 48.0	.71 .67	8 10	2.69*
Anomalopus gowi Woodstock, Qld; 30 September 1984	♂♂ ♀♀	52-53 54-56	54.6 52.8	.70 .45	5 10	5.20***
Eulepis trilineata Esperance, WA; 19 October 1976	♂♂ ♀♀	30-31 33-34	30.9 33.1	.39 .33	9 9	12.97***
Morethia taeniopleura Townsville, Qld; 20, 28 May, 2 June 1976	♂♂ ♀♀	28-30 30-31	29.0 30.3	.50 .48	17 10	6.65***
Nannoscincus maccoyi Bondi State Forest, NSW 23 December 1980	♂♂ ♀♀	34-35 34-36	34.4 35.0	.52 .41	10 13	3.08***
Ophioscincus truncatus Red Scrub Flora Reserve, NSW 21 April 1976	♂♂ ♀♀	45-47 47-50	46.6 48.0	.63 1.09	16 6	5.04***
Saiphos equalis Spring Creek, NSW 9 March 1976	♂♂ ♀♀	38 38-40	38.0 38.6	.00 .63	6 14	2.28*

Table 5. Relative lung length (as % of SVL) in four Australian lineages of scincid lizards with reduced limbs.

Taxon	Left Lung			Right Lung			N
	Range	Mean	SD	Range	Mean	SD	
1) *Anomalopus*							
Anomalopus							
leuckartii	19-27	22.3	4.16	18-25	21.0	3.61	3
verreauxii	21-24	22.3	1.50	18-21	19.5	1.73	4
Vermiseps							
swansoni	19-29	23.2	3.70	10-21	16.4	4.22	5
pluto	22	22.0	—	38	38.0	—	1
gowi	20-26	22.7	2.75	13-17	14.5	1.91	4
brevicollis	18-26	23.5	3.79	14-18	15.7	1.71	4
2) *Ophioscincus*							
ophioscincus	8-18	11.7	5.50	25-29	26.7	2.08	3
truncatus	12-16	13.9	1.60	25-30	27.4	2.30	5
3) *Coeranoscincus*							
frontalis	12	12.0	—	39	39.0	—	1
reticulatus	29	29.0	—	24	24.0	—	1
4) *Saiphos*							
equalis	18-25	22.4	2.70	18-23	21.2	1.92	5

Table 6. Primary habits of the species of *Egernia*. Species known to make their own burrows are marked with an asterisk. Data from literature, R. Sadlier and G. Shea (pers. comms), and pers. obs.

Species	Saxicolous	Terrestrial	Semi-arboreal
E. coventryi		+	
E. cunninghami	+		
E. depressa	+		+
E. formosa	+		+
E. frerei		+	
E. hosmeri	+		
*E. inornata**		+	
E. kingii		+	
*E. kintorei**		+	
E. luctuosa		+	
E. major		+	
E. margaretae	+		
*E. modesta**		+	
*E. multiscutata**		+	
E. napoleonis	+		+
E. pilbarensis	+		
E. pulchra	+	+	
E. richardii	+		+
E. rugosa		+	
E. saxatilis	+		+
E. slateri		+	
E. stokesii	+	+	+
*E. striata**		+	
E. striolata	+		+
*E. whitii**		+	

Table 7. References to feeding habits of Australian skinks both in the wild (W) and in captivity (C).

Taxon	Reference
Carinascincus	
ocellata	Hewer and Mollison 1974 (W)
pretiosa	Hewer and Mollison 1974 (W); Rounsevell *et al.* 1985 (W)
metallicum	Hewer and Mollison 1974 (W,C)
Carlia	
amax	James *et al.* 1984 (W)
gracilis	James *et al.* 1984 (W)
munda	James *et al.* 1984 (W, as *C. foliorum*)
triacantha	James *et al.* 1984 (W)
rhomboidalis	Wilhoft 1963a (W)
Claireascincus	
entrecasteauxii form A	Hewer and Mollison 1974 (W)
Coeranoscincus	
reticulatus	McDonald 1977 (W)
Cryptoblepharus	
plagiocephalus	James *et al.* 1984 (W)
virgatus	Rose 1974 (W, as *C. boutonii)*; Webb 1983 (W)
Ctenotus	
ariadnae	Pianka 1969b (W)
atlas	Pianka 1969b (W)
brooksi	Pianka 1969b (W)
calurus	Pianka 1969b (W)
colletti	Pianka 1969b (W)
dux	Pianka 1969b (W)
essingtoni	James *et al.* 1984 (W)
fallens	Smith 1976 (W)
grandis	Pianka 1969b (W); Smith 1976 (W)
helenae	Pianka 1969b (W)
hilli	James *et al.* 1984 (W)
leae	Pianka 1969b (W)
leonhardii	Pianka 1969b (W)
pantherinus	Pianka 1969b (W); Smith 1976 (W); Dell and Chapman 1979a (W); Chapman and Dell 1980a (W)
piankai	Pianka 1969b (W); Smith 1976 (W)
quattuordecimlineatus	Pianka 1969b (W)
schomburgkii	Pianka 1969b (W); Dell and Chapman 1979a (W); Chapman and Dell 1980a (W)
taeniolatus	Rose 1974 (W); Webb 1983 (W); Taylor 1986 (W)
uber	Dell and Chapman 1979a (W)
vertebralis	James *et al.* 1984 (W)
Egernia[1]	
cunninghami	Longley 1947c (C); Honegger 1964 (C); Barwick 1965 (W); Matz 1968 (C); Shine 1971 (C); Rankin 1972 (C); Münsch 1981 (C); Niekisch 1975 (C); Bartlett 1981 (C); Brown 1983 (W); Rose 1985 (C)
frerei	Banks 1985 (C)
inornata	Pianka and Giles 1982 (W)
kingii	Arena 1986 (W)
major	Longman 1918 (C, as *E. bungana*)
rugosa	Adler 1958 (C, as *E. dorsalis*)
saxatilis	Brown 1983 (W)
striata	Pianka and Giles 1982 (W)
striolata	Bustard 1970e (W); Brown 1983 (W)
whitii	Hewer 1948 (?); Adler 1958 (C); Hickman 1960 (W,C); Brown 1983 (W)
Emoia	
atrocostata	Cogger *et al.* 1983 (W)
Eremiascincus	
fasciolatus	Greer 1979a (C)
richardsonii	Smith 1976 (W)
Eulamprus	
heatwolei	Brown 1983 (W)
quoyii	Veron 1969b (W); Rose 1974 (W); Popper 1980 (C, not seen, as quoted in Daniels 1987); Daniels 1987 (W)
tenuis	Antenor 1973 (C); Rose 1974 (W); Webb 1983 (W)
tympanum	Brown 1983 (W)
Eulepis	
duperreyi	Hewer and Mollison 1974 (W, as *Leiolopisma trilineatum*)
Glaphyromorphus	
douglasi	James *et al.* 1984 (W)
isolepis	Smith 1976 (W)

Table 7 continued

Taxon	Reference
Harrisoniascincus	
zia	Ingram and Ehmann 1981 (W)
Hemiergis	
decresiensis	Robertson 1976, 1981; Crome 1981 (W)
peroni	Smyth 1968 (W)
Lampropholis	
delicata	Hewer and Mollison 1974 (W); Rose 1974 (W); Crome 1981 (W)
guichenoti	Crome 1981 (W)
Lerista	
bipes	Smith 1976 (W)
bougainvillii	Hewer and Mollison 1974 (W)
muelleri	Smith 1976 (W)
Lygisaurus	
burnettii	Webb 1983 (W)
Morethia	
boulengeri	Smyth and Smith 1974 (W)
obscura	Chapman and Dell 1980a (W)
ruficauda	Smith 1976 (W, as *M. taeniopleura*)
Menetia	
greyii	Smyth and Smith 1974 (W); Chapman and Dell 1980a (W)
Nannoscincus	
maccoyi	Robertson 1976, 1981 (W); Brown 1983 (W)
Notoscincus	
ornatus	Smith 1976 (W)
Proablepharus	
reginae	Smith 1976 (W)
Pseudemoia	
palfreymani	Rounsevell *et al.* 1985 (W)
spenceri	Brown 1983, 1986 (W)
Saiphos	
equalis	Rose 1974 (W)
Saproscincus	
mustelina	Rose 1974 (W)
Tiliqua[1]	
branchialis	Smith 1976 (W)
casuarinae	Hewer in Morrison 1948 (C); Hewer and Mollison 1974 (W); Mebs 1974 (C)
gerrardii	Longley 1938 (C); Adler 1958 (C); Wilhoft 1960 (C); Morris *et al.* 1963 (W,C); Miles 1973 (C); Mudrack 1974 (C); Stephenson 1977 (C); Field 1980 (C); Robertson 1980 (C)
multifasciata	Adler 1958 (C)
occipitalis	Dell and Chapman 1977, 1979a; Chapman and Dell 1979
nigrolutea	Longley 1941 (C); Adler 1958 (C); Hewer and Mollison 1974 (W); Bartlett 1984 (C); Giddings 1984 (C); Webb and Simpson 1986 (W)
rugosa	Cole 1930; Tubb 1938 (W); Coleman 1944 (C); Longley 1940 (C), 1944a-b (C); Matz 1968 (C); Dell and Chapman 1979a; Bamford 1980 (W); Belan 1980 (?, not seen); Chapman and Dell 1980a (W); Hitz 1983 (C); Shugg 1983 (W); Brown 1983 (W)
scincoides	Longley 1940 (C); Tschambers 1949 (C); Honegger 1964 (C); Matz 1968 (C); Rose 1974 (W); Vestjens 1977 (W); Schafer 1979 (C); Webb 1983 (W)

[1] References for these two genera are by no means exhaustive.

Table 8. Estimated age to maturity for various Australian skinks, expressed as season after hatching or birth when breeding first occurs. Most common age to maturity, in cases where variation occurs, is underlined.

Taxon	Age to Maturity		Reference
	♂♂	♀♀	
Carinascincus			
entrecasteauxii form A	?	2	Pengilley 1972
entrecasteauxii form B	1	2	Pengilley 1972
Ctenotus			
taeniolatus	1-<u>2</u>	1-<u>2</u>	Taylor 1985b
Egernia			
cunninghami	5	5	Barwick 1965
modesta	2	2	Milton 1987
striolata	2-<u>3</u>	2-<u>3</u>	Bustard 1970e
whitii	2	2	Milton 1987
Eulamprus			
heatwolei	2-<u>3</u>	2-<u>3</u>	Pengilley 1972
quoyi	1	1	Veron 1969a
Harrisoniascincus			
coventryi	2	2	Pengilley 1972 (as *Leiolopisma* sp. *x*)
Eulepis			
duperreyi	?	2?	Pengilley 1972 (as *Leiolopisma trilineatum*)
Hemiergis			
peroni	?	3	Smyth 1968
Lampropholis			
guichenoti	?	2	Pengilley 1972
	1	1	Simbotwe 1985
Menetia			
greyii	1	1	Smyth and Smith 1974
Morethia			
boulengeri	1-2	2	Smyth and Smith 1974
Pseudemoia			
spenceri	<u>2</u>-3	2	Pengilley 1972
Nannoscincus			
maccoyi	?	2?	Pengilley 1972
Tiliqua			
rugosa	3-4	3-4	Bamford 1980; Bull, in press

Table 9. Mode of reproduction, female size and brood size in Australian skinks. Letter in parentheses after species name indicates oviparous (O) or ovoviviparous (V) mode of reproduction; number in parentheses indicates mean if given as decimal or mode if whole number. Unreferenced entries are personal observations.

Taxon	SVL of gravid ♀♀ (mm)	Brood Size	N	Reference
Anomalopus				
gowi (O)	99-108(101.9)	1-3(2.1)	7	
leuckartii (O)	107-137(120.7)	3-4(3.8)	4	
swansoni (V)	96-107 (101.5)	3(3.0)	2	
verreauxii (O)	128-182(157.0)	3-11(8.0)	4	
Calyptotis				
lepidorostrum (O)	39-55 (48.5)	2-4(3.2)	10	
ruficauda (O)	42-52(48.5)	2-6(3.3)	10	
scutirostrum (O)	37-59(47.5)	1-5(3.3)	21	
temporalis (?)	36(36.0)	2(2.0)	1	
Carinascincus				
greeni (V)	—	3(3.0)	3	Rawlinson 1974a (as *Leiolopisma* n. sp.), 1975
	62-68(64.8)	2-3(2.3)	4	
metallicum (V)	—	1-7(3.9)	44	Rawlinson 1974a
	42-66(53.4)	2-8(3.9)	75	
ocellatum (V)	—	4(4.0)	2	Rawlinson 1974a
	58-71(64.4)	2-4(2.6)	11	Green and McGarvie 1971
pretiosum (V)	—	4(4.0)	1	Rawlinson 1974a
	—	2-4(2.9)	8	
	45-63(52.5)	1-4(2.2)	19	
Carlia				
amax (O)	36-39(37.5)	2(2.0)	4	
fusca (O)	52-55(53.7)	1-2(1.7)	3	
munda (O)	38-40(39.0)	2(2.0)	5	
tetradactyla (O)	49-60(52.1)	2(2.0)	22	
triacantha (O)	41(41.0)	2(2.0)	1	
rhomboidalis (O)	61(61.0)	2(2.0)	1	
Claireascincus				
baudini (V)	45-53(49.0)	3-4(3.5)	2	
entrecasteauxii form A (V)	—	2-6(3.1)	59	Pengilley 1972
entrecasteauxii form B (V)	—	1-6(3.4)	153	Pengilley 1972
	—	1-7(4.1)	56	Rawlinson 1974a
Coeranoscincus				
reticulatus (O)	100-192(154.0)	2-6(4.2)	5	McDonald 1977; pers. obs.
Cryptoblepharus				
carnabyi (O)	32.5-46(38.7)	1-2(1.9)	37	G. Shea, pers. comm.
litoralis (O)	39-46(43.5)	1-2(1.5)	12	
plagiocephalus (O)	35.5-47(40.3)	2(2.0)	27	G. Shea, pers. comm.
	41(41.0)	2(2.0)	1	Chapman and Dell 1979, 1985
	—	—(2.0)	?	Davidge 1980
virgatus (O)	36-41(38.3)	1-2(1.9)	17	
Ctenotus				
ariadnae (O)	64(64.0)	4(4.0)	1	Pianka 1969b, *in litt.*
arnhemicus (O)	53(53.0)	4(4.0)	1	James 1983
	54(54.0)	3(3.0)	1	R. Sadlier, pers. comm.
atlas (O)	64-?(?)	1-2(1.5)	2	Pianka 1969b, *in litt.*
calurus (O)	42-?(?)	2-4(2.7)	6	Pianka 1969b, *in litt.*
colletti (O)	45(45.0)	2(2.0)	1	Pianka 1969b, *in litt.*
essingtoni (O)	52-70(58.9)	(2.9)	48	James 1983
	61-64(61.4)	2-4(3.0)	7	R. Sadlier, pers. comm.
fallens (O)	—	—(4.0)	?	Murray 1980
gemmula (O)	56	2(2.0)	1	Chapman and Dell 1975
helenae (O)	80-?(?)	1-6(3.5)	19	Pianka 1969b, *in litt.*
impar (O)	63(63.0)	2(2.0)	1	Chapman and Dell 1977, 1985
labillardieri (O)		2-4	—	Chapman and Dell 1975
	59-71(67.7)	3-5(4.1)	14	
leae (O)	59-?(?)	3-4(3.7)	3	Pianka 1969b, *in litt.*
leonhardii (O)	62-?(?)	2-7(4.7)	7	Pianka 1969b, *in litt.*
mimetes (O)	80(80.0)	6(6.0)	1	Dell and Chapman 1981b; Chapman and Dell 1985
pantherinus (O)	76-?(?)	3-9(6.1)	11	Pianka 1969b, *in litt.*
	—	6(6.0)	1	Smith 1976
	83(83.0)	3(3.0)	1	Chapman and Dell 1985
	75(75.0)	3(3.0)	1	Chapman and Dell 1980b

Table 9 (continued)

Taxon	SVL of gravid ♀♀ (mm)	Brood Size	N	Reference
Ctenotus continued				
schomburgkii (O)	46-?(?)	2-4(3.0)	6	Pianka 1969b, *in litt.*
storri (O)	39(39.0)	2(2.0)	1	R. Sadlier, pers. comm.
taeniolatus (O)	—	1-7(3.7)	?	Taylor 1985b
Cyclodina				
lichenigera (O)	62(62.0)	3(3.0)	1	
Egernia				
cunninghami (V)	—	4(4.0)	2	Matz 1968 and Petit therein
	—	1-7(4.5)	15	Niekisch 1980 and references therein
	—	2-7	?	Bartlett 1981
depressa (V)	—	2(2.0)	1	Stirling and Zietz 1893
	116(116.0)	2(2.0)	1	Day 1980
kingii (V)	—	2-8	?	Arena 1986
inornata (V)	—	3	1	Miller 1978
	74(74.0)	3(3.0)	1	Chapman and Dell 1979, 1985
	—	1-4(2.1)	32	Pianka and Giles 1982
mcpheei (V)	—	3	1	Wells and Wellington 1984
modesta (V)	—	1-5(2.7)	42	Milton 1987
multiscutata (V)	82(82.0)	1(1.0)	1	Chapman and Dell 1977, 1985
napoleonis (V)	—	1-2	?	Chapman and Dell 1975
	91(91.0)	4(4.0)	1	Dell and Chapman 1977
stokesi (V)	—	1(1.0)	1	Nankivell 1976
striata (V)	—	1-4(2.6)	19	Pianka and Giles 1982
striolata (V)	—	2-6(4.3)	6	Bustard 1970e
whitii (V)	—	3(3.0)	1	Adler 1958
	—	1-5(2.8)	27	Rawlinson 1974a
	—	1-5(2.8)	70	Milton 1987
Emoia				
atrocostata (O)	67-90(74.6)	1-2(1.9)	10	
cyanogaster (O)	83-92(88.7)	2(2.0)	4	
Eremiascincus				
richardsonii (O)	79-116(94.6)	3-7(4.6)	5	
Eugongylus				
rufescens (O)	121-124(122.5)	2-4(3.3)	4	
Eulamprus				
amplus (V)	—	5(5.0)	1	Covacevich and McDonald 1980
heatwolei (V)	78-98(87.5)	2-5(3.1)	73	Pengilley 1972
	—	3-5(4.0)	2	Millar 1978
kosciuskoi (V)	58-76(69.1)	1-6(3.1)	19	
leuraensis (?)	—	2(2.0)	1	Shea and Peterson 1985
luteilateralis (V)	80-90(85.0)	3-4(3.5)	2	
	—	5(5.0)	3	Covacevich and McDonald 1980
murrayi (V)	59-102(87.8)	2-5(3.6)	5	
quoyii (V)	86-108(100.8)	2-7(4.8)	21	data from Veron 1969a
tenuis (V)	—	7(7.0)	1	Antenor 1973
tympanum (V)	—	1-8(3.7)	67	Rawlinson 1974a
Eulepis				
duperreyi (O)	—	4-8(5.5)	11	Rawlinson 1974a (as *Leiolopisma trilineatum*)
	50-55(52.5)	5(5.0)	2	Morley 1985 (as *L. trilineatum*)
	54-71(60.1)	3-7(4.8)	12	
platynota (O)	65-80(71.6)	3-9(5.4)	5	
trilineata (O)	51-66(59.6)	3-6(4.5)	18	
	—	3(3.0)	1	Chapman and Dell 1975
Glaphyromorphus				
crassicaudus (O)	47-55(50.6)	1-4(3.0)	5	
darwiniensis (O)	52(52.0)	5(5.0)	1	James 1983 (as *Sphenomorphus crassicaudus*)
douglasi (O)	70(70.0)	5(5.0)	1	
fuscicaudis (O)	60-90(77.0)	2-4(3.0)	4	
gracilipes (V)	65-80(70.8)	2-6(3.8)	11	
isolepis (O)	—	4(4.0)	1	Loveridge 1949
	51-64(58.8)	3-8(4.8)	4	
pardalis (O)	67-68(67.5)	3-6(4.5)	2	
pumilus (O)	54-55(54.5)	3(3.0)	2	
Gnypetoscincus				
queenslandiae (V)	61-82(70.1)	2-6(3.2)	28	
Harrisoniascincus				
coventryi (V)	37-54(42.5)	1-4(2.5)	41	
	—	1-7(3.0)	15	Rawlinson 1975

Table 9 (continued)

Taxon	SVL of gravid ♀♀ (mm)	Brood Size	N	Reference
Harrisoniascincus (con.)				
zia (O)	54-59(56.7)	5-6(5.7)	3	
Hemiergis				
decresiensis (V)	—	2	1	Armstrong 1979a
	63-79(69.0)	1-8(4.5)	15	
initialis (V)	36-47(40.7)	1-2(1.7)	11	
millewae (V)	50-55(53.3)	2(2.0)	3	
peroni (V)	55-71(64.8)	2-4(3.1)	34	Smyth 1968
	—	2(2.0)	3	Chapman and Dell 1975
quadrilineatum (V)	53-66(58.4)	3-6(4.0)	5	
Lampropholis				
amicula (O)	29(29.0)	2(2.0)	1	
delicata (O)	—	1-4(3.0)	34	Rawlinson 1974a
	34-46	1-7(3.9)	60	Baker 1979 (introduced Hawaiian population)
	35-42	1-4(2.8)	6	Shea 1986
guichenoti (O)	38-49(44.1)	1-5(3.0)	32	Pengilley 1972
	—	1-3(2.3)	31	Simbotwe 1985
Lerista				
allanae (O)	73-89 (80.0)	2-3(2.3)	3	
bipes (O)	50-62(54.9)	2(2.0)	10	
bougainvillii				
mainland (O)	—	2-4(2.8)	6	Smyth and Smith 1974
	52-70(59.6)	1-4(2.8)	29	
Bass St., Tas. (V)	55-74(68.2)	2-4(2.7)	11	
	—	3(3.0)	3	Rawlinson 1974a
Kangaroo I. (V)	54-71(59.5)	1-3(2.0)	11	
borealis (O)	58-63(59.7)	3(3.0)	3	
christinae (O)	37(37.0)	2(2.0)	1	
cinerea (O)	65-72(68.0)	2-3(2.4)	5	
connivens (?)	78-86(81.7)	5-7(6.0)	3	
desertorum (O)	86(86.0)	3(3.0)	1	
distinguenda (O)	44-51(47.5)	2-3(2.5)	2	
	—	1(1.0)	1	Chapman and Dell 1975
elegans (O)	39-43(41.0)	2-4(3.0)	2	
	—	2(2.0)	?	Davidge 1980
fragilis (?)	47(47.0)	2(2.0)	1	
frosti (O)	49-68(56.1)	2-3(2.3)	6	
gerrardi (?)	80(80.0)	2(2.0)	1	
griffini (O)	58-65(61.6)	1-3(2.3)	7	
humphriesi (?)	58(58.0)	2(2.0)	1	
ips (O)	65(65.0)	2(2.0)	1	
karlschmidti (O)	—(63.6)	1-3(1.6)	5	James 1983
labialis (O)	51-54(52.7)	2(2.0)	3	
lineata (O)	43-50(47.4)	2-3(2.3)	9	
	—	2-3(3)	?	Davidge 1980
lineopunctulata (O)	94(94.0)	3(3.0)	1	
microtis (V)	45-53(50.8)	2-3(2.6)	5	
muelleri (O)	40-46(43.4)	1-2(1.9)	18	
picturata (O)	79-84(81.5)	2-4(3.0)	4	
planiventralis (?)	54-56(55.0)	2-4(3.0)	2	
punctatovittata (O)	85-97(90.5)	2-4(3.3)	7	
stylis (O)	62-75(70.0)	2-3(2.2)	9	
uniduo (O)	51-56(53.3)	2-3(2.3)	3	
vittata (?)	61-76(67.3)	2-3(2.6)	10	
walkeri (O)	62(62.0)	2(2.0)	1	
wilkinsi (?)	61-74(66.2)	1-3(1.9)	19	
xanthura (?)	46(46.0)	2(2.0)	1	
Lygisaurus				
burnettii (O)	30-34(31.4)	2(2.0)	12	
sp. 1 (O)	31(31.0)	2(2.0)	1	
sp. 2 (O)	29-36(32.3)	2(2.0)	37	
Menetia				
greyii (O)	—	1-3(2.0)	9	Smyth and Smith 1974
	—	—(2.0)	?	Murray 1980
	27-33(?)	1-2(1.9)	10	Chapman and Dell 1980a
	30-33(?)	1-2(1.7)	3	Chapman and Dell 1980b
	—	(2.0)	?	Davidge 1980
	28-30(?)	1(1.0)	3	Dell and Chapman 1981a
	30-37(34.0)	2-3(2.1)	8	

Table 9 (continued)

Taxon	SVL of gravid ♀♀ (mm)	Brood Size	N	Reference
Morethia				
adelaidensis (O)	44-58(49.0)	2-6(3.2)	33	
boulengeri (O)	48-57(?)	2-5(3.7)	28	Smyth and Smith 1974
	—	3-5(3.5)	11	Rawlinson 1976
	—	2-3(2.5)	2	Armstrong 1979b
	40-54(47.5)	1-6(3.2)	74	
butleri (O)	49-57(53.3)	2-5(3.8)	11	
lineoocellata (O)	38-57(46.1)	2-5(2.8)	32	
	57(57.0)	3(3.0)	1	Dell and Chapman 1977
obscura (O)	39-55(47.4)	1-5(3.1)	52	
	—	2-3	?	Chapman and Dell 1975
	47(47.0)	4(4.0)	1	Dell and Chapman 1979b
	41(41.0)	2(2.0)	1	Chapman and Dell 1980a
	—	—(3.0)	4	Chapman and Dell 1985
ruficauda (O)	31-42(36.2)	1-3(2.0)	12	
storri (O)	35-38(36.5)	2-3(2.1)	8	
taeniopleura (O)	36-44(39.2)	2-4(2.4)	14	
Nannoscincus				
maccoyi (O)	41-55(50.0)	1-9(3.5)	28	
Notoscincus				
wotjulum (O)	31-36(34.0)	2-4(2.8)	9	
sp. nov. (O)	34-41(37.5)	4(4.0)	2	
Ophioscincus				
cooloolensis (?)	67(67.0)	3(3.0)	1	
truncatus (O)	77-79(78.0)	2-3(2.5)	2	
ophioscincus (O)	68-97(88.7)	2-3(2.5)	4	
Proablepharus				
tenuis (O)	31-32(31.5)	1-2(1.5)	2	
reginae (?)	—	2(2.0)	1	Smith 1976
Pseudemoia				
spenceri (V)	—	1-3(1.9)	29	Rawlinson 1974b
	52-63(57.2)	1-4(2.5)	13	
Saiphos				
equalis (V)	58-87(74.8)	1-7(3.1)	49	
Saproscincus				
basiliscus (O)	42-46(44.0)	2(2.0)	2	
challengeri (O)	48-55(52.0)	3-4(3.5)	4	
czechurae (O)	33-36(34.5)	2(2.0)	2	
mustelina (O)	39-62(48.1)	2-7(3.7)	15	
tetradactyla (O)	31-33(32.2)	2(2.0)	5	
Techmarscincus				
jigurru (O)	62(62.0)	4(4.0)	1	Shea 1987b
Tiliqua				
branchialis (V)	—	5(5.0)	1	Smith 1976
	90(90.0)	4(4.0)	1	Dell and Chapman 1977
casuarinae (V)	—	19	1	Bustard 1970d
	—	2-7(4)	15	Rawlinson 1974a
	158(158.0)	13(13.0)	1	Timms 1977
	98-158(119.7)	4-14(6.8)	8	
gerrardii (V)	—	—(5.3)	3	Longley 1940
	—	18(18.0)	1	Longley 1938, 1941
	195(195.0)	33(33.0)	1	Wilhoft 1960
	—	13-22(16.7)	3	Morris *et al.* 1963
	—	17	1	Swan 1972
	195(195.0)	15(15.0)	1	Miles 1973
	—	5	1	Mudrack 1974
	—	21-49(35.0)	?	Anonymous 1977
	—	12-26(16.7)	6	Stephenson 1977
	—	23-53(38.0)	2	Barnett 1977a
	—	14-17(15.5)	2	Field 1980
	—	20-27(23.5)	2	Robertson 1980
	—	—(18.0)	?	Bartlett 1984
multifasciata (V)	—	2-7(4)	?	Christian 1977a
	—	3-6(4.4)		Barnett 1977b
nigrolutea (V)	—	3(3.0)	1	Fleay 1931
	—	8(8.0)	1	Green and McGarvie 1971
	—	3-8(5.6)	8	Rawlinson 1974a
	—	4-5(4.5)	2	Fowler 1978
	—	2-3(2.5)	2	Bartlett 1984
	—	3(3.0)	1	Green 1984
occipitalis (V)	—	7(7.0)	1	Chapman and Dell 1975

Table 9 (continued)

Taxon	SVL of gravid FF (mm)	Brood size	N	Reference
Tiliqua (continued)				
rugosa (V)	244-282(260.3)	1-3(2.0)	8	
	—	1-2(1.3)	20	Bamford 1980
	230(230.0)	1(1.0)		Chapman and Dell 1980a
	200(200.0)	1(1.0)		Chapman and Dell 1980b
	—	1-3(2.2)	9	Bull, in press
	—	21	1	Schafer 1979
scincoides (V)	227-308(266.5)	5-10(7.3)	4	Shea 1981 and data cited therein
	—	1-18(11.6)	17	Shea 1981 and data cited therein
	—	—(15.5)	?	Bartlett 1984
	—	6	1	Phillips 1986

Table 10. A very conservative estimate of the Australian skink lineages in which live-bearing habits have evolved.

Taxon	Specific Live-bearing Taxa
1. *Anomalopus*	*Anomalopus (Vermiceps) swansoni*
2. *Calyptotis* + *Saiphos*	*Saiphos*
3. *Carinascincus*	all
4. *Claireascincus* + *Pseudemoia*	all
5. *Cyclodina*	all New Zealand species
6. *Egernia* + *Tiliqua*	all
7. *Eulamprus* + *Gnypetoscincus*	all
8. *Glaphyromorphus*	*G. nigricaudis* (some)
9. *Harrisoniascincus*	*H. coventryi*
10. *Hemiergis*	all
11. *Lerista*	*Lerista bougainvillii, L. microtis*

Table 11. Relative clutch mass for Australian skinks.

Taxon	RCM 1			RCM 2			RCM 3			RCM 4			Reference
	X̄	R	SD	X̄	R	SD	X̄	R	SD	X̄	R	SD	
Carlia													
amax	—			—			—			.15		.02	James 1983 (N=5)
gracilis	—			—			—			.12		.04	James 1983 (N=18)
munda	—			—			—			.13		.03	James 1983 (N=10; as *C. foliorum*)
tetradactyla	.29			—			—			—			Pers. obs. (N=2)
	—			—			—			.13		.02	James 1983 (N=16)
triacantha	—			—			—			.13		.03	James 1983 (N=4)
Cryptoblepharus													
plagiocephalus	—			—			—			.19		.05	James 1983 (N=18)
	—			—			.22		.06	—			Pianka 1986 (N=7)
Ctenotus													
colletti	—			—			.19			—			Pianka 1986 (N=1)
essingtoni	—			.45		.20	—			.19		.03	James 1983 (N=4)
robustus													
tropical	—			—			—			.21		.02	James 1983 (N=3)
temperate	—			—			—			.22		.02	James 1983 (N=2)
pantherinus	—			—			.17		.04	—			Pianka 1986d (N=10)
quatturodecim-lineatus	—			—			.04		.01	—			Pianka 1986 (N=3)
saxatilis	—			—			—			.15		—	James 1983 (N=1)
taeniolatus	—			—			—			.23		.04	James 1983 (N=5)
Egernia													
inornata	—			—			.13		.05	—			Pianka 1986 (N=21)
striata	—			—			.09		.04	—			Pianka 1986 (N=18)
Eremiascincus													
richardsonii	—			—			.11			—			Pianka 1986 (N=1)
Eulepis													
platynota	.46			—			—			—			Pers. obs. (N=1)
Glaphyromorphus													
darwiniensis	—			—			—			.24			James 1983 (N=1; as *Sphenomorphus crassicaudus*)
douglasi	—			—			—			.15		.00	James 1983 (N=3)
isolepis	—			—			—			.17			James 1983 (N=6)
Lampropholis													
delicata	.41			—			—			—			Pers. obs. (N=1)
guichenoti	.36			—			—			—			Pers. obs. (N=1)
	—			—			—			.27		.05	James 1983 (N=3; after Pengilley 1972)
Lerista													
bipes	—			—			.18		.05	—			Pianka 1986 (N=3)
karlschmidti	—			—			—			.11		.03	James 1983 (N=4)
Lygisaurus													
burnettii	—			—			—			.22			James 1983 (N=1)
Morethia													
boulengeri	.46	.42-.50	.06	—			—			—			Pers. obs. (N=2)
storri	—			—			—			.17			James 1983 (N=1)
Tiliqua													
branchialis	—			—			—			.19			Pianka 1986 (N=1)

Table 12. Incubation times and temperatures for eggs of various species of Australian skinks. Means in parentheses.

Taxon	Incubation Period (days)	Incubation Temperature (°C)	Reference
Ctenotus *taeniolatus*	40	30	R. Shine in Taylor 1985b
Cyclodina *lichenigera*	68	25(avg.)	Cogger *et al.* 1983
Eulepis *duperreyi*	34-35(34.8)	26	Shine 1983 (as *Leiolopisma trilineatum*)
	29(29.0)	30	Shine 1983 (as *L. trilineatum*)
	37-38(37.9)	22-24	Morley 1985 (as *L. trilineatum*)
Lampropholis *delicata*	64-66(65.3)	20	Shine 1983
	35-42(36.8)	26	Shine 1983
guichenoti	150(150.0)	15	Shine 1983
	66-73(72.0)	20	Shine 1983
	35-45(38.3)	26	Shine 1983
	26-29(27.8)	30	Shine 1983
Menetia *greyii*	46-49(47.0)	28±4	Bush 1983
	104(104.0)	20(avg)	Bush 1983
Nannoscincus *maccoyi*	94-98(96.0)	15	Shine 1983
	58-61(59.5)	20	Shine 1983
	30-35(32.8)	26	Shine 1983

Table 13. Body temperatures (°C; FBT) for Australian skinks active in the field; heliothermic, surface active species.

Taxon	Mean	Range	SD or SE	N	Reference
Carinascincus					
greeni	27.4	24.3-32.3	2.50	16	Greer 1982
pretiosa	27.3	22.6-31.8	2.06	30	Greer 1982
Carlia					
longipes	30.8	?	±0.32	?	Wilhoft 1961 (as *C. fusca*)
rhomboidalis	28.9	—	±0.58		Wilhoft 1961
Claireascincus					
entrecasteauxii					
form B	29.8	22.9-34.3	3.77	21	Shine 1983 and pers. comm.
Ctenotus					
ariadnae	35.8	—	0.92	4	Pianka 1969b
atlas	34.5	—	3.96	23	Pianka 1969b
brooksi	30.6	—	2.81	61	Pianka 1969b
calurus	35.6	—	1.93	29	Pianka 1969b
colletti	35.9	—	1.83	4	Pianka 1969b
dux	32.0	—	2.32	45	Pianka 1969b
grandis	34.2	—	3.32	43	Pianka 1969b
helenae	32.8	—	4.13	64	Pianka 1969b
leae	37.7	—	1.72	18	Pianka 1969b
leonhardii	38.0	—	1.91	92	Pianka 1969b
piankai	35.5	—	3.31	10	Pianka 1969b
quattuordecim lineatus	35.8	—	2.67	79	Pianka 1969b
regius	36.4	35.7-37.1	±0.41	3	Bennett and John-Alder 1986
robustus	35.0	34.4-36.3	±0.67	3	Bennett and John-Alder 1986
schomburgkii	33.3	—	3.38	77	Pianka 1969b
taeniolatus	25.0	—	—	?	Heatwole 1976
Egernia					
cunninghami	30.1	28.3-32.5	±0.84	19	Barwick 1965
	31.6	—	—	6	Horton 1968
	32.9	—	1.70	19	Wilson and Lee 1974
kintorei	25.2	—	—	—	Pianka and Pianka 1970
modesta	27.7	—	—	5	Horton 1968
stokesii	36.0	—	—	1	Horton 1968
striolata	30.2	—	—	6	Horton 1968
whitii	31.9	—	—	7	Horton 1968
	33.1	30.8-33.5	—	—	Heatwole 1976
	34.0	31.1-35.6	±0.37	74	Johnson 1977
Eulamprus					
heatwolei	29.4	18.2-36.5	—	59	Spellerberg 1972c (as *Sphenomorphus tympanum* warm temperate form)
	29.0	20.8-32.3	2.58	20	Shine 1983 and pers. comm. (as *S. tympanum*)
kosciuskoi	28.2	25.8-29.6	±0.82	4	Bennett and John-Alder 1986
	30.9	16.9-36.0	—	37	Spellerberg 1972c
quoyii	28.1	22.2-33.9	—	112	Veron and Heatwole 1970
	30.0	17.4-34.2	—	42	Spellerberg 1972c
tympanum	29.3	25.2-33.0	±0.74	12	Bennett and John-Alder 1986
	30.2	16.3-34.5	—	83	Spellerberg 1972c
Eulepis					
duperreyi	28.3	23.5-31.7	—	?	Spellerberg 1972a and Heatwole 1976
	26.2	23.2-31.8	—	3	Shine 1983 and pers. comm. (as *Leiolopisma trilineatum*)
Harrisoniascincus					
coventryi	27.3	24.4-30.6	—	3	Shine 1983 and pers. comm.

Table 13 (continued)

Taxon	Mean	Range	SD or SE	N	Reference
Lampropholis *guichenoti*	26.8	21.1-34.7	4.00	9	Shine 1983 and pers. comm.
Pseudemoia *spenceri*	31.6	29.4-33.6	±0.45	12	Shine 1983 and pers. comm.
Tiliqua *occipitalis*	33.7	30.5-35.5	1.93	8	Licht *et al.* 1966a
rugosa	32.7	25.0-35.5	2.63	72	Licht *et al.* 1966a
	34.3	32.4-35.9	±0.34	10	Bennett and John-Alder 1986

Table 14. Body temperatures (°C; FBT) for Australian skinks active in the field; shade active, crepuscular to nocturnal, or cryptozoic to fossorial species.

Taxon	Mean	Range	SD or SE	N	Reference
Ctenotus *pantherinus*	33.1	—	3.11	67	Pianka 1969b
Cyclodina *lichenigera*	25.9	24.6-26.5	—	4	Cogger *et al.* 1983
Hemiergis *decresiensis*	20.3	17.3-23.0	±0.70	8	Bennett and John-Alder 1986
peroni	23.6	19.2-27.0	±0.68	12	Bennett and John-Alder 1986
Nannoscincus *maccoyi*	17.6	9.1-25.1	4.84	42	Shine 1983 and pers. comm.

Table 15. Preferred body temperatures (°C; PBT) for Australian skinks in a laboratory thermal gradient; heliothermic, surface active species.

Taxon	Mean	Range	SD or SE	N/n	Reference
Carinascincus					
greeni	28.9	21.5-37.3	—	6/1009	Rawlinson 1974a, 1975
metallicum	29.0	22.4-35.1	—	8	Rawlinson 1974a
ocellatum	30.7	23.2-37.3	—	8	Rawlinson 1974a
pretiosum	29.1	21.5-37.2	—	8	Rawlinson 1974a
Claireascincus					
entrecasteauxii form A	32.5	30.2-34.1	±1.03	3	Bennett and John-Alder 1986
entrecasteauxii form B	33.9	33.0-35.2	±0.78	4	Bennett and John-Alder 1986
Ctenotus					
atlas	35.5	34.5-36.6	—	1	Bennett and John-Alder 1986
labillardieri	32.2	28.5-35.6	3.36	11	Licht *et al.* 1966a
	33.9	—	±1.81	8	Williams 1965
lesueurii	32.6	28.1-35.8	1.97	—	Licht *et al.* 1966a
	35.2	—	±1.43	6	Williams 1965
regius	35.6	33.5-37.3	±0.46	13	Bennett and John-Alder 1986
robustus	34.4	33.4-35.1	±0.75	6	Bennett and John-Alder 1986
taeniolatus	35.3	33.4-37.8	±0.67	6	Bennett and John-Alder 1986
uber	35.3	33.2-37.2	±0.53	6	Bennett and John-Alder 1986
Egernia					
cunninghami	33.3	32.1-35.1	0.30	5	Licht *et al.* 1966a
	35.3	32.6-37.6	±0.14	14/97	Johnson 1977
	33.0	32.3-34.2	—	13	Johnson 1977
	32.5	—	±0.17	10/1000	Wilson and Lee 1974
	34.3	33.5-35.1	—	2/?	Horton 1968
depressa	34.0	30.8-36.3	1.45	4	Licht *et al.* 1966a
frerei	32.9	31.6-34.6	±0.10	1/64	Johnson 1977 (as *E. major*)
	34.4	32.3-36.1	—	1/?	Horton 1968 (as *E. major*)
major	34.7	32.5-36.4	±0.15	1/67	Johnson 1977 (as *E. bungana*)
	31.7	32.7-36.4	—	1/?	Horton 1968 (as *E. bungana*)
modesta	33.6	33.1-34.2	—	2/?	Horton 1968
napoleonis	34.7	32.7-36.7	0.64	5	Licht *et al.* 1966a (as *E. carinata*)
saxatilis	33.0	32.0-34.0	—	2/?	Horton 1968 (but most of the specimens probably *E. striolata*, G. Shea, pers. comm.)
stokesi	32.6	29.7-35.3	2.16	12	Licht *et al.* 1966a
striolata	32.7	31.3-33.1	±1.08	6	Bennett and John-Alder 1986
	34.8	32.0-37.3	—	9/?	Horton 1968
whitii	33.6	29.5-37.3	—	12	Rawlinson 1974c
	34.1	32.9-34.9	±0.67	5	Bennett and John-Alder 1986
	33.6	31.8-35.9	—	4/?	Horton 1968
Eulamprus					
heatwolei	29.0	19.6-35.3	—	10	Spellerberg 1972c (as *Sphenomorphus tympanum* warm temperate form)

Table 15 (continued)

Taxon	Mean	Range	SD or SE	N	Reference
Eulamprus (continued)					
kosciuskoi	29.8	27.2-32.1	±1.08	8	Bennett and John-Alder 1986
	30.8	21.6-35.1	—	7	Spellerberg 1972c
quoyii	28.8	28.3-29.7	±0.40	7	Bennett and John-Alder 1986
	29.6	22.7-34.4	—	12	Spellerberg 1972c
tympanum	29.1	—	—	16	Rawlinson 1974a
	29.6	27.4-32.0	±0.72	8	Bennett and John-Alder 1986
	29.8	19.4-35.0	—	16	Spellerberg 1972c
Harrisoniascincus					
coventryi	30.1	20.2-37.8	—	7/1136	Rawlinson 1975
Lampropholis					
delicata	26.1	18.6-33.6	—	4	Rawlinson 1974a
guichenoti	33.7	32.0-34.8	±0.36	10	Fraser 1980; Fraser and Grigg 1984
Lerista					
bougainvillii	30.7	21.6-38.0	—	4	Rawlinson 1974a
Morethia					
boulengeri	34.1	29.9-39.3	—	?	Rawlinson 1976
Pseudemoia					
spenceri	31.9	25.3-40.3	—	?/1106	Rawlinson 1974b
Tiliqua					
casuarinae	31.2	27.8-33.0	—	1	Bennett and John-Alder 1986
	32.6	27.1-39.0	—	6	Rawlinson 1974a
occipitalis	32.9	29.2-35.3	0.21	4	Licht *et al.* 1966a
nigrolutea	34.8	31.9-37.7	—	6	Rawlinson 1974a
rugosa	33.8	31.2-36.0	1.10	13	Licht *et al.* 1966a
	32.6	—	±0.62	10	Wilson 1974
	33.0	29.5-36.0	±0.82	9	Edwards 1978
	31.9	29.4-34.3	±0.75	10/1000	Bennett and John-Alder 1986
scincoides	33.1		2.1		Phillips 1986[1,3,5]
	30.8		1.3		Phillips 1986[1,3,4]
	27.4		1.2		Phillips 1986[1,4]
	37.5		1.6		Phillips 1986[2,3,5]
	37.1				Phillips 1986[2,3,6]
	31.1				Phillips 1986[2,4,5]
	28.9				Phillips 1986[2,4,6]

[1]Adults, [2]Juveniles, [3]Photophase, [4]Scotophase, [5]Fed, [6]Unfed.

Table 16. Preferred body temperatures (°C; PBT) for Australian skinks in a laboratory thermal gradient; shade active, crepuscular to nocturnal, or cryptozoic to fossorial species.

Taxon	Mean	Range	SD or SE	N	Reference
Eremiascincus					
fasciolatus	21.2[1]	18.0-23.9	±1.17	10	Bennett and John-Alder 1986
	24.9[2]	19.5-29.4	±1.61	10	Bennett and John-Alder 1986
richardsonii	24.6[1]	22.4-27.5	—	1	Bennett and John-Alder 1986
	27.3[2]	26.3-29.1	—	1	Bennett and John-Alder 1986
Hemiergis					
decresiensis	26.4	22.0-32.3	±1.9	44	Robertson 1981
	17.6[1]	—	±0.58	7	Bennett and John-Alder 1986
	24.8[2]	—	±1.37	7	Bennett and John-Alder 1986
peroni	20.3[1]	—	±1.57	10	Bennett and John-Alder 1986
	23.5[2]	—	±1.40	10	Bennett and John-Alder 1986
quadrilineatum	27.0	25.3-31.1	1.15	4	Licht *et al.* 1966a
Nannoscincus					
maccoyi	21.1	9.5-25.2	±4.6	21	Robertson 1981

[1]Morning and mid-day temperatures. [2]Late afternoon and early evening temperatures.

Table 17. Critical thermal minimum temperatures (°C; CTMin) for Australian skinks; heliothermic, surface active species.

Taxon	Mean	Range	SD or SE	N	Reference
Carinascincus					
metallicum	2.7	2.2- 3.4	—	9	Spellerberg 1972a
ocellatum	3.0	—	—	1	Spellerberg 1972a
Claireascincus					
entrecasteauxii	2.5	2.2- 2.8	—	7	Spellerberg 1972a
Ctenotus					
regius	8.7	—	±0.35	3	Bennett and John-Alder 1986
taeniolatus	11.4	—	±0.12	2	Bennett and John-Alder 1986
uber	9.1	—	±0.53	3	Bennett and John-Alder 1986
Egernia					
cunninghami	4.7	—	—	1	Spellerberg 1972a
inornata	8.9	—	—	1(juv.)	Spellerberg 1972a
	9.8	9.5-10.1	—	2(ad.)	Spellerberg 1972a
saxatilis	5.9	—	—	1(juv.)	Spellerberg 1972a
	4.5	4.0- 5.9	—	6(ad.)	Spellerberg 1972a
striolata	6.1	—	±0.47	3	Bennett and John-Alder 1986
whitii	4.0	3.4- 4.6	—	2(juv.)	Spellerberg 1972a
Eulamprus					
kosciuskoi	2.5	2.0- 3.2	—	36	Spellerberg 1972a
heatwolei	6.1	5.7- 7.2	—	42	Spellerberg 1972a
	4.5	4.0- 5.9	—	10	Spellerberg 1972a
quoyii	6.0	5.8- 6.4	—	52	Spellerberg 1972a
tympanum	2.9	2.2- 3.8	—	65	Spellerberg 1972a
	2.2	1.5- 2.6	—	16	Spellerberg 1972a
Eulepis					
duperreyi	3.0	2.9- 3.1	—	2	Spellerberg 1972a (as E. trilineatum)
Harrisoniascincus					
coventryi	2.8	2.8	—	2	Spellerberg 1972a (as Pseudemoia spenceri, fide Rawlinson 1975)
	1.9	1.5- 2.1	—	6	Spellerberg 1972a (as P. spenceri)
Lampropholis					
delicata	4.7	4.6- 5.1	—	4	Spellerberg 1972a
guichenoti	3.2	3.0- 3.6	—	11	Spellerberg 1972a
	1.9	1.4- 2.1	—	7	Spellerberg 1972a
Tiliqua					
nigrolutea	5.2	4.7- 5.8	—	3	Spellerberg 1972a
rugosa	4.5	4.0- 5.0	—	3	Spellerberg 1972a
	3.5	—	±0.18	3	Bennett and John-Alder 1986

Table 18. Critical thermal minimum temperatures (°C; CTMin) for Australian skinks; shade active, crepuscular to nocturnal, or cryptozoic to fossorial species.

Taxon	Mean	Range	SD or SE	N	Reference
Eremiascincus					
fasciolatus	9.0	—	±0.50	3	Bennett and John-Alder 1986
richardsonii	7.3	—	—	1	Bennett and John-Alder 1986
Hemiergis					
decresiensis	6.8	5.9-7.4	—	3	Spellerberg 1972a
peroni	9.6	—	±0.12	2	Bennett and John-Alder 1986
Nannoscincus					
maccoyi	4.0	3.4-4.6	—	2	Spellerberg 1972a
Saproscincus					
mustelina	6.5	6.0-7.1	—	2	Spellerberg 1972a
	5.9	—	—	2	Spellerberg 1972a

Table 19. Critical thermal maximum temperatures (°C; CTMax) for Australian skinks; heliothermic, surface active species. Note that where there are two sets of values for a single species in Spellerberg's (1972a) data the top row pertains to animals caught in summer and the bottom to animals caught in winter; where there is only one row the animals were caught in summer, except for *Carinascincus metallicum* which was caught in winter.

Taxon	Mean	Range	SD or SE	N	Reference
Carinascincus					
metallicum	39.6	39.2-40.0	—	7	Spellerberg 1972a
ocellatum	40.1	40.1	—	1	Spellerberg 1972a
Carlia					
longipes	40.3	40.3	—	1	Greer 1980a (as *C. fusca*)
munda	42.4	41.2-43.6	0.91	9	Greer 1980a (as *C. melanopogon*)
pectoralis	41.3	40.0-42.1	0.59	15	Greer 1980a
rhomboidalis	41.2	39.2-42.2	1.04	7	Greer 1980a
schmeltzii	42.3	41.7-43.4	0.64	7	Greer 1980a
vivax	41.3	40.5-42.5	0.87	4	Greer 1980a
Claireascincus					
entrecasteauxii	42.8	42.2-43.6	0.51	10	Greer 1980a
Cryptoblepharus					
litoralis	41.4	40.2-42.4	1.11	3	Greer 1980a
plagiocephalus	43.1	42.1-43.7	0.59	5	Greer 1980a
virgatus	41.1	39.8-42.0	1.15	3	Greer 1980a
Ctenotus					
essingtoni	45.0	44.4-45.7	0.59	4	Greer 1980a (as *Ctenotus* sp. 1)
regius	45.1	—	±0.29	3	Bennett and John-Alder 1986
spaldingi	45.4	44.2-46.3	1.12	3	Greer 1980a
taeniolatus	42.5	40.8-44.0	1.36	5	Greer 1980a
	44.7	—	±0.24	3	Bennett and John-Alder 1986
uber	45.5	—	±0.18	3	Bennett and John-Alder 1986
Egernia					
cunninghami	41.9	41.9	—	1	Spellerberg 1972a
inornata	42.8	42.8	—	1	Spellerberg 1972a
saxatilis	41.9	41.3-42.5	—	2	Spellerberg 1972a
	41.8	40.3-43.8	—	3	Spellerberg 1972a
striolata	44.2	—	±0.13	3	Bennett and John-Alder 1986
whitii	41.2	40.6-41.6	—	3	Spellerberg 1972a
	41.8	41.6-41.9	—	2	Spellerberg 1972a
	39.5	37.1-41.9	±1.3	21(ad.)	Johnson 1977
	38.2	36.3-40.6	±1.2	13(juv.)	Johnson 1977
	42.8	42.0-43.6	0.65	6	Greer 1980a
Eulamprus					
kosciuskoi	40.2	39.8-40.8	—	36	Spellerberg 1972a
quoyii	39.8	39.2-40.6	—	52	Spellerberg 1972a
	40.8	—	—	50	Veron and Heatwole 1970
	37.1	37.1	—	1	Brattstrom 1971b
heatwolei	40.0	39.4-40.6	—	42	Spellerberg 1972a
	39.0	38.6-39.4	—	10	Spellerberg 1972a
tympanum	39.8	39.0-40.6	—	65	Spellerberg 1972a
	38.3	38.2-38.5	—	14	Spellerberg 1972a
Eulepis					
duperreyi	43.5	43.5	—	2	Spellerberg 1972a (as *E. trilineatum*)
platynota	43.6	43.2-44.0	0.29	6	Greer 1980a
Harrisoniascincus					
coventryi	40.2	39.4-41.0	0.55	7	Greer 1980a
	42.3	42.0-42.5	—	2	Spellerberg 1972a (as *Pseudemoia spenceri*; Rawlinson 1975)
	41.8	41.3-42.5	—	6	Spellerberg 1972a (as *P. spenceri*)

Table 19 (continued)

Taxon	Mean	Range	SD or SE	N	Reference
Lampropholis					
delicata	40.8	40.6-41.3	—	7	Spellerberg 1972a
guichenoti	41.0	40.7-41.7	0.58	3	Greer 1980a
	42.0	41.3-42.1	—	17	Spellerberg 1972a
	40.7	40.6-41.3	—	4	Spellerberg 1972a
n. sp.	37.3	36.9-37.6	0.49	2	Greer 1980a (as *Lampropholis* sp. 2)
n. sp.	38.5	38.2-39.1	0.35	5	Greer 1980a (as *Lampropholis* sp.6)
Lygisaurus					
burnettii	38.7	38.7	—	1	Greer 1980a
Morethia					
butleri	44.6	43.6-45.2	0.69	7	Greer 1980a
lineoocellata	44.6	44.0-44.8	0.51	4	Greer 1980a
obscura	42.3	41.0-42.9	0.59	11	Greer 1980a
taeniopleura	42.0	41.2-43.7	0.77	8	Greer 1980a
Tiliqua					
branchialis	41.5	?	—	?	Warburg 1965b
	43.6	43.6	—	1	Greer 1980a
nigrolutea	42.5	41.0-43.0	—	3	Spellerberg 1972a
rugosa	>41.7	>41.7	—	1	Warburg 1965a[1]
	43.0	—	±0.26	3	Bennett and John-Alder 1986
scincoides	43.9	43.9	—	1	Heatwole *et al.* 1973

[1]Based on the observation that one animal survived for 2 h at a body temperature of 41.7°C.

Table 20. Critical thermal maximum temperatures (°C; CTMax) for Australian skinks; shade active, crepuscular to nocturnal, or cryptozoic to fossorial species.

Taxon	Mean	Range	SD or SE	N	Reference
Anomalopus					
brevicollis	36.3	35.2-37.0	0.65	5	Greer 1980a (as *Anomalopus* sp. 1)
leuckartii	35.8	35.8	—	1	Greer 1980a (as *Anomalopus lentiginosus*)
Calyptotis					
temporalis	34.5	34.5	—	1	Greer 1980a (as *Sphenomorphus* sp. 1)
Ctenotus					
pantherinus	36.9	36.9	—	1	Heatwole 1976
Eremiascincus					
fasciolatus	41.6	41.6	—	1	Greer 1980a
	41.2	41.2	±0.32	3	Bennett and John-Alder 1986
richardsonii	42.0	42.0	—	1	Bennett and John-Alder 1986
Eulamprus					
murrayi	35.1	35.1	—	?	Heatwole 1976
	36.4	35.9-36.9	0.36	6	Greer 1980a
tenuis	39.2	38.3-40.1	0.64	6	Greer 1980a
Glaphyromorphus					
punctulatus	38.6	37.7-39.7	0.68	14	Greer 1980a
Gnypetoscincus					
queenslandiae	35.9	34.4-37.9	1.22	6	Greer 1980a
Hemiergis					
decresiensis	32.4	32.0-33.5	—	3	Brattstrom 1971b
	39.3	39.2-39.4	—	2	Spellerberg 1972a
	38.6	—	±0.34	4	Bennett and John-Alder 1986
peroni	38.6	37.0-40.2	0.76	28	Greer 1980a
quadrilineatus	38.1	37.9-38.5	0.32	3	Greer 1980a
Lerista					
bougainvillii	43.8	42.4-45.0	0.67	12	Greer 1980a
frosti	42.9	42.9	—	1	Greer 1980a
neander	43.1	43.1	—	1	Greer 1980a
orientalis	40.9	40.9	—	1	Greer 1980a
terdigitata	45.1	43.3-47.0	1.57	5	Greer 1980a
Nannoscincus					
maccoyi	37.0	36.4-38.0	—	3	Spellerberg 1972a
Saiphos					
equalis	37.3	36.1-39.0	0.90	9	Greer 1980a
Saproscincus					
basiliscus	35.5	34.7-37.0	0.72	7	Greer 1980a (as *Lampropholis* sp. 4)
mustelina	38.6	38.2-38.8	—	3	Spellerberg 1972a
	36.9	35.7-37.5	—	5	Spellerberg 1972a
tetradactyla	35.3	34.5-36.2	0.74	5	Greer 1980a (as *Lampropholis* sp. 5)

Table 21. Data on reproductively active females in seven, far-northern *Lerista* species. Sample size for each species pertains to all the females as large or larger than the smallest female with either yolking follicles or oviducal eggs, except for *L. orientalis* where all females in the sample were counted.

Species and Locality	Date	N	% Reproductively Active		
			Inactive	Ovarian	Oviducal
L. orientalis					
Qld	24 June-5 July	39	100	—	—
N.T.	30-31 July	10	100	—	—
L. borealis	22-26 July	4	25	25	50
L. griffini	25-26 August	10	30	40	30
	2-4 October	4	75	25	—
L. karlschmidti[1]	August-September[2]	5	—	+	+
L. simillima	7 September	4	75	25	—
L. greeri	5, 9 September	6	50	50	—
L. bipes	3-15 September	19	32	26	42

[1]Data from James (1983). [2]A 24 July specimen in the Northern Territory Museum, Darwin is "gravid" (pers. obs.).

CHAPTER 6

Varanidae — Goannas

VARANIDS are medium to large-sized lizards with well-developed limbs and tails, lanky bodies, long necks, snake-like tongues and small, relatively uniform-sized scales which render the skin tough but supple. They are active, largely diurnal animals with predatory and scavenging habits. There are terrestrial, saxicolous, arboreal and semi-aquatic species, and they occur in a variety of habitats ranging from mangroves and closed forests to arid sand plains. Varanids are perhaps the most distinctive and easily recognizable group of lizards, both popularly and technically (Table 1), in the world. In Australia all varanids are called goannas, a corruption of iguana, a Spanish and English word for certain large species of a primarily New World family of lizards, the Iguanidae (p. 9). Another common name, widely used elsewhere but occasionally also in Australia, is monitor.

There are approximately 35 species of varanids living today and all are placed in the genus *Varanus*. The family occurs in Africa, southern Asia, the Philippine and Indonesian Archipelagos, New Guinea, the Solomon Islands and Australia. The number of species increases from west to east with the largest number by far occurring in Australia. Thus there are only two species in Africa, 13 in Asia, the Philippines, the Indonesian Archipelago, New Guinea, the Solomons and the neighbouring Pacific Islands, and 24 in Australia, of which 20 occur only in Australia (Mertens 1942a-c, 1957, 1958b, 1966; Storr 1966b, 1980a). The usual reason given for the large number of species in the Australian Region is the dearth of small to medium-sized, carnivorous native mammals such as cats, weasels and dogs, which has allowed the more or less uninhibited evolution of this group of carnivorous lizards (Storr 1964b; Pianka 1973, 1981). However, this "explanation" may need review in light of increasing knowledge of the now extinct, mid to late Tertiary carnivorous mammal fauna (also see p. 6).

Varanids vary greatly in size (Table 2). The smallest living species is probably *Varanus brevicauda* from the Australia deserts; at its largest it measures 118 mm for the snout-vent length and 230 mm for the total length, and weighs 15 gm. The largest living Australian varanid is *Varanus giganteus*, the Perentie, also from the central deserts; it attains a snout-vent length of 750 mm, a total length of 1750 mm (Mertens 1958b), and a weight of 17 kg (Butler 1970). The largest living varanid, *V. komodoensis* from Komodo Island in the Indonesian Archipelago, reaches a total length of approximately 3 m and weight of 250 kg (Auffenberg 1981). However, this species was dwarfed by the extinct *Megalania* from the Pleistocene of Australia, which, by extrapolation from proportions of parts, e.g., limb bones, to the whole, has been calculated to have reached a snout-vent length of 4.5-5.0 m, a total length of close to 7 m (Hecht 1975) and a weight of 2200 kg (Auffenberg 1981). Yet despite these huge size differences, all varanids look more or less alike, differing only in details such as relative lengths of snouts and tails, shape and spininess of the tail, and position of the nostrils on the snout. Clearly the basic varanid body plan is functional across a remarkable size range.

Osteologically varanids are also rather uniform. Variation in the Australian species is confined to premaxillary tooth number (nine in most species, seven in *Varanus acanthurus* and *V. storri*), state of fusion of nasal bones (Table 3), variation in presacral and postsacral vertebral number (Table 4), and variation in the degree of expression of the osteoderms on the head and tail (poorly or not at all developed in most species, heaviest in *Varanus acanthurus* and *V. storri*).

Phylogenetic relationships within varanids remain poorly known despite having been assessed from a number of different approaches: morphological (Mertens 1942a; Branch 1982);

behavioural (Horn 1985); karyotypic (King and King 1975; King *et al.* 1982), and biochemical (Holmes *et al.* 1975). The morphological and behavioural approaches have fallen short, because the characters have not been thoroughly or comprehensively analysed in terms of primitive and derived character states, and the karyotypic and biochemical approaches are lacking because all attempts to date to infer the polarity of character states are based, at least in my view, on unrealistic assumptions[1]. Therefore, in the absence of a phylogenetic arrangement for goannas, an ecological one based on habitat may be most useful.

The habitats of Australian goannas fall into four broad categories recognizable by general physical features that transcend the usual botanical or geographical characterizations of habitat type (Cogger 1959). First, there is the arboreal/aquatic habitat, that is, trees or shrubs in and around water. Here the goannas often shelter and bask in the vegetation, feed in and around the water (consequently their diet features many aquatic animals) and will often flee into the water when disturbed. These species include *Varanus indicus*, *V. mertensi* Schürer and Horn 1976; Shine 1986a and b; Fig. 90), *V. mitchelli* (Worrell 1963; Peters 1968, 1969b; Shine 1986a and b), and *V. semiremex* (Peters 1969a,d).

The second general goanna habitat in Australia is the rocky outcrop. Here the goannas both shelter and forage in and around the crevices, caves, boulders and slabs that characterize this structural feature. These species include *Varanus acanthurus*, (Cogger 1959; Thompson and Hosmer 1963; Swanson 1979; King *et al.* 1982; pers. obs.), *V. glebopalma* (Mitchell 1955; Christian 1977; Horn and Schürer 1978; Swanson 1979; Fig. 84), *V. glauerti* (Storr 1980a), *V. kingorum* (Storr 1980a; Weigel 1985, as *Varanus* n. sp.), *V. pilbarensis* (Storr 1980a; Johnstone 1983), *V. primordius*, and *V. storri* (Peters 1969d, 1973a-b; Fig. 85). Here too we might include *V. giganteus* (White in Zietz 1914; Cogger 1959, 1965; Martin 1975; Brunn 1981, 1982; Pianka 1982) which, although

Fig. 84. *Varanus glebopalma,* a rather thin, lanky species of rock goanna. Photo: P. R. Rankin.

it may forage widely on nearby sand-plains, appears to shelter in and around rocky habitats (but see Pianka 1982), perhaps the only place it can find retreats suitable for its large size.

Third is the arboreal habitat. Here goannas rely heavily on trees and large shrubs as places to forage and shelter. They spend much time on the ground, although they seem to be most at home off the ground in vegetation. In a sense this is similar to the arboreal/aquatic habitat minus the aquatic component. The goannas in this group include *Varanus caudolineatus* (Schmida 1975; Storr and Harold 1980), *V. gilleni* (White in Zietz 1914; Mertens 1958; Thomson and Hosmer 1963; Cogger 1965; Martin 1975; Schmida 1975; Brunn 1980b; Delean 1980; Roberts 1985), *V. prasinus* (Czechura 1980; Carlzen 1982), *V. timorensis* (Peters 1969; Schmida 1971; Cogger 1973; Cogger and Lindner 1974; Christian 1981; Shine 1986b), *V. tristis* (White in Zietz 1914; Pianka 1982; Stammer 1983; Fitzgerald 1985; Shine 1986b), and *V. varius* (Irvine 1957; Cogger 1959; Worrell 1963; Bustard 1968e; Horn 1980; Brunn 1982; Fig. 86).

[1]The assumptions regarding the karyotypes were as follows: 1) ". . . the karyotype which is common to the greatest number of taxa, and that from which the most simple derivative karyotypes can be produced, is the primordial form."; 2) ". . . that the direction of change has been from metacentricity to acrocentricity in all large chromosomes . . .", and ". . . that less weight has been applied to the phylogenetic significance of the microchromosomes because of their poor resolution . . . and that changes in these elements have been from acrocentricity to metacentricity" (King and King 1975, and virtually repeated in Holmes *et al.* 1975). The first assumption is not valid because the most common form of a trait could have resulted from speciation in that part of the lineage showing the trait as a derived condition within the lineage, or as King (1981a) has remarked "it is important to realize that a predominant karyotypic form within a lineage or group of lineages is not necessarily the ancestral form . . . ." The second and third assumptions seem to contradict each other because it is difficult to see why large chromosomes should change in one direction and smaller ones in the opposite direction. Furthermore, with regard to the second assumption, i.e., change from metacentricity to acrocentricity, King (1981a) has opined elsewhere that "the most parsimonious model for chromosome evolution in lizards argues for an acrocentric ancestral karyotype with evolution primarily by fusion . . ." i.e., acrocentricity as primitive instead of derived.

Fig. 85. *Varanus storri,* a small species of rock goanna. Note the spiny tail.

Fig. 86. *Varanus varius*, a species of arboreal goanna, basking in late afternoon sunshine.

Crevices and holes in trees can often be limited in number and hence contested by species that use them. Usually this competition is most severe between members of the same species because of close similarity of size, shape and habits. However, it can also occur occasionally between distantly related species as the following anecdote told by H. G. Cogger as a result of his field work on the Coburg Peninsula in the far north of the Northern Territory. "I had a common but interesting experience while collecting in the vicinity of this and other swamps. Suddenly the air would be rent by a piercing and fearful scream, which, the first time I heard it, was only a few feet from my ears and gave me a terrific fright. On each occasion the screaming was found to emanate from the limb of a nearby tree, where, after some effort with an axe, the culprits were always found to be a tree monitor or goanna, *Varanus timorensis*, and one or more Green Tree Frogs *(Litoria caerulea)*. The goanna, on entering the hollow limb, would find its way blocked by one, or more sleeping tree frogs. Unwilling, or sometimes unable, to reverse in the confined hollow, the goanna would try to force its way past the loudly protesting frogs. This was so common that by the end of a month I regarded it as an almost daily event. We never found any evidence that the goannas might be feeding on the frogs" (Cogger 1973).

Finally, there is the ground habitat. Here goannas spend most of their foraging time on the ground, and for retreats they use either burrows of their own or another animal's making, or hollow logs on the ground. The goannas in this group are *Varanus brevicauda* (Schmida 1975), *V. eremius* (Pianka 1982), *V. gouldii* (White in Zietz 1914; Cogger 1959 and 1965; Thomson and Hosmer 1963; Martin 1975; Brunn 1980c; King 1980; Bush 1981; ianka 1982; Shine 1986a-b), *V. panoptes* (Storr 1980a; Shine 1986a-b; pers. obs.), *V. rosenbergi*

(Tubb 1938; King and Green 1979; Brunn 1980c; King 1980; Bush 1981) and *V. spenceri* (Pengilley 1981).

In none of the species of the preceding group is it known to what degree the animals dig their own burrows. Actual observations of burrow construction have been made only in *Varanus brevicauda* (Schmida 1975) and *V. storri* (Bartlett 1982). Burrows thought to have been constructed by *V. rosenbergi* have been described as "quite regular in size and arrangement of passages". They are 9 to 9.3 m long and about 15 cm in diameter. "The main gallery is L-shaped, with a second L-shaped gallery opening near the angle" of the first. The roofs of the burrows examined were never more than 30 cm below the ground (usually 15-22 cm) (Tubb 1938). The regularity and distinctive shape of these burrows is intriguing and would be worth confirming in this species and others that dig their own burrows. A *V. gouldii* burrow on a sand plain was about 1.3 m in length, ran about 10 cm below the surface for its entire length, and was ∼-shaped when viewed from above. The burrow ended just below the surface so that the goanna could "pop" out if approached from behind via the main entrance (pers. obs.)

Not all species adhere strictly to their usual or general habitats as given above. Some of the rocky outcrop species may become ground dwelling in areas lacking rock outcrops (e.g., *Varanus acanthurus* — Swanson 1979; King *et al.* 1982) while certain arboreal species may also occur on rocks (e.g., *V. caudolineatus* — Pianka 1969c and *V. tristis* — White in Zietz 1914; Pianka 1971b; Low 1978; Christian 1981; Brunn 1982; Fitzgerald 1985) or on the ground (e.g. *V. tristis* — Christian 1981). However, in general, the habitat associations described above appear to be remarkably tight.

Some ground goannas appear to be associated with certain soil types. For example, *Varanus panoptes* is said to occur largely on hard-packed soils whereas *V. gouldii* occurs largely on looser, sandy soils (Storr 1980a; Shine 1986b), and *V. spenceri* apparently occurs only on grass-covered, black cracking clay soils (Pengilley 1981).

Certain morphological and physiological features of goannas appear to be directly related to their habitat type. Some aquatic species such as *Varanus mertensi* and *V. mitchelli* have the nostrils placed dorsally, allowing the animal to breathe while the rest of the head and body is underwater (Cogger 1959); they may also have the tail strongly laterally compressed to aid in swimming. Some arboreal species such as *V. gilleni* (Bustard 1970d) and *V. prasinus* (Czechura 1980) have a prehensile tail and *V.*

gilleni is said to have "more strongly curved" claws than *V. eremius*, a similar-sized ground dweller (Pianka 1969c). Some of the rock dwelling species, such as *V. glebopalma* (Mitchell 1955: fig. 3; Swanson 1979) and some of the arboreal species, such as *V. prasinus* (Czechura 1980; Greene 1986) have a very dark substance on the soles of the feet which is thought to give increased traction to these climbing species. Another rock dwelling species, *V. acanthurus*, is said to use its spiny tail to help keep itself wedged in crevices when grasped by a predator (Swanson 1979).

Goannas living in habitats lacking fresh water also show a physiological adaptation to dealing with one consequence of this shortage. The reptile kidney can not secrete urine more concentrated than the blood plasma, hence in situations where fresh water, for flushing the salt, is hard to find, for example during periods of drought or life in or around the sea (where the water is more saline than blood), reptiles may have a problem with the increasing levels of salt in the body. As alluded to above (p. 26), certain arid zone agamids simply tolerate increasing body salt concentrations until the rains come. Arid zone goannas such as *Varanus gouldii* (Green 1972a-b), and mangrove inhabiting goannas, such as *V. semiremex* (Dunson 1974), have another method of coping: a nasal salt gland. This gland excretes salt in concentrations higher than the blood plasma and requires no fresh water to function. Certain other lizards, notably all from arid or littoral environments, where potable fresh water is scarce, also have such a gland, e.g., *Tiliqua rugosa*.

The nasal gland has a distinctive, "striated" ultrastructure which makes it readily identifiable, and this feature has been found in various degrees of development in different goanna species: little developed — *Varanus giganteus*, moderately — *V. rosenbergi*, and well — *V. acanthurus* and *V. gouldii* (Saint-Girons *et al.* 1981). Although this is strong circumstantial evidence for a salt gland in these species, an actual excretory ability has yet to be proved. The salt gland is lacking in one overseas varanid that lives along the rivers and hence, presumably, has no problem in obtaining fresh water. From this, one would expect the fresh water goannas such as *V. mitchelli* and *V. mertensi* to also lack the gland. In the case of the mangrove-inhabiting *V. semiremex*, the ability to get rid of "excess salt" is so great that the animals can live indefinitely without drinking water, the "desalinated water" processed from the prey being adequate to sustain them (Dunson 1974).

Goannas are basically diurnal. To judge from the best studied species such as *Varanus gouldii* and *V. varius*, they emerge from their retreats only after sunrise, bask either at or near the entrance to their retreats to raise their body temperature and once the desired temperature is attained, set off on their activities until sometime well before sunset when they re-enter their retreats for the night. This basic routine is varied with later emergences and earlier retreats, or with no emergence at all, during winter, and with a period of mid-day retreat during summer. During activity, they maintain body temperatures at remarkably uniform levels, usually averaging about the mid-30's (°C) (Stebbins and Barwick 1968; D. King 1980; Pianka 1982: table 5; Tables 10-11). Basking *V. tristis* may curl the tail over the back (Christian 1981), but the significance of this behaviour is unknown. The fact that *V. tristis* may also hold the tail in the same way when moving across flat terrain (Fyfe 1979, 1980) suggests it may have some other function than thermoregulation; perhaps it provides protection from flying predators.

Although goannas are primarily active during the day, a few have been reported active after dark. *Varanus glebopalma* is said to forage primarily "during the first couple of hours after sunset" (Christian 1977b; see also Armstrong 1981) and *V. tristis* has been reported to feed at night around outside building lights as late as midnight (Fyfe 1979, 1980). In the north, goannas are active year around, i.e., they are seen on the surface, although they are more commonly seen during the summer wet than during the winter dry (Shine 1986b). However, in the south only the smaller species and the island species, *V. rosenbergi*, are active throughout the year; the larger mainland species appear to become dormant during the winter. A *V. varius* (origin not stated) kept under warm conditions during a Sydney winter, lost interest in food, indicating a possible circannual rhythm in physiology (Peters 1970a). The smaller species in the south can apparently remain active during the winter because their small body size allows them to heat up in the limited amount of time available for basking in the winter, and the island populations possibly remain active due to the milder maritime climate and perhaps to their darker colouration which facilitates heating. During the summer, goannas in the south may avoid the mid-day heat by entering their retreats (Green 1972b) or in captivity, their water dishes (e.g., *V. varius* — Peters 1970a). *V. rosenbergi* on Kangaroo Island has been shown to shift its activity areas seasonally; in summer the animals occur in more open exposed areas while in winter they shift to more heavily vegetated or better drained areas (Green and King 1978).

There is evidence to suggest that at least some species of goannas have relatively large home areas which they know well and in which they have one or more specific retreats. For example, some *Varanus varius* seem to have a large home area which they "tour", spending a few days in each section and sheltering in a specific tree hollow each night (Stebbins and Barwick 1968). That these home areas may be long standing is suggested by the fact that one individual apparently inhabited the same tree hollow for at least three years (Frauca 1966). In *V. rosenbergi* the home areas often overlap and different animals may use the same burrows on different nights or, even occasionally, the same night (Green and King 1978).

Distances covered in a day by a foraging goanna can be quite large. For example, *Varanus eremius*, has been tracked for distances of "up to half a mile" (Pianka 1968), *V. varius* for three-quaters of a mile (Stebbins and Barwick 1968), *V. gouldii flavirufus* for "over a mile" (Pianka 1970b) and *V. tristis* "nearly a mile" (Pianka 1971b).

When active, goannas usually move slowly, but steadily and methodically, with the long, narrow tongue flicking in and out of the mouth sensing the environment. They investigate crevices and disturbed areas, enter burrows, hollow logs and rock crevices, stick their snouts here and there, and scratch and dig. They have a well-developed sense of smell and use it to find buried prey or food, such as insect larvae (pers. obs.), scorpions, spiders and geckos in burrows (Koch 1970; H. Ehmann pers. comm.), reptile eggs in nests (Cogger 1973), and buried carcasses and corpses. Occasionally, at least some species of intermediate-sized goannas will stand up on their rear legs and look around, e.g., *Varanus gouldii* (Glazebrook 1977; pers. obs.) and *V. panoptes* (Schmida 1985:75 as *V. gouldii*).

The food eaten by goannas has been determined from studies of both captive and free ranging (Table 5) specimens. Two generalizations can be made from these studies: goannas are largely carnivorous, and they are ultimately opportunistic, i.e., they will eat almost any animal they can catch and overcome. Actual prey of wild animals runs heavily to medium and large sized invertebrates and small vertebrates. Not surprisingly, perhaps, larger species of goannas take, on average, larger prey (Pianka 1982). Some goannas eat reptile (Butler 1970; Cogger and Lindner 1969, 1974; King and Green 1979) and bird eggs (King and Green 1979; Horn 1980), the latter habit often earning them the wrath of farmers and bird keepers. Some goannas also eat carrion (Kennerson 1980) which puts them at risk from highway

traffic and poisoned baits. Goannas are also occasionally cannibalistic both in confinement (Cogger 1959; Mertens 1963; Johnson 1976) and in nature (King and Green 1979; Shine 1986b). There is apparently one authenticated case of a goanna *(Varanus varius)* eating plant matter, specifically, berries (Vincent 1981).

The foraging habits of *Varanus mertensi* are distinctive and in keeping with its strongly aquatic habits. Apparently when undisturbed they walk underwater along the bottom just as other species would on land (Swanson 1976) and in shallow water they "corral" fish by encircling them with the body and tail and then turning the head into the corral to feed (Hermes 1981). In captivity, they hunt fish with the head under water, the eyes open and the tongue flicking in and out (Schürer and Horn 1976); they also lunge at fish from above the water.

The only clear case of prey selectivity to come to notice to date among goannas is with *Varanus brevicauda*. We have noticed in the lab that whereas this species readily devours cockroaches, mealworms, skinks and geckos (see also Schmida 1975), it refuses hairless pink baby mice. When presented with "pinkies" it becomes alert and rushes over to investigate, but after tonguing them a couple of times, backs off and loses interest. This may be the only goanna that will not eat mammalian flesh (see also Pianka 1970a; Schmida 1974); certainly other small goannas such as *V. acanthurus* (Husband 1979b), *V. gilleni* (pers. obs.) and *V. storri* (Peters 1973a; Stirnberg and Horn 1981; Bartlett 1982) will eat baby mice.

Goannas appear to have prodigious stomach capacities. For example, upon capture a 1.2 kg *Varanus varius* regurgitated a 500 gm rabbit carcass representing 42% of the lizard's weight (Weavers 1983) and a 20 kg specimen of the same species contained four fox cubs, three young rabbits and three large blue-tongue skinks (Fleay 1950).

Despite their seemingly "eat almost anything" habits there is apparently at least one "natural" prey item they can not eat, the introduced cane toad *(Bufo marinus)*. A captive *Varanus gouldii* died in agony 20 minutes after having bitten and mouthed a toad (Stammer 1981).

The teeth in most varanids are laterally compressed, sharp-pointed, recurved and, in the larger species at least, finely serrated along the anterior and posterior edges (Figs 87-88). This would appear to be a cutting and severing tooth instead of a crushing tooth as seen in the peg-like teeth of scincids and certain pygopodids. Varanids subdue their prey not only by lacerating

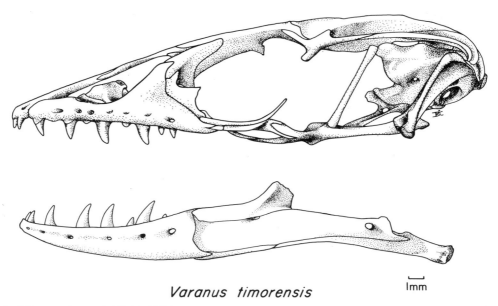

Varanus timorensis

⊢ 1mm

Fig. 87. Skull of *Varanus timorensis* (AM R 12371).

it with their teeth, but also by bashing it against the substrate. Prey held in the mouth is smashed and scraped against the ground, rock or tree first on one side and then on the other.

Varanids have a method of orienting prey in the mouth prior to swallowing it that is unusual in lizards. It is known by the slightly imposing name of inertial feeding. In this the food in the mouth is either held under a loosened grip or actually given a slight backward (toward the throat) acceleration and then the head "thrown" forward over the food against the weight of the stationery or posteriorly moving food (Longley 1947a; Gans 1961, 1969; Smith 1982). The behaviour is the same in principle, if not in appearance, to a person lunging with an open

Fig. 88. Tooth from the middle of the right maxilla of *Varanus varius* (AM R 3699) to show the serrated edges. Scale=2 mm. Photo: G. Avern.

mouth to catch falling grapes or peanuts that have been tossed into the air. The reasons for this peculiar feeding behaviour in varanids may be related to the relatively small size of the tongue. Most other lizards have a large, fleshy tongue with which they push food through the mouth and back into the throat. However, the highly modified, slender tongue of varanids is too "weak" to serve this role, hence the development of the alternate method of feeding. Once prey is past the mouth, it is pushed down the long throat by sinuous, posteriorly moving contractions of the neck muscles ("throat kinking"). This is similar to the swallowing behaviour of the pygopod *Lialis* (p. 107).

The inguinal, or visceral, fat bodies of three desert species of goannas — *Varanus giganteus* (Stirling 1912), *V. gouldii* (Pianka 1982), and *V. tristis* (Pianka 1982) — have been commented upon for their large size, and Pianka (1982) has suggested they serve as a useful energy reserve in the unpredictable environments of the interior. This interesting idea would be worth pursuing with further anatomical observations and perhaps some experimental work.

Goannas are usually encountered by themselves which leads to the impression that they are rather solitary animals. However, certain species have been found together in relatively large numbers (e.g., *Varanus gilleni* — pers. obs., *V. glebopalma* — Christian 1977; *V. storri* — Bustard 1970d (as *V. acanthurus brachyurus*), raising the question of the nature of the aggregations. They are probably the result of either short-term associations formed during courtship or of several animals independently finding and settling in a particularly favourable part of the habitat. However, should the opportunity ever arise, long term study of a locally dense population of goannas might reveal more complex and stable network of interactions than herefore suspected.

The sex ratio among "random" samples of wild-caught goannas is often biased in favour of males (e.g., *Varanus gouldii* — King and Green 1979; Pengilley 1981; pers. obs.; *V. panoptes* — Shine 1986a-b; *V. storri* — Peters 1969d, 1973a; *V. varius* — Horn 1980). The reason for this is unknown but is generally assumed to be due to different activity patterns (King and Green 1979; King and Rhodes 1982). This seems like a plausible explanation for animals first seen and pursued in the open, say as a result of driving along a road, but it becomes less plausible when it turns out to be true for animals caught in their overnight retreats (Peters 1973a). There is one case of a sex ratio running heavily in favour of females (*V. spenceri* — Pengilley 1981), but in this instance the observation was made around fresh piles of soil that were apparently being

investigated by gravid females looking for a place to dig their nesting burrows. There are also some cases where the sex ratio is not significantly different from 1:1, (e.g., in museum specimens of *V. acanthurus* (King and Rhodes 1982) and in wild-caught specimens of *V. gouldii*, *V. mertensi* and *V. mitchelli* (Shine 1986a-b). These cases support the notion that biased sex-ratios are the result of different activity patterns in the two sexes — an interesting notion in its own right and worthy of further investigation.

Unfortunately for field studies there are no obvious sexual differences in goannas other than internal ones (gonads and their associated ducts and, in males, the hemipenes and their associated bones — Shea and Reddacliff 1986; Fig. 89).

There is marked sexual dimorphism in size in some goanna species (e.g., *Varanus gouldii* and *V. panoptes*) but not others (e.g., *V. mertensi* and *V. mitchelli*). In the dimorphic species, males are much larger than females, often not attaining sexual maturity until near the maximum size of females and going on to exceed them in length by a third (Shine 1986b).

Two Australian varanids, e.g., *Varanus acanthurus* (King *et al.* 1982) and *V. varius* (King and King 1975) have a heteromorphic pair (nine) of chromosomes in the female, but other Australian species, e.g., *V. tristis,* lack these sex chromosomes. The same kind of heteromorphism seen

Fig. 89. X-ray of a male (left; AM R 41796) and female (right; AM R 60200) *Varanus gilleni* to show the hemipenial bone (arrow) of male goannas (bone on right side hidden from view).

in the Australian species also seems to occur in two African species but in no other non-Australian species investigated to date (King and King 1975). The phylogenetic significance of this "shared" sex chromosome in varanids is unclear.

Males of at least some, perhaps all, goanna species perform a ritual combat (reviewed in Horn 1985). In its basic features this involves two males pressing their bodies together in some form of upright posture in an effort to topple the opponent. In the larger species the posture is a bipedal, "dancing" embrace, but in the smaller species, it begins as a horizontal, head-to-head, embrace which is then briefly carried into an unstable arch rooted at the heads and tails. A single combat may involve numerous bouts before a clear victor emerges. The combats are ritualized in that they follow a general pattern and rarely lead to any serious physical injury. The purpose of these combats is unknown. Amongst Australian goannas the vertical embrace has been reported in *Varanus mertensi* — photo of R. Braithwaite, reproduced here as Fig. 90, *V. spenceri* (Waite 1929 as *V. giganteus*, reidentified by Horn 1981), and *V. varius* (Worrell 1963) and the horizontal, arching embrace in *V. gilleni* (Murphy and Mitchell 1974; Carpenter *et al.* 1976), *V. semiremex* (Horn 1985) and possibly *V. timorensis* (Horn 1985). In large, semi-natural, outdoor enclosures, male *V. storri* established territories centred over discrete rock piles (Bartlett 1982), but just how territorial goannas are in nature is unknown.

The males of many snake groups also perform a ritual combat which involves up-raised and intertwining bodies attempting to push each other to the ground. This male combat is very similar to that in varanids, and adds to the notion, originally based on morphology, that varanids and their close relatives may be the closest living relatives of snakes (p. 119-120).

Courtship and mating have been recorded only occasionally for Australian goannas (i.e., *Varanus acanthurus* — Murphy 1972; *V. gilleni* — Schmida 1975; *V. prasinus* Carlzen 1982; *V. rosenbergi* — King and Green 1979; *V. storri* — Peters 1969d; Eidenmüller and Horn 1985; *V. varius* — Murphy 1972; Tasoulis 1985) and details are yet too sketchy to see any coherent patterns. The mouth mating grip of the male, when applied, appears to be always on the neck or shoulder (Murphy and Mitchell 1974). The former is the primitive grip amongst lizards. A pair of goannas may copulate several times over a short period (e.g., *V. rosenbergi* — King and Green 1979).

Information on the egg laying period in goannas comes from two sources: the dates of collection for females with oviducal eggs (i.e., shelled eggs

Fig. 90. Varanus mertensi in male combat embrace; 17 km E of Cooinda in Kakadu National Park, Northern Territory; March 1981. Photo: R. Braithwaite.

about to be laid) and actual dates of laying for wild or wild-caught females (Table 6). The available information is skimpy but indicates that southern goannas lay their eggs anytime between early spring (mid-September) to late summer (February) whereas northern species lay during the wet if terrestrial (e.g., northern populations of *Varanus gouldii*) and during the early dry if aquatic (e.g., *V. mertensi* and *V. mitchelli*).

All varanids are oviparous. Clutch sizes in Australian species range 2-35 (Table 7) and RCMs (3) range .14-.16 (Table 8). A positive correlation between female size and egg number has been demonstrated for some species (e.g., *Varanus acanthurus* — King and Rhodes 1982; *V. spenceri* — Pengilley — 1981) but not others (e.g., *V. panoptes* — Shine 1986b). There is no evidence that goannas lay more than one clutch per season.

Most goannas usually lay their eggs in a hole dug in the ground or sometimes at the end of a burrow (e.g., *Varanus gilleni* — Schmida 1975; *V. spenceri* — Christian 1979) or in the case of certain arboreal species such as *V. varius* in the decaying wood of a rotting stump. Occasionally, vegetable matter may be included in the nest (*V. varius* — Cogger 1967; *V. mertensi* — Swanson 1976), but whether this is intended or accidental is unclear.

Three goannas, *Varanus gouldii* (the distinctive Sydney area form; Cogger 1959; Wells and Wellington 1985), *V. rosenbergi* (King and Green 1979) and *V. varius* (Longley 1945a; Fleay 1953; Worrell 1963; Cogger 1959, 1960b, 1967; Swanson 1976; Weavers 1983; Mertens 1986) are known to occasionally lay their eggs in active termite nests which may be either on the ground or in a tree. The female excavates a hole in a nest, lays her eggs and then covers them superficially before departing. Afterwards, the termites repair the nest and solidly seal in the clutch. Termite nests make perfect incubators, of course, because not only do they protect the eggs from predation (the termites themselves seem to consider the eggs as just another physically inert inconvenience that nature has thrust upon them, seal them off within the nest itself, and go on about their business) but they also provide nearly perfect conditions of temperature and humidity for the development of reptile eggs. Varanids in the other parts of the world also use termite nests for egg laying. The trait is well known in the African *V. niloticus* (Cowles 1930), and it probably also occurs in at least some populations of *V. prasinus* (Carlzen 1982; R. G. Allen in Greene 1986).

There are as yet no direct observations of how the young break out of these solidly constructed incubators, and it has been conjectured that the female may return to the nest at the time of hatching and help free the young (Cogger 1967) — as do crocodilians. Although this is possible, it seems unlikely because female goannas appear not to attend their nest as do crocodilians nor do goannas have a voice with which to alert the mother as do crocodilians (although perhaps the smell of the newly hatched young or their scratching could attract the female). Clearly, however, this small problem in goanna nesting biology could be solved once and for all by a single astute field observation.

An even bigger problem in goanna reproductive biology is the variation in the incubation times between species. These vary from the more usual periods of two to three months up to the very unusual 10 months or more (Table 9). The reasons for the longer incubation periods are not known, but for high latitude species they suggest that the developing young may have to "overwinter" in the nests and hatch out the following spring.

The relationship between mass and size has been studied in three goanna species and the results suggest that the rate of weight increase is inversely proportional to species size, i.e., smaller species gain weight more rapidly per unit length than do larger species (*Varanus mitchelli* > *V. mertensi* > *V. panoptes*) (Shine 1986b).

Colour patterns in goannas vary with age, within populations, and geographically. Variation with age is most marked in species with distinctly cross-banded young, e.g., *Varanus spenceri* — Peters 1969c; *V. varius* — pers. obs. These patterns almost always become much muted with age. In contrast, young *V. tristis* of the interior only acquire their characteristic dark heads and neck with age (Fyfe 1979, 1980).

The best known case of distinct colour morphs within a population is in *Varanus varius*. Adults in the southern part of the range occur in two forms or phases: the more drably patterned form found throughout the range, and the distinctly and beautifully cross-banded Bell's form (Fig. 91). The latter colour pattern was thought to occur only in some, but not all, males (Horn 1980), but a female "Bell's" has recently been found (J. Montgomery, pers. comm.). The genetic and/or developmental basis of this colour pattern polymorphism is unknown.

Some goannas also show obvious geographic variation in colour pattern. For example, the Sydney area population of *Varanus gouldii* has the young "brightly marked with vivid orange to pink" and the adults darker than elsewhere (Worrell 1963), and the interior, arid-region population is said to be more brightly coloured than elsewhere (Storr *et al.* 1983). The inland and western population of *V. tristis* has a dark head and neck (giving rise to the species name *tristis* — sad), which is usually lacking in the more peripheral population (*orientalis*) (Christian 1981). In addition to these specific patterns of variation and in direct contradiction to the latter one, there also seems to be a general trend for species or populations from the more mesic/peripheral

Fig. 91. The plain (top) and the Bell's (bottom) colour phases of *Varanus varius.* Photos: S. Wilson.

parts of Australia to be darker than those from the more xeric/central areas (Mertens 1957a).

Most goannas appear to have lost the ability to change colour over short time periods, i.e., seconds or minutes (Bartholomew and Tucker 1964). However, one species, *Varanus storri* is said to retain this ability, going from a greyish colour to vivid rusty-orange on the back with rising temperatures or intense social interactions (Sprackland 1980).

Goannas have a variety of strategies for defence. The most common of course is crypsis, that is, to flatten themselves against the substrate, become absolutely quiet, and trust their colour pattern to camouflage them. In fact, most goannas will "hold" until an intruder is quite close, so much do they trust in crypsis to conceal them. However, when it becomes apparent that they have been "spotted", they then rely on flight, generally to some refuge, usually a burrow or crevice for the ground and rock dwellers, the water for arboreal/aquatic species or a tree for aboreal species. Many of the larger, terrestrial species will begin their flight on four legs but as they accelerate, raise themselves into a bipedal run on rear legs (e.g., *Varanus giganteus* — Ayliffe, in Stirling 1912; *V. gouldii* — Bustard 1970d; *V. panoptes* — pers. obs.). Why they do this is unknown; perhaps it is a faster or more efficient mode of locomotion, but this has yet to be demonstrated. Aquatic monitors such as *V. mitchelli* will swim away under the surface with strong lateral undulations of the body and tail while holding the limbs tucked back along the sides of the body and tail (Schürer and Horn 1976). *V. gouldii* has a burrow with the end near the surface permitting it to burst out if pursued within the burrow (Cogger 1967; Swanson 1976; H. Ehmann, pers. comm.; pers. obs.). Arboreal species tend to keep the trunk or a branch between themselves and the intruder (Cogger 1959; Schmida 1971).

When startled or cornered, goannas often turn sidewise to the intruder, arch the back, bend the neck downward, inflate the throat, hiss, and hold the tail ready for lashing (Irvine 1957; Cogger 1959; Bustard 1968e; Johnson 1976; Murphy and Lamoreaux 1978; Stirnberg and Horn 1981; Delean 1981). some of the smaller species such as *Varanus storri* (Sprackland 1980) will rapidly vibrate the tip of the tail creating a buzzing noise if done against resonant objects. Sometimes if strongly provoked the larger species will raise themselves up on their hind legs and hiss (e.g., *V. giganteus* — Johnson 1976; *V. gouldii* — Longley 1947a; Johnson 1976; photo by Weigel in Grigg *et al.* 1985: vii; *V. panoptes* — Barbour 1943, as *V. gouldii; V.*

spenceri — Johnson 1976, *V. varius* — Johnson 1976). This display presumably startles the intruder and buys the goanna time to edge slowly toward cover, or it may even serve as a deterrent. If approached too closely or actually grasped, goannas will rub or lash with the tail, bite and scratch. A wound from a tooth or claw is nasty enough in its own right, but it may also become infected, especially if the animal has been feeding on carrion.

Some goannas seem to have only a slight fear of people. For example, *Varanus giganteus* and *V. spenceri* will sometimes allow a person to move up next to them (Peters 1968; Martin 1975; Brunn 1978; White 1979; pers. obs.), while *V. rosenbergi* on Kangaroo Island and *V. gouldii* on Lizard Island are both well-known for their "tameness". The large size and hence relatively few predators of adult *V. giganteus* may explain its boldness, while the long isolation from Aborigines and the later restraint of Europeans may explain the animals' reduced fear of people on the two islands. In captivity, *V. semiremex* is said to be usually "tame" in demeanor and disinclined to bite (Peters 1969a).

Like other large lizards, goannas are relatively long-lived, at least in captivity. The following longevity records have been established: *Varanus gilleni* — 4½ years (pers. obs.); *V. gouldii* — 6 years, 10 months (Flower 1937); *V. varius* — 10 years (Bredl and Schwaner 1985), and at least 15 years, 7 months (period of captivity of a specimen mature when caught; Kennerson 1979).

Goannas have attracted a good deal of interest from physiologists because their large size make them easy to handle and to measure. In one aspect of their physiology — metabolism — goannas differ notably from most other lizards: stimulated to maximum activity at body temperatures that are similar to their normal activity temperatures (i.e., in the mid-30's °C) goannas have levels of aerobic metabolism that are about twice those of other similar-sized lizards (Bartholomew and Tucker 1964). At high levels of activity other lizards are using anaerobic metabolism, a kind of incomplete metabolism of carbohydrate that leads to a build-up of lactic acid (a breakdown product of carbohydrate) in the muscle which soon leads to fatigue. In contrast, goannas use to a greater degree aerobic metabolism, a process that breaks down carbohydrate completely to carbon dioxide and oxygen and allows a more sustained level of activity (Bennett 1972).

For efficient aerobic metabolism, large quantities of oxygen are required, and in order to get the oxygen from the air into the cells in the muscles where most of the metabolism is occurring, goannas have undergone at least three modifications to the delivery system for oxygen going from the lung to the muscles. First, the lung is more finely subdividual internally and more heavily vascularized. This increased area of vascularization results in greater exchange of gases (oxygen in and carbon dioxide out) (refs in Bennett 1973a). Second, goannas have a buffering system in the blood that prevents the breakdown products of metabolism from making the blood too acid and thereby making it less able to carry oxygen (Bennett 1973b). And third, goannas have more myoglobin (an oxygen transport molecule in the muscles like haemoglobin in the blood) which facilitates use of oxygen in the tissues to which it is being transmitted. All of these features, and perhaps others yet to be discovered, allow goannas to use a more efficient and sustainable metabolism that allows for prolonged periods of activity such as their long distance dashes to escape predators, long distance foraging, active pursuit of mammalian prey (see below), exploratory digging for food, the excavation of nests in termite nests, and perhaps, even, the escape of the young from nests.

Body temperatures of goannas "active in the field" are fairly high, ranging 34.2–38.2°C (Table 10), but perhaps not surprisingly so, given their active, diurnal habits. Body temperatures preferred by goannas placed in a thermal gradient are lower than those for animals active in the field but show similar high values, ranging 34.7–37°C (Table 11). The near lethal temperature has been determined for only one species of goanna, Varanus varius, at 43.7°C (Heatwole 1976).

At stressfully high temperatures goannas often pump the gular area, a movement similar to the gular flutter of geckos but much slower (approximately 1/sec). Goannas apparently do not pant, and in this they are similar to skinks (Heatwole et al. 1973). The reason for the loss of the panting response is unknown.

A relatively high level of intelligence is frequently attributed to goannas by people who have studied them intensively. For example, the very experienced lizard ecologist Eric Pianka (1982) has said "my experiences with all Varanus, particularly with these perenties, have greatly impressed me with their mammalian-level intelligence". In view of these claims, it would be interesting to have some objective tests by which to make comparisons between not only goannas and other lizards but also goannas and other vertebrates.

Goannas may also be noted for exceptional capacities for what might be called, for want of a better term, persistence, as the following brief account shows: "a large Varanus varius was seen chasing a pair of Brushtailed Possums (Trichosurus vulpecula) for an hour in a clump of red Gums . . . The possums escaped by jumping from the top branches to the ground then climbing another tree" (Brooker and Wombey 1986).

Goannas have been used by people for a variety of purposes. Aborigines traditionally ate them (recipe: break neck, throw onto coals, eat when skin begins to peel — Worrell 1963). They also incorporated them into their mythology. The people of New Guinea and the Solomon Islands ate them, the fat bodies being especially favoured in the latter place (M. McCoy, pers. comm.), and in both areas their skins were used for drum heads. The Japanese introduced them on to several Pacific Islands to control rats, and in India they are protected because they eat the crabs that riddle the rice paddy dikes. Europeans favour goannas for their fine leather, and consequently their skins are frequently rendered into coverings for small cases, wallets, boots, and belts; they are even made into trousers. Europeans also employ goannas in their own mythologies, witness the medicinal "goanna oil" advertised in the Queensland press (for a brief history and the current status of this curious substance, see Harris 1985 and Anonymous 1987). In South Australia, Varanus rosenbergi may have been introduced on to Reevesby Island and perhaps other small islands, in order to reduce the number of snakes (Robinson et al. 1985 and references therein).

Commercial exploitation and habitat destruction worldwide have caused all varanid species to be classed as endangered by the Committee for the International Trade in Endangered Species (CITES) and hence banned from all international trade between countries signatory to this group's recommendations. Australia is a signatory.

Table 1. Derived character states for Varanidae in comparison to all other lizards and snakes.

External morphology
 Head scales fragmented

Skull, mandible and hyoid
 Premaxillae fused, at least dorsally
 Premaxillary process long, extending to frontal
 Narial slit extends posteriorly to separate completely nasal from maxilla and
  prefrontal
 Frontals encircle forebrain
 Parietals fused
 Lacrimal foramen divided
 Prefrontal and jugal widely separated by lacrimal
 Alar process of prootic overlaps descending process of parietal
 Foramen between premaxilla and maxilla in palate
 Maxilla contacts vomer behind opening to Jacobson's organ
 Vomers lack lateral shelves
 Ectopterygoid contacts palatine to exclude maxilla from infraorbital vacuity
 Palatines lack choanal grooves
 Palatal foramen entirely within palatine
 Teeth pointed, recurved, basally fluted
 Lower jaw hinged medially
 Second branchial arch absent

Postcranial skeleton
 Presacral vertebrae $\geq$ 27
 Lumbar vertebrae $\geq$ 1
 Vertebrae with zygosphenes
 Postcaudal vertebrae lack autotomy septa; tail does not regenerate
 Centra of postcaudal vertebrae with pedicels for caudal haemapophyses
 First rib on fifth presacral vertebra
 Inscriptional ribs incomplete
 Clavicles rod-shaped, medially separated
 Interclavicle anchor-shaped
 Mesosternum absent
 Intermedium absent
 Fifth metatarsal separated from fourth metatarsal
 Hemipenis contains two bones

Soft anatomy
 Tongue long and thin, deeply forked apically
 Intrapulmonary bronchi cartilage-lined
 Lungs with extensive alveolae

Behaviour
 Males perform ritual combat "dance"

Table 2. List of the Australian varanid species with the maximum snout-vent length (mm) known for each.

Taxon	Maximum SVL	Reference
V. primordius	114 (N=6)	Mertens 1942c; Storr 1966b; pers. obs.
V. brevicauda	118 (N=21)	Storr 1980a
V. kingorum	120 (N=5+)	Storr 1980a; Weigel 1985
V. storri	132 (N=10)	Storr 1980a
V. caudolineatus	132 (N=110)	Storr 1980a
V. eremius	164 (N=36)	Storr 1980a
V. pilbarensis	169 (N=9)	Storr 1980a
V. gilleni	186 (N=48)	Storr 1980a; Pianka 1982
V. glauerti	227 (N=34)	Storr 1980a
V. acanthurus	237 (N=111)	Storr 1980a
V. timorensis	253 (N=69)	Storr 1980a
V. semiremex	263 (N=5)	Kinghorn 1923 (as *V. boulengeri*); Mertens 1942c, 1958b
V. tristis	280 (N=110)	Storr 1980a
V. prasinus	287 (N=26)	Mertens 1942b
V. mitchelli	346 (N=54)	Peters 1968; Storr 1980a; Shine 1986b
V. glebopalma	355 (N=30)	Storr 1980a
V. rosenbergi	395 (N=46)	Storr 1980a
V. mertensi	475 (N=33)	Storr 1980a
V. panoptes	500 (N=18)	Storr *et al.* 1983
V. spenceri	500 (N=13)	Pengilley 1981
V. indicus	530 (N=?)	Boulenger 1885
V. gouldii	655 (N=147)	Storr 1980a
V. varius	765 (N=94)	Brattstrom 1973; Weavers 1983
V. giganteus	795 (N=17)	Stirling 1912; Mertens 1958b; Storr 1980a

Table 3. State of fusion of the nasal bones in Australian varanids (pers. obs.).

Distinct	Fused	Unknown
V. acanthurus	*V. giganteus*	*V. kingorum*
V. brevicauda	*V. glauerti*	
V. caudolineatus	*V. glebopalma*	
V. eremius	*V. gouldii*	
V. gilleni	*V. indicus*	
V. mitchelli	*V. mertensi*	
V. primordius	*V. panoptes*	
V. semiremex	*V. pilbarensis*	
V. storri	*V. prasinus*	
V. timorensis	*V. spenceri*	
V. tristis	*V. rosenbergi*	
	V. varius	

Table 4. The number of vertebrae in Australian varanids. Sexes have been combined as there appears to be no sexual dimorphism, at least in presacral vertebrae number, based on the largest sample available from a single population: *Varanus gilleni:* ♂♂ 29-30, 29.1, 7 vs ♀♀ 29-30, 29.5, 8 for ranges, mean and N, respectively.

| | Vertebrae | | | | | |
| | Presacral | | | Postsacral | | |
Taxon	Range	Mean	N	Range	Mean	N
Varanus						
acanthurus	28-29	28.3	12	81-95	87.0	5
brevicauda	31-35	32.3	7	47-66.5	56.5	3
caudolineatus	28	—	1	66	—	1
eremius	29	29.0	4	110	—	1
giganteus	30	—	1	130	—	1
gilleni	29-30	29.3	16	70-82	75.2	6
gouldii	29-30	29.2	5	109-115	112.3	3
indicus	29	29.0	5	119-128	124.0	3
mertensi	29	29.0	3	114-118	116.0	2
mitchelli	30-31	30.3	4	122-123	122.3	3
pilbarensis	29	—	1	102+	—	1
panoptes	29	29.0	2	123	—	1
primordius	29-30	29.3	4	74-81	77.5	2
semiremex	29-32	30.5	2	100	—	1
spenceri	29	—	1	—	—	—
storri	27-28	27.2	9	73-78	75.5	5
timorensis	29-31	29.6	13	102-107	104.6	5
tristis	29	29.0	9	105-117	112.3	6
varius	29	29.0	9	137-143	139.4	8

Table 5. References to food eaten by Australian varanids both in the wild (W) and in captivity (C).

Taxon	Reference
V. acanthurus	Murphy 1972 (C); Husband 1979b (C); King and Rhodes 1982 (W)
V. brevicauda	Pianka 1970a (W); Schmida 1974 (C), 1975 (C), 1985 (C)
V. caudolineatus	Pianka 1969c (W)
V. eremius	Pianka 1968 (W), 1982 (W)
V. giganteus	Murphy 1972 (C); Smith 1976 (W); Barnett 1977a (C)
V. gilleni	Pianka 1969c (W), 1982 (W); Peters 1970a (C); Murphy 1972 (C); Schmida 1975 (C); Barnett 1981a (C); Roberts 1985 (W); Bickler and Anderson 1986 (C)
V. glebopalma	Christian 1977b (C); Horn and Schürer 1978 (C)
V. gouldii	Longley 1947a (C); Irvine 1957 (C); Douglas and Ride 1962 (W); Koch 1970 (W); Pianka 1970b (W), 1982 (W); Murphy 1972 (C); Pengilley 1981 (W); Shine 1986a (W) and b (W); Irwin 1986 (C)
V. indicus	Dryden and Taylor 1969 (W)
V. mertensi	Worrell 1956 (C), as *V. bulliwallah;* Brotzler 1965(c); Schürer and Horn 1976 (C); Murphy 1972 (C); Hermes 1981 (W); Shine 1986a (W) and b (W)
V. mitchelli	Peters 1968 (C), 1969b (C), 1971 (C); Murphy 1972 (C); Shine 1986a (W) and b (W)
V. pilbarensis	Johnstone 1983 (W)
V. panoptes	Shine 1986a (W) and b (W)
V. prasinus	Murphy 1972 (C); Carlzen 1982 (C); Greene 1986 (W, C)
V. rosenbergi	Braysher and Green 1970 (C, as *V. gouldii*); King and Green 1979 (W)
V. semiremex	Peters 1969a, d (C)
V. spenceri	Peters 1968 (C), 1969c, e (C); Christian 1979 (C); Pengilley 1981 (W)
V. storri	Mertens 1963 (C, as *V. acanthurus primordius*); Peters 1969d (C); Sprackland 1980 (C); Stirnberg and Horn 1981 (C); Bartlett 1982 (C); Eidenmüller and Horn 1985 (C)
V. timorensis	Peters 1969d (C), 1970a (C); Schmida 1971 (C); Murphy 1972 (C)
V. tristis	Pianka 1971b (W), 1982 (W); Murphy 1972 (C); Brunn 1980a (W)
V. varius	Krefft 1866 (W); Longley 1945a (C); Fleay 1950 (W); Irvine 1957 (C); Peters 1970a (C); Murphy 1972 (C); Rose 1974 (W); Vestjens 1977 (W); Kennerson 1979 (C); Vincent 1981 (W); Weavers 1983 (W)

Table 6. Egg laying season for various species of Australian varanids as indicated by date of collection of females gravid with oviducal eggs or by actual date of laying.

Taxon	Date of collection	Date of laying	References
V. acanthurus	Aug.-Nov.	—	King and Rhodes 1982
V. eremius	Jan.	—	Pianka 1968
V. gilleni	—	18 Sept.	Gow 1982
V. glebopalma	—	mid-Nov.	Barnett 1977b (captive specimen kept in Melbourne)
V. gouldii	22 Nov., 9 Dec., 11 Dec.	—	Pianka 1970b
	—	29-30 Nov.	Barnett 1979
	Jan.-Feb.	—	Shine 1986a-b
V. mertensi	April, June	—	Shine 1986a-b
V. mitchelli	April-June	—	Shine 1986a-b
V. rosenbergi	Feb.	—	King and Green 1979
V. spenceri	—	11 Nov.	Peters 1969c, e, 1971a
	—	mid-Sept., 25 Sept.	Christian 1979
	late Sept. to early Oct.	—	Pengilley 1981
V. storri	Sept.	—	Bustard 1970d (as *V. acanthurus brachyurus*)
V. tristis	28 Oct.-19 Nov.	—	Pianka 1982
V. varius	—	Dec.-Jan.	Fleay 1950
	—	11 Dec.	Bredl and Schwaner 1985

Table 7. Summary of clutch size in Australian varanids. Species arranged in order of increasing size.

Taxon	Clutch Size			Reference
	Range	Mean	N	
V. brevicauda	2	—	1	Pianka 1970a
	8	—	1	Schmida 1975 (more than one clutch?)
V. caudolineatus	4-5	—	?	Pianka 1969c
	3	—	1	Schmida 1975
V. storri	4	—	1	Mudrack 1969, as quoted by Eidenmüller and Horn 1985
	5-7	6.0	2	Bustard 1970d (as *V. acanthurus brachyurus*)
	2-3	2.3	4	Eidenmüller and Horn 1985
V. eremius	3-6	4.1	8	Pianka 1982
V. gilleni	3	—	1	Schmida 1975
	4	—	1	Horn 1978
	4-7	5	3	Gow 1982
V. acanthurus	8	—	1	Husband 1979b (young in nest)
	4	—	1	Murphy 1972
	2-11	—	8	King and Rhodes 1982
V. tristis				
orientalis	8-12	10.0	2	Mitchell 1955
tristis	7	—	1	Christian 1981
	5-17	10.2	19	Pianka 1982
V. prasinus	3-7	5.1	9	Carlzen 1982
V. mitchelli	10-12	11.0	3	Shine 1986a-b
V. glebopalma	7	—	1	Christian 1977b
V. rosenbergi	11-19	14.0	6	King and Green 1979; gravid female
	10-12	11.0	2	King and Green 1979; nests
V. mertensi	10-14	12.3	3	Bustard 1970d
	13	—	1	Brotzler 1965
	3-11	6.9	10	Shine 1986a-b
	14	—	2	Irwin 1986
V. panoptes	9-11	11.0	3	Shine 1986a-b
V. spenceri	20-35	27.5	2	Christian 1979
	18	—	1	Peters 1969c, e, 1971a
	11-31	18.7	13	Pengilley 1981
V. gouldii	8	—	1	Barnett 1979
	7	—	1	Brooker and Wombey 1978
	3-11	6.0	3	Shine 1986a-b
	7	—	1	Irwin 1986
V. gouldii				
flavirufus	4-8	6.4	11	Pianka 1982
	6	—	1	King and Green 1979
V. varius	ca. 12	—	—	Waite 1929
	9	—	—	Irvine 1957
	8-10	9.0	2	Frauca 1966
	6-9	—	—	Fleay 1950, 1953
	6	—	—	Swanson 1976
	5	—	1	Horn 1980
	4	—	1	Peters 1970a
	7-14	9.7	3	Weavers 1983
	7	—	1	Bredl and Schwaner 1985

Table 8. Relative clutch mass for Australian varanid lizards.

Taxon	RCM 3			Reference
	x̄	R	SD	
Varanus				
eremius	.16	—	.02	Pianka, 1986 (N= 2)
gouldii	.14	—	.03	Pianka, 1986 (N= 2)
tristis	.16	—	.03	Pianka, 1986 (N=11)

Table 9. Incubation times and temperatures (°C) for eggs of various species of Australian varanids. Abbreviations: day (d), weeks (w) and months (m).

Taxon	Incubation Period	Incubation Temperature	Reference
V. brevicauda	ca. 70-84 d	18-25	Schmida 1974
	10 w	?	Schmida 1975
V. gilleni	92 ± 3 d	29-30	Horn 1978
	103-104 d	?	Gow 1982
V. gouldii	169-172 d	29.5-32	Barnett 1979
	208	ca 24-25 for 70 d, then 29.5-32 for remainder	Barnett 1979, 1981b
	265 d	30-32	Irwin 1986
V. mertensi	182-217 d	29-30	Brotzler 1965
	269-281 d	30-32	Irwin 1986
V. prasinus	57-70 d	27-33	Carlzen 1982
V. rosenbergi	6+ m	27	King and Green 1979
V. spenceri	123-130 d	29.5	Peters 1969c, e, 1971a
	98-101 d		Christian 1979
V. storri	85 d	?	Barnett 1978
	80+ d	28.5-30	Bartlett 1982
	72 d	?	Rese 1984
	105-109 d	28 ± 1	Eidenmüller and Horn 1985
	102-103 d	?	Eidenmüller and Horn 1985
V. timorensis similis	139-140 d	28-31	Rüegg 1974
V. tristis orientalis	93 d	?	Barnett 1978
V. varius	6 w	?	Fleay 1950
	6-7 m	?	Horn 1980
	317 d	? variable	Markwell 1985
	6 m	?	Peters 1970a
	ab. 5 m	?	Bredl and Schwaner 1985

Table 10. Body temperatures (°C; FBT) of Australian varanids active in the field.

Taxon	Mean	Range	SD or SE	N	Reference
V. caudolineatus	37.8	—	3.45	10	Pianka 1982
V. eremius	37.5	—	3.04	53	Pianka 1982
V. giganteus	36.8	—	—	1	Heatwole 1976
	38.2	—	—	2	Pianka 1982
V. gilleni	37.4	36.4-38.4	—	2	Pianka 1969c, 1982
V. gouldii	37.2	36.1-38.3	2.49	2	Bartholomew and Tucker 1964
	37.1	34.4-36.2	1.87	6	Licht *et al.* 1966a — but note discrepancy between mean and upper end of range
	37.5	—	3.46	67	Pianka 1982
V. tristis	36.5	34-38.9	3.46	2	Bartholomew and Tucker 1964 (as *V. punctatus*)
	34.8	—	2.72	38	Pianka 1982
V. varius	34.2	28.9-37	2.97	7	Bartholomew and Tucker 1964
	35.5	35.0-36.2	—	1	Stebbins and Barwick 1968, Spellerberg 1972a

Table 11. Preferred body temperatures (°C; PBT) of Australian varanids in a laboratory thermal gradient.

Taxon	Mean	Range	SD or SE	N	Reference
V. gilleni	37	—	—	?	Bickler and Anderson 1986
V. gouldii	34.7	32.0-38.8	±0.57	?	Johnson 1972

CHAPTER 7

Carnivorous Legless Lizards — Snakes

THERE is, of course, another group of lizards that we have not even considered in this book but is nonetheless well-represented in Australia: snakes. Thinking of snakes as lizards, or, more appropriately, as nearly "legless" lizards, is useful for several reasons. First, it is almost certain that at some point in their evolution snakes did have an exclusive common ancestor that anybody — herpetologist or otherwise — would call a lizard if it were alive today. Second, lizards have repeatedly evolved a "snake-like" form and true snakes are simply another, albeit very successful in terms of species, example of this. Third, there are two parts to the problem of the origin of snakes — their actual phylogenetic relationships and the ecological/behavioural associations of the first snakes, and comparison of snakes with other limb-reduced lizards may shed light on at least this latter question. In this last chapter we pursue the second part of the question of snake origins, their ecological/behavioural associations, by comparing snakes with other limb-reduced and legless lizards, drawing especially on examples of Australian lizards, and we conclude with a brief look at the first part of the question, their phylogenetic relationships, a problem that may involve a group well-represented in Australia, the varanids.

There are three current theories as to the original[1] ecological/behavioural characteristics of the first snakes that will help in our comparison between snakes and other limb-reduced lizards: 1. aquatic forms (Nopsca 1923); 2. diurnal, surface dwelling forms living primarily on the ground and perhaps in low vegetation such as grass and shrubs (Janesch 1906; Camp 1923), and 3. cryptozoic/nocturnal burrowers (Mahendra 1938;

Walls 1940, 1942; Brock 1941; Bellairs and Underwood 1951; Bellairs 1972; Senn and Northcutt 1973). As there are no living, fully aquatic limb-reduced lizards, the first theory can not be examined by comparison with such a group. However, there are a few diurnal, surface dwelling, limb-reduced lizards, e.g., *Pygopus*, and many cryptozoic/nocturnal burrowers, e.g., *Aprasia* and most limb-reduced skinks. Perhaps even this frequency distribution of the three ecological/behavioural types is telling us something important.

To begin the comparison, consider first the two most obvious attributes of snakes amongst their many diagnostic characters (Table 1): elongation and extreme limb reduction.

Just how elongate snakes are in comparison to lizards is difficult to judge because the relevant measurement, e.g., snout-vent length/head length, is not widely available for many lizards or snakes. However, the stockiest Australian snakes, e.g., the rather highly modified *Acanthophis*, *Echiopsis* and certain *Simoselaps*, have SVL/HL ratios, very generously estimated, of about 12.3-14.4. Most other snakes would have much higher ratios. For example, the primitive genus *Cylindrophis* has a SVL/HL on the order of 20. In comparison, Australian lizards with well-developed limbs have SVL/HL ratios of 3.1-5.7 and limb-reduced species ratios of 5.4-21.7 (Table 2). However, it is only the burrowers that well exceed the lower end of the snake range. Further, in terms of the number of presacral vertebrae, or the number of precloacal vertebrae in snakes (which lack an obvious sacrum), the range is 120-320 in living snakes and up to 457 in extinct snakes but only 22-145 in lizards (Hoffstetter and Gasc 1979). However, the lizards that range up to the lowest value for snakes are all burrowers, e.g., *Aprasia* with 137 (Parker 1956).

Snakes have lost completely the front limbs, pectoral girdle and sternum. Loss of the front limbs is common in other lizards, but only the

---

[1]It is important to emphasize that we are interested here primarily in the characteristics of the ancestor of all living snakes, not in the many modifications they have undergone in their subsequent evolution. The consensus is that among living forms the tropical anilioids are most similar to the ancestor of snakes (Savitsky 1980) and among fossil forms the Cretaceous *Dinilysia* (Estes *et al.* 1970; Rage 1977) is the most similar. Both of these taxa weigh heavily in the characterization of the ancestor of snakes hypothesized here.

legless burrowers have greatly reduced pectoral girdles and sternum. Snakes also have the rear limb reduced to a femur and claw, the pelvic girdle separated medially and only loosely attached to the vertebral column, and a dedifferentiated sacrum. Loss of the rear limbs and separation of the two halves of the pelvic girdle are again common in other lizards, but only the legless burrowers, such as *Lerista apoda*, have dedifferentiated the sacrum. Overall the front limb has been reduced ahead of the hind limb in snakes, as in most other limb-reduced lizard lineages in which there is a disparity between front and rear limbs, e.g., pygopodids and *Lerista*. Those snakes that retain rear limbs use them in courtship (the male rubs or vibrates the claw against the female; Murphy *et al.* 1978) but as yet no non-locomotory role is known for the reduced rear limb in any limb-reduced Australian lizard.

Snakes probably have, as a primitive feature, a single longitudinal row of enlarged ventral scales, although one group of possibly early-diverging (but nonetheless highly specialized) snakes, the blind snakes or scolecophidians, represented in Australia by the blind snakes of the family Typhlopidae, have all the scales of uniform size. Many lizard groups have the ventral scales larger than the dorsal scales, e.g., pygopodids and varanids amongst Australian lizards, but none has only a single row of enlarged scales. Pygopodids perhaps come the closest to this condition in usually having only two rows of enlarged ventral scales.

Snakes are well-known for having kinetic skulls, that is the skull bones, or at least those of the snout, palate, jaw and jaw suspension, are individually moveable. Although certain "advanced" snakes have highly kinetic skulls, this appears to have been a development within snakes, and primitively snakes probably had only a moderately kinetic skull. Lizard skulls are also kinetic in some of the same ways snake skulls are, e.g., in the jaw suspension, but not in others; for example, they lack independent movement of the left and right sides of the upper and lower jaws, and their snout movement occurs at the parietal/frontal joint instead of the nasal/frontal joint (p. 107). The increased kineticism of the snake skull may be related to the need to get the relatively small head around prey large enough to nourish the relatively large body and perhaps also to the fact that snake teeth are relatively ineffective in reducing the size of the prey through crushing or cutting prior to swallowing it.

Snake teeth are long, pointed, rounded in basal cross-section, and recurved. Only a few lizards have similar teeth, in the Australian fauna — *Lialis* amongst pygopodids and *Coeranoscincus* amongst scincids (Fig. 88). These and other lizards with similar teeth are elongate lizards feeding on large, "slippery" prey, skinks in the case of *Lialis* and perhaps earthworms in the case of *Coeranoscincus*. The functional connection here is unclear, but it is interesting to note that at least one group of snakes thought to be the most generally primitive among living forms, the aniliids, eat elongate, "slippery" prey such as eels, frogs, caecilians, skinks and snakes (Greene 1983). Goanna teeth are also long, pointed and recurved, but compressed instead of rounded. Their dentition seems more geared to severing instead of grasping as in snakes and the lizards *Lialis* and *Coeranoscincus*.

The carnivorous habits of snakes are in strong contrast to the general arthropod feeding habits of lizards. To be sure some snakes, like the highly specialized blind snakes eat arthropods, as do occasionally more ordinary snakes, and conversely some lizards eat fleshy prey. However, most snakes, as probably did the first snakes, eat fleshy prey — basically vertebrates — and avoid arthropods — basically chitinous prey. Why this should be so is unclear, unless snakes made a basic feeding switch such that today, like *Lialis*, they are apparently no longer "wired" neurologically to recognize arthropods as prey. However, this would only be the proximate cause; the ultimate cause of such a switch remains unknown.

Primitively the hyoid apparatus of snakes, i.e., the skeleton of the tongue and throat, was a vastly simplified-shaped structure, devoid of most of the branches seen in the primitive lizard (Langebartel 1968). A similarly reduced hyoid is rare in other lizards, being found only in strongly limb-reduced forms that spend almost their entire lives underground, e.g., some African skinks (*Typhlosaurus*) and dibamids (Rieppel 1981). That the simplified throat skeleton is not simply due to the long, thin tongue is evident from the fact that goannas which have a similar tongue have a well-developed throat skeleton. It is interesting to note that no Australian lizard, not even the most fossorial, e.g., *Anomalopus*, *Aprasia*, *Coeranoscincus*, *Lerista*, *Ophidiocephalus*, and *Ophioscincus* have an appreciably reduced hyoid apparatus.

Snakes show one derived behaviour, related to feeding, which apparently does not occur in any other lizard: constriction. All the generally primitive snakes constrict their prey (Greene 1983), but as yet no lizard is known to possess this ability. Perhaps the greater motor skills

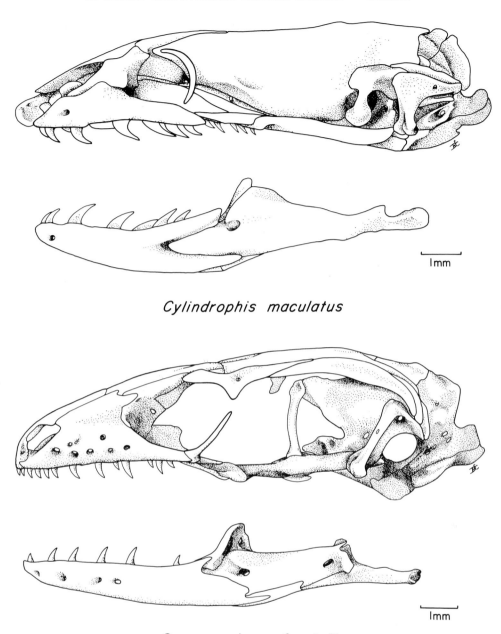

Cylindrophis maculatus

Coeranoscincus frontalis

Fig. 92. A primitive snake *Cylindrophis maculatus* (AM 5579) in comparison with a burrowing lizard, the skink *Coeranoscincus frontalis* (AM R 3823).

inherent in a very elongate body are prerequisites for constriction; alternately perhaps prey has to be of a relatively large size to facilitate constriction as a method of subjugation.

In contrast to the generally kinetic skull and mandible of snakes is the solid brain case, that part of the skull immediately surrounding the brain. These bones are thick and have extended forward to encase completely in bone that area of the brain that is covered in thick fascia in other lizards. A similar brain case is seen in other lizards only in amphisbaenians, where it appears to have evolved slightly differently and is thought to strengthen the brain case during

burrowing. It may have had a similar function in snakes. Alternatively, there is a suggestion that it evolved in snakes to add protection during the overcoming and swallowing of relatively large prey. However, in this context, it is relevant that *Lialis* which also regularly swallows large, sometimes still living prey, has the standard open brain case.

Snakes have lost the upper temporal arch in the skull through diminution of the bones[1,2] and

[1]Snakes retain the anterior elements of the upper temporal arch, the postfrontal and postorbital, but in a much reduced state and have lost the posterior element, the squamosal.

[2]Recall that one of the features attendant to the evolution of lizards and snakes as a group was the loss of the lower temporal arch.

the post-temporal arch through apposition with the parietal (Fig. 88). The upper arch has been lost in several burrowing groups but in only one group of strongly limbed lizards, geckos + pygopodids. Hence its loss in snakes is consistent with, but not conclusive of, a burrowing ancestry. Loss of the post-temporal arch is a feature seen only in burrowers. The loss of both these arches in burrowers appears to be related to a streamlining and strengthening of the skull.

Perhaps as a development following at least in part from the loss of the upper temporal arch, snakes have the jaw muscles extending up along the side of the skull all the way to the midline. The only other lizards with similarly expanded jaw muscles are in two groups of burrowers: amphisbaenians and dibamids.

Snakes also lack an epipterygoid, a finger-like bone projecting up along the side of the brain case in most squamates. This bone has been lost in only three other groups of squamates: amphisbaenians (most species), chameleonids, and dibamids (one of the two genera). The chameleons, of course, are arboreal but the other two are burrowers.

Snakes show several modifications of sensory structures that appear to be related to life under ground. For example, both the eye and the visual centres in the brain show a number of modifications that are seen in other lizards only in small-eyed forms living under conditions of reduced light intensity — a morphology and microhabitat restricted almost entirely to burrowers. Changes associated with a small eye are the loss of scleral cartilages and ossicles and the reduction and loss of the ciliary body and annular pad, while changes associated with life in diminished light are loss of the oil droplets and paraboloids in the retinal cells. The only surface-dwelling lizard group which has small eyes and lives under low light intensity is the nocturnal *Heloderma*. Surface-active species living under low light intensities, e.g., geckos, are usually large-eyed. Another sensory structure lost in snakes and many other lizards living under low light intensity, such as geckos, is the parietal eye.

The ear and its associated structures have also been profoundly modified in snakes. The external ear canal and tympanum have been lost as has also the eustachian tube. The stapes which formerly attached to the tympanum now attaches to the quadrate bone, and the conch of the quadrate, which formerly formed a rim of attachment for the tympanum has been lost, leaving the quadrate a simple, rod-like strut (Fig. 88). The only other lizards to have undergone the same suite of modifications are all burrowers. Among Australian species, the earless *Aprasia* approach the snake condition most closely in having lost the external ear opening and the eustachian tube. However, their much-reduced stapes has, at most, only a thin ligamentous attachment to the quadrate.

The internal organs of snakes have been variously changed in shape and relative position, largely as a consequence of the elongation and narrowing of the body cavity (Underwood 1976b). One of the most striking adjustments is the reduced left lung, which at its largest and presumably most primitive is about a 15% reduction of the right (Underwood 1976a) but in many taxa is reduced to a small diverticulum or absent. A large reduction of one lung over the other occurs in a variety of both surface-dwelling and burrowing limb-reduced, elongate squamates (pp. 113, 172).

Snakes breathe by contracting uniformly the entire part of the body over the lung area. Recall that legless lizards employ only this circumferential method, e.g., pygopodids, or ventral displacement, e.g., *Lerista*, the latter method seemingly being a special adaptation amongst sand-swimmers to prevent the sand from caving in the free space created by the contracting body wall. Unfortunately too little is known about breathing in non-Australian legless lizards to say how general the mode of uniform contraction is or whether other modes exist as well.

We can conclude this chapter and the book by a brief look at the first part of the question of snake origins: their actual phylogenetic relationships. This is one of the most intractable problems in herpetology and has engaged some of the best minds in the field in the twentieth century.

There are three current theories regarding the relationships of snakes: 1) they are one half of an early split in the squamate lineage, the other half being the lizards (Underwood 1970); 2) they are related to some living burrowing group (Senn and Northcutt 1973; Northcutt 1978; Rage 1982), and 3) they are related to the varanoids — varanids and their closest living relatives, the lanthanotids and helodermatids (Nopsca 1908, 1923; Fejervary 1918; Camp 1923; McDowell and Bogert 1954; Bellairs 1972 and McDowell 1972).

Support for the first hypothesis is of two kinds: sentiment and fact. The sentiment is that snakes are so different from other lizards and are so unsatifactorily linked with any other lizard group, that they can only be related at the most basic level. The facts are a few derived characters which appear to unite all other "lizards" to the exclusion of snakes. These are the extensive fusion of the paired trabecular cartilages, the development of a moveable joint between the frontal and parietal bones, the loss of the rods in the retina, and a derived pathway through the middle ear cavity for a branch of the hyomandibular nerve (Rieppel 1980).

Support for the second hypothesis — or actually class of hypotheses — is a long list of characters shared between snakes and any one of several lineages of elongate, limb-reduced squamates, e.g., amphisbaenians (Rage 1982), dibamids (Senn and Northcutt 1973), pygopodids (Underwood 1957, as a heuristic argument only). The problem with these theories is that squamates have undergone elongation and limb-reduction on numerous independent occasions and there is a strong tendency to develop many of the same modifications each time this occurs. This leads to the possibility that similarity due to convergence will be mistaken for similarity due to common ancestry. One can really only hope to get out of this impasse by having derived characters that seem not to arise routinely in the burrowing mode, but as yet there are none that are particularly convincing.

Support for the third hypothesis, relationship with the varanoids, is a list of characters (Table 1) that is impressive in the context of trying to tie a highly modified group of lizards, i.e., snakes, to a relatively unmodified one. If you believe snakes may not be the sister group of all other lizards, and if you believe the similarity between snakes and limb-reduced lizards is likely to be due to convergence, then this is the best of the alternative hypotheses. Under this hypothesis snakes could be seen as limb-reduced, primitively burrowing, carnivorous relatives of the varanoids. In structure, but not habits, snakes would represent the stretched version of varanoids just as pygopodids are the stretched version of geckos.

Table 1. Derived characteristics of snakes in comparison to all other lizards and snakes. Those characters shared with varanoid lizards are indicated with a "V".

External features
    Form long and attenuate
    Dorsal crest absent

Skull, mandible and hyoid
    Frontal encircles fore brain (V)
    Parietal extends anteroventrally to meet frontals and parasphenoid
    Upper temporal arch absent
    Squamosal absent
    Exoccipitals meet above foramen magnum
    Epipterygoid absent
    Teeth long, pointed and recurved and replaced from medioposterior aspect (V)
    Stapes imperforate (V)
    Mandible hinged at splenial/angular suture (V)
    Ceratobranchial II absent (V)

Postcrannial skeleton
    Postcloacal vertebrae ≥120
    Sacrum dedifferentiated
    Sternum, pectoral girdle and forelimbs absent
    Pelvic girdle separated medially, only loosely attached to vertebral column
    Rear limb reduced to femur with claw
    Autotomy planes absent (V)

Soft anatomy
    Spectacle present
    Tongue long, narrow, deeply forked anteriorly, and retractable into glottal
        sheath
    Parietal eye absent
    Left lung at least 20% shorter than right
    Kidneys placed forward in body cavity
    Tympanum absent
    Eustachian tube absent
    Eye lacks scleral ossicles and cartilages
    Retinal elements lack parabaloids
    Ductus caroticus lacking (V)
    Tail does not regenerate (V)

Behaviour
    Male combat dance (V)
    Short-term colour change ability lost (V)
    Constriction of prey
    Carnivorous (V)

Table 2. Ratio of snout-vent length to head length (SVL/HL), a measure of elongation, in Australian lizards.

Taxon	SVL/HL
Agamidae	3.1-4.3
Gekkonidae	3.1-5.6
Scincidae *Egernia*	4.4-5.6
Varanidae	4.7-5.7
Scincidae (limb-reduced)	5.4-18.1
Pygopodidae	8.8-21.7
Snakes	≥12.3

References

ADLER, K. K., 1958. Observations on the Australian genera *Egernia* and *Tiliqua* in captivity. *Ohio Herp. Soc.* **1**(3): 9-12.

ALCALA, A. C. AND BROWN, W. C., 1967. Population ecology of the tropical scincoid lizard, *Emoia atrocostata*, in the Philippines. *Copeia* **1967**(3): 596-604.

ALEXANDER, W. B., 1914. The history of zoology in Western Australia. Part I. — Discoveries in the 17th century. *J. Nat. Hist. Sci. Soc. West. Aust.* **5**: 49-64.

ALEXANDER, W. B., 1922. The vertebrate fauna of Houtman's Abrolhos (Abrolhos Islands), Western Australia. *J. Linn Soc. Zool.* **34**(230): 457-86.

ANNABLE, T., 1983. Some observations on vocalization and the use of limb flaps in the pygopodid lizard, *Delma inornata* Kluge. *Herpetofauna* **14**(2): 80-82.

ANONYMOUS, 1926. Birth of blue-tongued lizards. *Aust. Zool.* **4**(5): 293.

ANONYMOUS, 1944. Two snake stories . . . not what it seems. *Wildlife* **6**(7): 206.

ANONYMOUS, 1954. For your information. *Reptilia.* **1**(2): 5.

ANONYMOUS, 1973a. Snakes and ladders. *Herpetofauna* **6**(1): 25-26.

ANONYMOUS, 1973b. News. Sydney. *Roy. Zool. Soc. Bull. Herp.* **1**(2): 3.

ANONYMOUS, 1976. Observations on the eastern water dragon *Physignathus lesueurii* in the natural state and in captivity. *Herpetofauna* **8**(2): 20-22.

ANONYMOUS, 1977. Letter and news from Peter Richardson, Bundaberg, Q'land. *Newsletter Vict. Herp. Soc.* **3**: 7-9.

ANONYMOUS, 1987. Goanna oil to soothe American ills. Sydney Morning Herald, 26 January, 1987.

ANSTIS, M. AND PETERSON, M., 1973. Results of collecting on the Wilcannia field trip. *Roy. Zool. Soc. Bull. Herpet.* **1**(2): 6-7.

ANTENOR, A., 1973. Notes on *Sphenomorphus tenuis. Roy. Zool. Soc. Bull. Herp.* **1**(2): 10.

ANTENOR, A., 1974. Notes on *Amphibolurus pictus* (Painted Dragon). *Roy. Zool. Soc. Bull of Herp.* **1**(3): 17.

ARCHER, M. AND WADE, M., 1976. Results of the Ray E. Lemley Expeditions, Part 1. The Allingham Formation and a new Pliocene vertebrate fauna from northern Queensland. *Mem. Qld Mus.* **17**(3): 379-97.

ARENA, P. C., 1986. Aspects of the biology of the King's skink *Egernia kingii* (Gray). Unpublished B.Sc. Honours Thesis; Murdock University, Murdock, Western Australia. Abstract only seen.

ARMSTRONG, G., 1978. A tenacious gecko. *South Aust. Herp. Group Newsletter* 1978 (end of year?): 7.

ARMSTRONG, G., 1979a. Reptiles of the Golden Grove areas in the Adelaide Hills. *South Aust. Herp. Group Newsletter* 1979(March): 5-8.

ARMSTRONG, G., 1979b. Brief notes on egg-laying in *Phyllodactylus marmoratus* and *Morethia boulengeri. South Aust. Herp. Group Newsletter* 1979(March): 9.

ARMSTRONG, G., 1981. Northern Territory field trip report: Victoria River & Mataranka areas; 17th-19th October, 1980. *South Aust. Herp. Group Newsletter* 1981 (March): 3-4.

ARNOLD, E. N., 1973. Relationships of the Palaearctic lizards assigned to the genera *Lacerta, Algyroides* and *Psammodromus* (Reptilia: Lacertidae). *Bull. Brit. Mus. (Nat. Hist.) Zool.* **25**(8): 291-366.

ARNOLD, E. N., 1984. Evolutionary aspects of tail shedding in lizards and their relatives. *J. Nat. Hist.* **18**: 127-69.

AUFFENBERG, W., 1981. The behavioural ecology of the Komodo monitor. University Presses of Florida, Gainesville; 406 pp.

AYALA, F. J., 1986. On the virtues and pitfalls of the molecular evolutionary clock. *J. Heredity* **77**: 226-35.

BADHAM, J. A., 1971. Albumin formation in eggs of the agamid *Amphibolurus barbatus barbatus. Copeia* **1971**(3): 543-45.

BADHAM, J. A., 1976. The *Amphibolurus barbatus* species-group (Lacertilia: Agamidae). *Aust. J. Zool.* **24**: 423-43.

BAEHR, M., 1976. Beobachtungen zur bipeden Fortbewegung bei der australischen Agame *Physignathus longirostris* (Boulenger). *Stuttgarter Beitr. Naturk.* Ser. A (Biol.), No. 291; 7pp.

BAKER, J. K., 1979. The rainbow skink, *Lampropholis delicata*, in Hawaii. Pacific Science **33**(2): 207-12.

BAMFORD, M. J., 1980. Aspects of the population biology of the bobtail skink, *Tiliqua rugosa* (Gray). Unpublished B.Sc. Honours Thesis, Murdock University; 178 pp.

BANKS, C., 1980. Keeping reptiles and amphibians as pets. Thomas Nelson Australia Pty Ltd, West Melbourne; 129 pp.

BANKS, C., 1985. Notes on growth of the major skink *(Egernia frerei)* in captivity. *Herpetofauna* **15**(1): 5-6.

BARBOUR, T., 1943. Defense posture of *Varanus gouldii. Copeia* **1943**(1): 56.

BARNETT, B., 1977a. An untitled series of news items. *Newsletter Vict. Herp. Soc.* No. 1: 3-7.

BARNETT, B., 1977b. Additional notes on new-born Centralian Bluetongues *(Tiliqua Multifasciata). Newsletter Vict. Herp. Soc.* No. 1: 10.

BARNETT, B., 1977c. Hatching of the frill-neck lizard *(Chlamydosaurus kingii). Newsletter Vict. Herp. Soc.* No. 2: 7-8.

BARNETT, B., 1978. Incubation periods. *Newsletter Vict. Herp. Soc.* **8:** 10.

BARNETT, B., 1979. Incubation of sand goanna *(Varanus gouldii)* eggs. *Herpetofauna* **11**(1): 21-22.

BARNETT, B., 1981a. Observations of fish feeding in reptiles. *Herpetofauna* **13**(1): 11-13.

BARNETT, B., 1981b. Artificial incubation of snake eggs. *Monitor* **1**(2): 31-39.

BARRETT, C., 1931. The Gippsland water dragon. *Vict. Nat.* **47**(10): 162-65.

BARTHOLOMEW, G. A. AND TUCKER, V. A., 1963. Control of changes in body temperature, metabolism, and circulation by the agamid lizard, *Amphibolurus barbatus. Physiol. Zoöl.* **36**(3): 199-218.

BARTHOLOMEW, G. A. AND TUCKER, V. A., 1964. Size, body temperature, thermal conductance, oxygen consumption, and heart rate in Australian varanid lizards. *Physiol. Zoöl.* **37**(4): 341-54.

BARTLETT, R. D., 1981. Notes on *Egernia cunninghami kreffti,* an Australian skink. *Brit. Herp. Soc. Bull.* **4:** 36-37.

BARTLETT, R. D., 1982. Initial observations on the captive reproduction of *Varanus storri,* Mertens. *Herpetofauna* **13**(2): 6-7.

BARTLETT, R. D., 1984. Note on the captive reproduction of the Australian skink, *Tiliqua nigrolutea. Brit. Herp. Soc. Bull.* **10:** 34-35.

BARWICK, R. E., 1965. Studies on the scincid lizard *Egernia cunninghami* (Gray 1832). Unpublished Ph.D. Thesis, Australian National University, Canberra; 177 pp.

BAUER, A. M., 1986. Saltation in the pygopodid lizard, *Delma tincta. J. Herpetol.* **20**(3): 462-63.

BAVERSTOCK, P. R., 1975. Effect of variations in rate of growth on physiological parameters in the lizard *Amphibolurus ornatus. Comp. Biochem. Physiol.* **51**A: 619-31.

BAVERSTOCK, P. R., 1978. The probable basis of the relationship between growth rate and winter mortality in the lizard *Amphibolurus ornatus* (Agamidae). *Oecologia* **37**(1): 101-7.

BAVERSTOCK, P. R., 1979. A three year study of the mammals and lizards of Billiat Conservation Park in the Murray mallee, South Australia. *South Aust. Nat.* **53**(4): 52-58.

BAVERSTOCK, P. R. AND BRADSHAW, S. D., 1975. Variation in rate of growth and adrenal corticosteroidogenesis in field and laboratory populations of the lizard *Amphibolurus ornatus. Comp. Biochem. Physiol.* **52**A: 557-65.

BEDDARD, F. E., 1905. A contribution to the anatomy of the frilled lizard *(Chlamydosaurus kingi)* and some other Agamidae. *Proc. Zool. Soc. Lond.* **1**(1): 9-22.

BELAN, I., 1980. The role of olfaction and vision in food-seeking behaviour of the scincid lizard *Trachydosaurus rugosus.* Unpublished M.Sc. (prelim.) Thesis, University of New England. Not seen.

BELLAIRS, A. d'A., 1972. Comments on the evolution and affinities of snakes. Pp. 157-72 *in* Studies in vertebrate evolution ed by K. A. Joysey and T. S. Kemp. Oliver and Boyd, Edinburgh; 284 pp.

BELLAIRS, A. d'A. AND UNDERWOOD, G., 1951. The origin of snakes. *Biol. Rev.* **26**(2): 193-237.

BELMONT, C. M., 1979. Resource partitioning and habitat selection in ground skinks. Unpublished Honours Thesis, University of New England, Armidale; 119 pp. Not seen.

BENNETT, A. F., 1972. The effect of activity on oxygen consumption, oxygen debt, and heart rate in the lizards *Varanus gouldii* and *Sauromalus hispidus. J. Comp. Physiol.* **79:** 259-80.

BENNETT, A. F., 1973a. Ventilation in two species of lizards during rest and activity. *Comp. Biochem. Physiol.* **46**A: 653-71.

BENNETT, A. F., 1973b. Blood physiology and oxygen transport during activity in two lizards, *Varanus gouldii* and *Sauromalus hispidus. Comp. Biochem. Physiol.* **46**A: 673-90.

BENNETT, A. F. AND JOHN-ALDER, H., 1986. Thermal relations of some Australian skinks (Sauria: Scincidae). *Copeia* **1986**(1): 57-64.

BENNETT, G., 1876. Notes on the Chlamydosaurus or frilled lizard of Queensland *(Chlamydosaurus Kingii,* Gray), and the discovery of a fossil species on the Darling Downs, Queensland. *Pap. Proc. Roy. Soc. Tasmania* **1875:** 56-58.

BENTLEY, P. J. AND BLUMER, W. F. C., 1962. Uptake of water by the lizard *Moloch horridus. Nature* **194:** 699-700.

BENTON, M. J., 1986. The demise of a living fossil? *Nature* **323:** 762.

BEVAN, J., 1983. A defensive reaction of *Gonocephalus spinipes* (Duméril). *Herpetofauna* **14**(2): 99.

BICKLER, P. E. AND ANDERSON, R. A., 1986. Ventilation, gas exchange, and aerobic scope in a small monitor lizard, *Varanus gilleni. Physiol. Zoöl.* **59**(1): 76-83.

BÖHME, W. AND BISCHOFF, W., 1976. Das Paarungsverhalten der kanarischen Eidechsen (Sauria, Lacertidae) als systematisches Merkmal. *Salamandra* **12**(3): 109-19.

BOOTH, P., 1981. Field trip (No. 12). *Newsletter Vict. Herp. Soc.* **22:** 16-19.

BOULENGER, G. A., 1885. Catalogue of the lizards in the British Museum (Natural History). Vol. 1. British Museum (Natural History), London; 436 pp.

BOULENGER, G. A., 1887. Catalogue of the lizards in the British Museum (Natural History). Vol. 3. British Museum (Natural History), London; 575 pp.

BOURNE, A. R., STEWART, B. J. AND WATSON, T. G., 1986. Changes in blood progesterone concentration during pregnancy in the lizard *Tiliqua (Trachydosaurus) rugosa. Comp. Biochem. Physiol.* **84**A(3): 581-83.

BOURNE, G., 1932. The origin of the liquid appearing from the soft spines and the tail of the lizard — *Diplodactylus spinigerus.* — Gray. *J. Roy. Soc. West. Aust.* **19:** 9-11.

BOWLER, J. K., 1977. Longevity of reptiles and amphibians in North American collections as of 1 November, 1975. Society for the Study of Amphibians and Reptiles. *Misc. Publs., Herp. Circ.* No. 6; 32 pp.

BOYLAN, T., 1970. Thorny devil. *Animal Kingdom,* Feb. 1970: 25-27.

BRADSHAW, B. J., 1976. Discovery of *Proablepharus reginae* a small member of Scincidae at Ormiston Gorge Pound & Scenic Reserve. *Ranger Review,* May 1976: 11.

BRADSHAW, S. D., 1965. The comparative ecology of lizards of the genus *Amphibolurus*. Unpublished Ph.D. Thesis, University of Western Australia, Nedlands. Not seen.

BRADSHAW, S. D., 1970. Seasonal changes in the water and electrolyte metabolism of *Amphibolurus* lizards in the field. *Comp. Biochem. Physiol.* **36:** 689-718.

BRADSHAW, S. D., 1971. Growth and mortality in a field population of *Amphibolurus* lizards exposed to seasonal cold and aridity. *J. Zool. Lond.* **165:** 1-25.

BRADSHAW, S. D., 1975. Osmoregulation and pituitary-adrenal function in desert reptiles. *Gen. Comp. Endocrin.* **25:** 230-48.

BRADSHAW, S. D., 1977a. Reptiles and their adaptation to arid environments. Pp. 145-60 *in* Australian animals and their environment ed by H. Messel and S. T. Butler. Shakespeare Head Press, Sydney; 367 pp.

BRADSHAW, S. D., 1977b. The regulation of water and electrolyte balance in desert lizards. Pp. 161-78 *in* Australian animals and their environment ed by H. Messel and S. T. Butler. Shakespeare Head Press, Sydney; 367 pp.

BRADSHAW, S. D., 1981. Ecophysiology of Australian desert lizards: studies on the genus *Amphibolurus*. Pp. 1393-434 *in* Ecological biogeography of Australia ed by A. Keast. W. Junk, The Hague; 2142 pp.

BRADSHAW, S. D., 1986. Ecophysiology of desert reptiles. Academic Press, Sydney; 324 pp.

BRADSHAW, S. D. AND MAIN, A. R., 1968. Behavioural attitudes and regulation of temperature in *Amphibolurus* lizards. *J. Zool.* **154:** 193-221.

BRADSHAW, S. D. AND SHOEMAKER, V. H., 1967. Aspects of water and electrolyte changes in a field population of *Amphibolurus* lizards. *Comp. Biochem. Physiol.* **20**(3): 855-65.

BRADSHAW, S. D., GANS, C. AND SAINT-GIRONS, H., 1980. Behavioural thermoregulation in a pygopodid lizard, *Lialis burtonis*. *Copeia* **1980**(4): 738-43.

BRADSHAW, S. D., TOM, J. A. AND BUNN, S. E., 1984. Corticosteroids and control of nasal salt gland function in the lizard *Tiliqua rugosa*. *Gen. Comp. Endocrin.* **54**(2): 308-13.

BRANCH, W. R., 1982. Hemipeneal morphology of platynotan lizards. *J. Herp.* **16**(1): 16-38.

BRATTSTROM, B. H., 1971a. Social and thermoregulatory behaviour of the bearded dragon, *Amphibolurus barbatus*. *Copeia* **1971**(3): 484-97.

BRATTSTROM, B. H., 1971b. Critical thermal maxima of some Australian skinks. *Copeia* **1971**(3): 554-57.

BRATTSTROM, B. H., 1973. Rate of heat loss by large Australian monitor lizards. *Bull. South. Calif. Acad. Sci.* **72**(1): 52-54.

BRAYSHER, M., 1971. The structure and function of the nasal salt gland from the Australian sleepy lizard *Trachydosaurus* (formerly *Tiliqua*) *rugosa*: Family Scincidae. *Physiol. Zoöl.* **44**(3): 129-36.

BRAYSHER, M. AND GREEN, B., 1970. Absorption of water and electrolytes from the cloaca of an Australian lizard, *Varanus gouldii* (Gray). *Comp. Biochem. Physiol.* **35:** 607-14.

BRAZENOR, C. W., 1932. A lizard not previously recorded from Victoria. *Vict. Nat.* **49**(7): 171.

BREDL, J. AND SCHWANER, T. D., 1985. First record of captive propagation of the lace monitor, *Varanus varius* (Sauria: Varanidae). *Herpetofauna* **15**(1): 20-21.

BRIGGS, R. J., 1973. Notes on *Delma fraseri*. *South Australian Herpetologist* 1973 (June): 20.

BROCK, G. T., 1941. The skull of *Acontias meleagris*, with a study of the affinities between lizards and snakes. *J. Linn. Soc. Lond. (Zool.)* **41**(277): 71-88.

BROOKER, M. G. AND WOMBEY, J. C., 1978. Some notes on the herpetofauna of the western Nullarbor Plain, Western Australia. *West. Aust. Nat.* **14**(2): 36-41.

BROOKER, M. AND WOMBEY, J., 1986. Some observations on the herpetofauna of the Macquarie Marshes region, N.S.W, with special reference to Chelidae. *Aust. Zool.* **23**(1): 1-4.

BROOM, R., 1897. On the lizards of the Chillagoe District, N. Queensland. *Proc. Linn. Soc. New South Wales* **22**(3): 639-45.

BROTZLER, A., 1965. Mertens — Wasserwarane züchteten in der Wilhelma. Freunde Kölner Zoo **8**(3): 89.

BROWN, A. M., 1985. Ultrasound in gecko distress calls (Reptilia: Gekkonidae). *Israel J. Zool.* **33**(3): 95-101.

BROWN, G. W., 1983. Comparative feeding ecology of south-eastern Australian scincids. Unpublished Ph.D. Thesis, La Trobe University, Bundoora; 382 pp.

BROWN, G. W., 1986. The diet of *Pseudemoia spenceri* (Lucas and Frost 1894) (Lacertilia: Scincidae), a species endemic to south-eastern Australia. *Vict. Nat.* **103**(2): 48-55.

BROWN, W. C. AND ALCALA, A. C., 1957. Viability of lizard eggs exposed to sea water. *Copeia* **1957**(1): 39-41.

BROWN, W. C. AND FALANRUW, M. V. C., 1972. A new lizard of the genus *Emoia* (Scincidae) from the Marianas Islands. *Proc. Calif. Acad. Sci.* (4)**39**(9): 105-10.

BROWN, W. C. AND McCoy, M., 1980. A new species of gecko of the genus *Cyrtodactylus* from Guadacanal Island, Solomon Islands. *Herpetologica* **36**(1): 66-69.

BROWN, W. C. AND PARKER, F., 1973. A new species of *Cyrtodactylus* (Gekkonidae) from New Guinea with a key to species from the island. *Breviora* No. 417; 7 pp.

BROWN, W. C. AND PARKER, F., 1977. Lizards of the genus *Lepidodactylus* (Gekkonidae) from the Indo-Australian Archipelago and the islands of the Pacific, with descriptions of new species. *Proc. Calif. Acad. Sci.* (4)**41**(8): 253-65.

BROWNE-COOPER, R., 1985. Notes on the reproduction of the bearded dragon *Pogona minor*. *Herpetofauna* **15**(2): 49.

BRUNER, H. L., 1907. On the cephalic veins and sinuses of reptiles, with description of a mechanism for raising the venous blood-pressure in the head. *Amer. J. Anat.* **7:** 1-117.

BRUNN, W., 1978. A trip to Alice Springs. *South Australian Herpetologist* 1978(April): 1-3.

BRUNN, W., 1980a. Beyond the banana curtin (sic). *South Aust. Herp. Group Newsletter* 1980(April): 2-7.

BRUNN, W., 1980b. Goannas caught on trips (part one). *South Aust. Herp. Group Newsletter* 1980(Oct.): 2-3.

BRUNN, W., 1980c. Some occasional observations, made by myself, and myself alone, on the lizards known to myself, and others, as goannas. *Slerpetofauna* **1**(1): 4-6.

BRUNN, W., 1981. Goannas caught on trips (cont. from previous issues). *South Aust. Herp. Group Newsletter* 1981 (Oct.): 4.

BRUNN, W., 1982. More of Werner Brunn's goannas caught on trips series. *South Aust. Herp. Group Newsletter* 1982(March): 6-9.

BULL, C. M., 1978. Dispersal of the Australian reptile tick *Aponomma hydrosauri* by host movement. *Aust. J. Zool.* **26**(4): 689-97.

BULL, C. M. In press. A population study of the viviparous Australian lizard *Trachydosaurus rugosus* (Scincidae). *Copeia.*

BULL, C. M. AND SATRAWAHA, R., 1981. Dispersal and social organization in *Trachydosaurus rugosus.* P. 24 *in* Proceedings of the Melbourne Herpetological Symposium ed by C. B. Banks and A. A. Martin. Zoological Board Victoria, Melbourne; 199 pp.

BUSH, B., 1981. Reptiles of the Kalgoorlie-Esperance region. Privately published (P.O. Box 192, Esperance, Western Australia 6450); 48 pp.

BUSH, B., 1983. Notes on reproduction in captive *Menetia greyii* (Lacertilia; Scincidae). *West. Aust. Nat.* **15**(6): 130-31.

BUSH, B., 1983. Diurnal or nocturnal? *South Aust. Herp. Group Newsletter* 1983(Sept.): 5.

BUSH, B., 1985. A record of reproduction in captive *Delma australis* and *D. fraseri* (Lacertilia: Pygopodidae). *Herpetofauna* **15**(1): 11-12.

BUSH, B., 1986. Seasonal aggregation behaviour in a mixed population of legless lizards, *Delma australis* and *D. fraseri. Herpetofauna* **16**(1): 1-6.

BUSTARD, R., 1963a. The marbled gecko *(Phyllodactylus porphyreus)* in captivity, with special reference to egg incubation. *Brit. J. Herp.* **3**(4): 76-78.

BUSTARD, H. R., 1963b. Gecko behavioural trait: tongue wiping spectacle. *Herpetologica* **19**(3): 217-18.

BUSTARD, H. R., 1964a. Defensive behaviour shown by Australian geckos, genus *Diplodactylus. Herpetologica* **20**(3): 198-200.

BUSTARD, H. R., 1964b. Reproduction in the Australian rainforest skinks, *Siaphos equalis* and *Sphenomorphus tryoni. Copeia* **1964**(4): 715-16.

BUSTARD, H. R., 1965a. The systematic status of the Australian geckos *Gehyra variegata* (Duméril and Bibron 1836) and *Gehyra australis* Gray 1845. *Herpetologica* **20**(4): 259-72.

BUSTARD, H. R., 1965b. The systematic status of the Australian gecko, *Gehyra variegata punctata* (Fry). *Herpetologica* **21**(2): 157-58.

BUSTARD, H. R., 1965c. Observations on Australian geckos. *Herpetologica* **21**(4): 294-302.

BUSTARD, H. R., 1966a. The *Oedura tryoni* complex: east Australian rock-dwelling geckos. (Reptilia: Gekkonidae). *Bull. Brit. Mus. (Nat. Hist.) Zool.* **14**(1): 1-14.

BUSTARD, H. R., 1966b. Notes on the eggs, incubation and young of the bearded dragon, *Amphibolurus barbatus barbatus* (Cuvier). *Brit. J. Herp.* **3**(10): 252-59.

BUSTARD, H. R., 1967a. Defensive display behaviour of the Australian gecko *Nephrurus asper. Herpetologica* **23**(2): 126-29.

BUSTARD, H. R., 1967b. Gekkonid lizards adapt fat storage to desert environments. *Science* **158**: 1197-198.

BUSTARD, H. R., 1967c. Activity cycle and thermoregulation in the Australian gecko *Gehyra variegata. Copeia* **1967**(4): 753-58.

BUSTARD, H. R., 1967d. Reproduction in the Australian gekkonid genus *Oedura* Gray 1842. *Herpetologica* **23**(4): 276-84.

BUSTARD, H. R., 1967e. A mechanism for greater predator survival during cold torpor in gekkonid lizards. *Brit. J. Herp.* **4**(1): 7-8.

BUSTARD, H. R., 1968a. The ecology of the Australian gecko, *Gehyra variegata,* in northern New South Wales. *J. Zool.* **154**: 113-38.

BUSTARD, H. R., 1968b. The egg-shell of gekkonid lizards: a taxonomic adjunct. *Copeia* **1968**(1): 162-64.

BUSTARD, H. R., 1968c. *Pygopus nigriceps* (Fischer): a lizard mimicking a venomous snake. *Brit. J. Herpetol.* **4**(2): 22-24.

BUSTARD, H. R., 1968d. Temperature dependent tail autotomy mechanism in gekkonid lizards. *Herpetologica* **24**(2): 127-30.

BUSTARD, H. R., 1968e. The reptiles of Merriwindi State Forest, Pilliga West, northern New South Wales, Australia. *Herpetologica* **24**(2): 131-40.

BUSTARD, H. R., 1968f. Temperature dependent activity in the Australian gecko *Diplodactylus vittatus. Copeia* **1968**(3): 606-12.

BUSTARD, H. R., 1968g. The ecology of the Australian gecko *Heteronotia binoei* in northern New South Wales. *J. Zool.* **156**: 483-97.

BUSTARD, H. R., 1969a. The population ecology of the gekkonid lizard (*Gehyra variegata* (Duméril and Bibron)) in exploited forests in northern New South Wales. *J. Anim. Ecol.* **38**: 35-51.

BUSTARD, H. R., 1969b. The micro-environment of a natural lizard nest. *Copeia* **1969**(3): 536-39.

BUSTARD, H. R., 1969c. The ecology of the Australian geckos *Diplodactylus williamsi* and *Gehyra australis* in northern New South Wales. I and II. *Proc. Koninkl. Nederl. Akad. Wetenschappen* Ser. C **72**(4): 451-77.

BUSTARD, H. R., 1970a. *Oedura marmorata* a complex of Australian geckos (Reptilia: Gekkonidae). *Senck. Biol.* **51**(1/2): 21-40.

BUSTARD, H. R., 1970b. The role of behavior in the natural regulation of numbers in the gekkonid lizard *Gehyra variegata. Ecology* **51**: 724-28.

BUSTARD, H. R., 1970c. Activity cycle of the tropical house gecko, *Hemidactylus frenatus. Copeia* **1970**(1): 173-76.

BUSTARD, R., 1970d. Australian lizards. Collins, Sydney; 162 pp.

BUSTARD, H. R., 1970e. A population study of the scincid lizard *Egernia striolata* in northern New South Wales. I and II. *Proc. Koninkl. Nederl. Akad. Wetenschap.* Ser. C **73**(2): 186-213.

BUSTARD, H. R., 1971. A population study of the eyed gecko, *Oedura ocellata* Boulenger, in northern New South Wales, Australia. *Copeia* **1971**(4): 658-69.

BUSTARD, H. R., 1979. Defensive mechanisms and predation in gekkonid lizards. *Brit. J. Herp.* **6**(1): 9-11.

BUSTARD, H. R. AND MADERSON, P. F. A., 1965. The eating of shed epidermal material in squamate reptiles. *Herpetologica* **21**(4): 306-08.

BUTLER, G. W., 1895. On the complete or partial suppression of the right lung in the Amphisbaenidae and the left lung in snakes and snake-like lizards and amphibians. *Proc. Zool. Soc. Lond.* **1895**(4): 691-712.

BUTLER, W. H., 1970. A summary of the vertebrate fauna of Barrow Island, Western Australia. *West. Aust. Nat.* **11**(7): 149-60.

CABANAC, M., HAMMEL, T. AND HARDY, J. D., 1967. *Tiliqua scincoides:* temperature-sensitive units in lizard brain. *Science* **158**: 1050-51.

CAGLE, F. F., 1946. A lizard population on Tinian. *Copeia* **1946**(1): 4-9.

CAMP, C. L., 1923. Classification of the lizards. *Bull. Amer. Mus. Nat. Hist.* **48**(11): 289-481.

CARLZEN, G., 1982. Breeding green tree monitors. *Herpetology Journal* **12**(2): 4-6.

CARPENTER, C. C., BADHAM, J. A. AND KIMBLE, B., 1970. Behavior patterns of three species of *Amphibolurus* (Agamidae). *Copeia* **1970**(3): 497-505.

CARPENTER, C. C., GILLINGHAM, J. C., MURPHY, J. B. AND MITCHELL, L. A., 1976. A further analysis of the combat ritual of the pygmy mulga monitor, *Varanus gilleni* (Reptilia: Varanidae). *Herpetologica* **32**(1): 35-40.

CARPENTER, C. C. AND MURPHY, J. B., 1978. Tongue display by the common bluetongue *(Tiliqua scincoides)* Reptilia, Lacertilia, Scincidae). *J. Herp.* **12**(3): 428-29.

CAUGHLEY, J., 1971. Discussion on the bearded dragon — *Amphibolurus, barbatus, barbatus. Herpetofauna* **3**(4): 19-22.

CHAPMAN, A. AND DELL, J., 1975. Reptiles, amphibians and fishes. *Rec. West. Aust. Mus.* Suppl. No. 1: 34-38.

CHAPMAN, A. AND DELL, J., 1977. Reptiles and frogs of Bendering and West Bendering Nature Reserves. *Rec. West. Aust. Mus.* Suppl. No. 5: 47-55 (note page numbers repeat in this No.).

CHAPMAN, A. AND DELL, J., 1978. Reptiles and frogs of Dongolocking Nature Reserve. *Rec. West. Aust. Mus.* Suppl. No. 6: 71-77.

CHAPMAN, A. AND DELL, J., 1979. Reptiles and frogs of Buntine and Nugadong Reserves. *Rec. West. Aust. Mus.* Suppl. No. 9: 117-25.

CHAPMAN, A. AND DELL, J., 1980a. Reptiles and frogs of Yorkrakine Rock, East Yorkrakine and North Bungulla Nature Reserves. *Rec. West. Aust. Mus.* Suppl. No. 12: 69-73 (note page numbers repeat in this No.).

CHAPMAN, A. AND DELL, J., 1980b. Reptiles and frogs of Badjaling Nature Reserve, South Badjaling Nature Reserve, Yoting Town Reserve and Yoting Water Reserve. *Rec. West. Aust. Mus.* Suppl. No. 12: 59-64. (note page numbers repeat in this No.).

CHAPMAN, A. AND DELL, J., 1985. Biology and zoogeography of the amphibians and reptiles of the Western Australian wheatbelt. *Rec. West. Aust. Mus.* **12**(1): 1-46.

CHAUMONT, F., 1963. Meine Beobachtungen bei der Geburt kleiner Blauzungenskinke. *Die Aquar. Terr. Zeit.* **16**(5): 151-52. Not seen.

CHISHOLM, E. C., 1923. The principal fauna found in district of Marrangaroo, County of Cook. N.S.W. *Aust. Zool.* **3**(3): 60-71.

CHISHOLM, E. C., 1924. The principal fauna of Katoomba and district, County of Cook. N.S.W. *Aust. Zool.* **3**(6): 206-14.

CHONG, G., HEATWOLE, H. AND FIRTH, B. T., 1973. Panting thresholds of lizards — II. Diel variation in the panting threshold of *Amphibolurus muricatus. Comp. Biochem. Physiol.* **46**A: 827-29.

CHOQUENOT, D. AND GREER, A. E. In press. Intrapopulational and interspecific variation in digital limb bones and presacral vertebrae of the genus *Hemiergis* (Lacertilia, Scincidae). *J. Herp.*

CHOU, L. M., 1974. Diet of the common Singapore house gecko, *Hemidactylus frenatus. J. Singapore Nat'l. Acad. Sci.* **4**(1): 11-13.

CHOU, L. M., 1979. Eggs and incubation period of three gekkonid lizards. *Copeia* **1979**(3): 552-54.

CHRISTIAN, T., 1977a. Notes on Centralian Bluetongues *(Tiliqua Multifasciata). Newsletter Vict. Herp. Soc.* **1**: 8-9.

CHRISTIAN, T., 1977b. Notes on *Varanus glebopalma. Vict. Herp. Soc. Newsletter* **6**: 11-13.

CHRISTIAN, T., 1979. Notes on Spencers monitor *(Varanus spenceri). Vict. Herp. Soc. Newsletter* **14**: 13-14.

CHRISTIAN, T., 1981. *Varanus tristis* — a variable monitor. *Herpetofauna* **12**(2): 7-12.

CHURCH, G., 1962. The reproductive cycles of the Javanese house geckos, *Cosymbotus platyurus, Hemidactylus frenatus,* and *Peropus mutilatus. Copeia* **1962**(2): 262-69.

CLARKE, C. J., 1965. A comparison between some Australian five-fingered lizards of the genus *Leiolopisma* Duméril & Bibron (Lacertilia: Scincidae). *Aust. J. Zool.* **13**: 577-92.

CLIFFORD, H. T. AND HAMLEY, T., 1982. Seed dispersal by water-dragons. *Qld Nat.* **23**(5-6): 49.

COGGER, H. G., 1957. Investigations in the gekkonid genus *Oedura* Gray. *Proc. Linn. Soc. New South Wales.* **82**(2): 167-79.

COGGER, H. G., 1958. Australian geckoes. *Aust. Mus. Mag.* **12**(11): 343-47.

COGGER, H. G., 1959. Australian goannas. *Aust. Mus. Mag.* **13**(3): 71-75.

COGGER, H. G., 1960a. The ecology, morphology, distribution and speciation of a new species and subspecies of the genus *Egernia* (Lacertilia: Scincidae). *Rec. Aust. Mus.* **25**(5): 95-105.

COGGER, H. G., 1960b. Snakes, lizards and chelonians. *Aust. Mus. Mag.* **13**(8): 250-53.

COGGER, H. G., 1961. An investigation of the Australian members of the family Agamidae (Lacertilia) and their phylogenetic relationships. Unpubl. M.Sc. Thesis, Sydney University; 129 pp. Not seen.

COGGER, H. G., 1964. The comparative osteology and systematic status of the gekkonid genera *Afroedura* Loveridge and *Oedura* Gray. *Proc. Linn. Soc. New South Wales* **89**(3): 364-72.

COGGER, H. G., 1965. Reptiles and frogs of Australia's arid regions. *Aust. Nat. Hist.* **15**(4): 128-31.

COGGER, H. G., 1967. Australian reptiles in colour. A. H. and A. W. Reed, Sydney; 112 pp.

COGGER, H. G., 1969. A study of the ecology and biology of the Mallee Dragon *(Amphibolurus fordi)* and its adaptations to survival in an arid environment. Unpublished Ph.D. Thesis, Macquarie University, Sydney, New South Wales; 298 pp.

COGGER, H. G., 1971. The reptiles of Lord Howe Island. *Proc. Linn. Soc. New South Wales.* **96**(1): 23-38.

COGGER, H. G., 1973. Reptiles and amphibians of Coburg Peninsula. *Aust. Nat. Hist.* **17**(9): 311-16.

COGGER, H. G., 1974. Thermal relations of the mallee dragon *Amphibolurus fordi* (Lacertilia: Agamidae). *Aust. J. Zool.* **22**: 319-39.

COGGER, H. G., 1975. New lizards of the genus *Pseudothecadatylus* (Lacertilia: Gekkonidae) from Arnhem Land and northwestern Australia. *Rec. Aus. Mus.* **30**(3): 87-97.

COGGER, H. G., 1978. Reproductive cycles, fat body cycles and socio-sexual behaviour in the mallee dragon, *Amphibolurus fordi* (Lacertilia: Agamidae). *Aust. J. Zool.* **26**: 653-72.

COGGER, H. G., 1984. Reptiles in the Australian arid zone. Pp. 235-52 in Arid Australia ed by H. G. Cogger and E. E. Cameron. Surrey Beatty and Sons Pty Ltd, Chipping Norton, New South Wales; 338 pp.

COGGER, H. G., 1986. Reptiles and amphibians of Australia. Reed Books Pty Ltd, Frenchs Forest, New South Wales; 688 pp.

COGGER, H. G. AND HEATWOLE, H., 1981. The Australian reptiles: origins, biogeography, distribution patterns and island evolution. Pp. 1333-73 in Ecological biogeography of Australia ed by A. Keast. W. Junk, The Hague; 2142 pp.

COGGER, H. G. AND LINDNER, D. A., 1969. Marine turtles in northern Australia. Aust. Zool. 15(2): 150-59.

COGGER, H. G. AND LINDNER, D. A., 1974. Frogs and reptiles. Pp. 63-107 in Fauna survey of the Port Essington District, Coburg Peninsula, Northern Territory of Australia ed by H. J. Frith and J. H. Calaby. Division of Wildlife Research Tech. Pap. No. 28, C.S.I.R.O.; 208 pp.

COGGER, H. G., CAMERON, E. E. AND COGGER, H. M., 1983. Zoological catalogue of Australia. Vol. 1. Amphibia and Reptilia. Bureau of Flora and Fauna, Canberra; 313 pp.

COGGER, H., SADLIER, R. AND CAMERON, E., 1983. The terrestrial reptiles of Australia's island territories. Aust. Nat'l Parks Wildlife Ser. Spec. Publ. 11; 80 pp.

COLE, C. E., 1930. Notes on the sleepy lizard. Trachysaurus rugosus (Gray). South Aust. Nat. 12(1): 2-4.

COLE, C. J., 1966. Femoral glands in lizards: a review. Herpetologica 22(3): 199-206.

COLEMAN, E., 1944. Lizards under domestication. Vict. Nat. 61(8): 137-38.

CONANT, R. AND HUDSON, R. G., 1949. Longevity records for reptiles and amphibians in the Philadelphia Zoological Garden. Herpetologica 5(1): 1-8.

COOK, R., 1973. The wall lizard Cryptoblepharus boutonii virgatus. Herpetofauna 6(2): 15-16.

COOPER, J. E. AND JACKSON, O. F. (eds), 1981. Diseases of the Reptilia. 2 Vols. Academic Press, London; 584 pp.

COOPER, W. E., 1986. Chromatic components of female secondary sexual colouration; influence on social behaviour of male keeled earless lizards (Holbrookia propinqua). Copeia 1986(4): 980-86.

COPLAND, S. J., 1946a. Geographic variation in the lizard Hemiergis decresiensis (Fitzinger). Proc. Linn. Soc. New South Wales 70(3-4): 62-92.

COPLAND, S. J., 1946b. Catalogue of reptiles in the Macleay Museum. Part I. Sphenomorphus pardalis pardalis (Macleay) and Sphenomorphus nigricaudis nigricaudis (Macleay). Proc. Linn. Soc. New South Wales. 70(5-6): 291-311.

COPLAND, S. J., 1947a. Catalogue of reptiles in the Macleay Museum. Part II. Sphenomorphus spaldingi (Macleay). Proc. Linn. Soc. New South Wales. 71(3-4): 136-44.

COPLAND, S. J., 1947b. Taxonomic notes on the genus Ablepharus (Sauria: Scincidae). I. A new species from the Darling River. Proc. Linn. Soc. New South Wales. 71(5-6): 282-86.

COPLAND, S. J., 1947c. Reptiles occurring above the winter snowline at Mt Kosciusko. Proc. Linn. Soc. New South Wales 72(1-2): 69-72.

COPLAND, S. J., 1949. Taxonomic notes on the genus Ablepharus (Sauria: Scincidae). II. The races of Ablepharus burnetti Oudemans. Proc. Linn. Soc. New South Wales. 73(5-6): 362-71.

COPLAND, S. J., 1952a. Taxonomic notes on the genus Ablepharus (Sauria: Scincidae). III. A new species from north-west Australia. Proc. Linn. Soc. New South Wales. 77(3-4): 121-25.

COPLAND, S. J., 1952b. A mainland race of the scincid lizard Lygosoma truncatum (Peters). Proc. Linn. Soc. New South Wales. 77(3-4): 126-31.

COPLAND, S., 1953. Presidential address. Recent Australian herpetology. Proc. Linn. Soc. New South Wales. 78 (1-2): v-xxxvii.

CORNWALL, G., 1965. Geckos. South Aust. Nat. 40(2): 27.

COURTICE, G. P., 1981a. Respiration in the eastern water dragon, Physignathus lesueurii (Agamidae). Comp. Biochem. Physiol. 68A: 429-36.

COURTICE, G. P., 1981b. The effect of temperature on bimodal gas exchange and the respiratory exchange ratios in the water dragon, Physignathus lesueurii. Comp. Biochem. Physiol. 68A: 437-41.

COURTICE, G. P., 1981c. Changes in skin perfusion in response to local changes in pCO2 in a diving lizard, Physignathus lesueurii. Comp. Biochem. Physiol. 69A: 805-07.

COURTICE, G. P., 1985. Effect of hypoxia on cardiac vagal action in a lizard Physignathus lesueurii, and its contribution to diving brachycardia. Pp. 373-77 in Biology of Australasian frogs and reptiles ed by G. Grigg, R. Shine and H. Ehmann. Surrey Beatty and Sons, Pty Ltd, Chipping Norton, New South Wales; 527 pp.

COVACEVICH, J., 1975. A review of the genus Phyllurus (Lacertilia: Gekkonidae). Mem. Qld Mus. 17(2): 293-303.

COVACEVICH, J., 1984. A biogeographically significant new species of Leiolopisma (Scincidae) from north eastern Queensland. Mem. Qld Mus. 21(2): 401-11.

COVACEVICH, J. AND INGRAM, G. J., 1978. An undescribed species of rock dwelling Cryptoblepharus (Lacertilia: Scincidae). Mem. Qld Mus. 18(2): 151-54.

COVACEVICH, J. AND MCDONALD, K. R., 1980. Two new species of skinks from mid-eastern Queensland rainforest. Mem. Qld Mus. 20(1): 95-101.

COVACEVICH, J., INGRAM, G. J. AND CZECHURA, G. V., 1982. Rare frogs and reptiles of Cape Peninsula, Australia. Biol. Conserv. 22: 283-94.

COVENTRY, A. J., 1976. A new species of Hemiergis (Scincidae: Lygosominae) from Victoria. Mem. Nat'l Mus. Vict. 37: 23-26.

COVENTRY, A. J. AND ROBERTSON, P., 1980. New records of scincid lizards from Victoria. Vict. Nat. 97(5): 190-93.

COWLES, R. B., 1930. The life history of Varanus niloticus (Linnaeus) as observed in Natal South Africa. J. Ent. Zool. 22(1): 1-32.

CREWS, D., 1987. Courtship in unisexual lizards: a model for brain evolution. Sci. Amer. 257(6): 72-77.

CROMBIE, R. I. AND DIXON, J. R. In press. The status of Mabuya deserticola (Sauria: Scincidae) with comments on the influence of Oceania on the neotropic herpetofauna. Herpetologica.

CROME, B., 1981. The diet of some ground-layer lizards in three woodlands of the New England Tableland of Australia. Herpetofauna 13(1): 4-11.

CUELLAR, O., 1984. Histocompatibility in Hawaiian and Polynesian populations of the parthenogenetic gecko Lepidodactylus lugubris. Evolution 38(1): 176-85.

CUELLAR, O. AND KLUGE, A. G., 1972. Natural parthenogensis in the gekkonid lizard Lepidodactylus lugubris. J. Genetics 61: 14-26.

CZECHURA, G. V., 1974. A new south-east locality for the skink *Anomalopus reticulatus*. *Herpetofauna* **7**(1): 24-25.

CZECHURA, G. V., 1980. The emerald monitor *Varanus prasinus* (Schegel): an addition to the Australian mainland herpetofauna. *Mem. Qld Mus.* **20**(1): 103-09.

CZECHURA, G. V., 1981. The rare scincid lizard, *Nannoscincus graciloides*: a reappraisal. *J. Herp.* **15**(3): 315-20.

CZECHURA, G. V., 1986. Skinks of the *Ctenotus schevilli* species group. *Mem. Qld Mus.* **22**(2): 289-97.

CZECHURA, G. AND MILES, J., 1983. Lizards. Pp. 80-99 *in* Wildlife of the Brisbane Area ed by W. Davies. Jacaranda Press, Milton; 216 pp.

CZECHURA, G. V. AND WOMBEY, J., 1982. Three new striped skinks, (*Ctenotus*, Lacertilia, Scincidae) from Queensland. *Mem. Qld Mus.* **20**(3): 639-45.

DAAN, S. AND BELTERMAN, T., 1968. Lateral bending in locomotion of some lower tetrapods. 1 & 2 *Proc. Koninkl. Nederl. Akad. Wetenschap.* Ser. C. **71**(3): 245-66.

DANIELS, C. B., 1983. Running: an escape strategy enhanced by autotomy. *Herpetologica* **39**(2): 162-65.

DANIELS, C. B., 1984. The importance of caudal lipid in the gecko *Phyllodactylus marmoratus*. *Herpetologica* **40**(3): 337-44.

DANIELS, C. B., 1987. Aspects of the aquatic feeding ecology of the riparian skink *Sphenomorphus quoyii*. *Aust. J. Zool.* **35**(3): 253-58.

DANIELS, C. B., FLAHERTY, S. P. AND SIMBOTWE, M. P., 1986. Tail size and effectiveness of autotomy in a lizard. *J. Herp.* **20**(1): 93-96.

DAREVSKY, I. S., KUPRIYANOVA, L. A. AND ROSHCHIN, V. V., 1984. A new all-female triploid species of gecko and karyological data on the bisexual *Hemidactylus frenatus* from Vietnam. *J. Herp.* **18**(3): 277-284.

DAVEY, H. W., 1923. The Moloch lizard, *Moloch horridus*, Gray. *Vict. Nat.* **40**(4): 58-60.

DAVEY, H. W., 1924. Marbled geckos hatched in captivity. *Vict. Nat.* **41**(3): 51.

DAVEY, H. W., 1944. Some lizards I have kept. *Vict. Nat.* **61**(5,6): 82-84, 101-03.

DAVEY, H. W., 1945. Lizard victims of a caged gecko. *Vict. Nat.* **61**(12): 216.

DAVEY, K., 1970. Australian lizards. Landsdowne Press, Melbourne; 111 pp.

DAVIDGE, C., 1979. A census of a community of small terrestrial vertebrates. *Aust. J. Ecol.* **4**(2): 165-70.

DAVIDGE, C., 1980. Reproduction in the herpetofaunal community of a *Banksia* woodland near Perth, Western Australia. *Aust. J. Zool.* **28**(3): 435-43.

DAWSON, W. R., 1960. Physiological responses to temperature in the lizard *Eumeces obsoletus*. *Physiol. Zoöl.* **33**(2): 87-103.

DAY, K., 1980. Notes on the pygmy spiny tailed skink, *Egernia depressa* (Gunther) in captivity. *Herpetofauna* **11**(2): 29.

DAY, K., 1982. A day down the gorge. Monitor **1**: 69-73.

DECKERT, K., 1940. *Pygopus nigriceps* Fischer, ein westaustralischer Flossenfuss. *Wschr. Aquar. -U. Terrarienk.* **37**: 269-70. Not seen.

DELEAN, S. 1980. A new record of the pygmy mulga monitor, *Varanus gilleni* (Lucas and Frost). *Herpetofauna* **12**(1): 35.

DELEAN, S., 1981. Notes on aggressive behaviour by Goulds goannas *(Varanus gouldii)* in captivity. *Herpetofauna* **12**(2): 31.

DELEAN, S. AND HARVEY, C., 1981. Some observations on the knob-tailed gecko, *Nephrurus laevissimus* in the wild. *Herpetofauna* **13**(1): 1-3.

DELEAN, S. AND HARVEY, C., 1984. Notes on the reproduction of *Nephrurus deleani* (Reptilia: Gekkonidae). *Trans. Roy. Soc. South Aust.* **108**(3/4): 221-22.

DE LISSA, G., 1981. Notes on the skink *Sphenomorphus tenuis*. *Herpetofauna* **13**(1): 33.

DELL, J., 1985. Arboreal geckos feeding on plant sap. *West. Aust. Nat.* **16**(4): 69-70.

DELL, J. AND CHAPMAN, A., 1977. Reptiles and frogs of Cockleshell Gully Reserve. *Rec. West. Aust. Mus.* Suppl. No. 4: 75-86.

DELL, J. AND CHAPMAN, A., 1978. Reptiles and frogs of Durokoppin and Kodj Kodjin Nature Reserves. *Rec. West. Aust. Mus.* Suppl. No. 7: 69-74.

DELL, J. AND CHAPMAN, A., 1979a. Reptiles and frogs of Wilroy Nature Reserve. *Rec. West. Aust. Mus.* Suppl. No. 8: 47-51 (note page numbers repeat in this No.).

DELL, J. AND CHAPMAN, A., 1979b. Reptiles and frogs of Marchagee Nature Reserve. *Rec. West. Aust. Mus.* Suppl. No. 8: 43-48 (note page numbers repeat in this No.).

DELL, J. AND CHAPMAN, A., 1981a. Reptiles and frogs of Billyacatting Hill Nature Reserve. *Rec. West. Aust. Mus.* Suppl. No. 13: 49-51 (note page numbers repeat in this No.).

DELL, J. AND CHAPMAN, A., 1981b. Reptiles and frogs of East Yuna and Bindoo Hill Nature Reserves. *Rec. West. Aust. Mus.* Suppl. No. 13: 95-102 (note page numbers repeat in this No.).

DELL, J. AND HAROLD, G., 1979. Reptiles and frogs of Yornaning Nature Reserve. *Rec. West. Aust. Mus.* Suppl. No. 8: 43-47 (note page numbers repeat in this No.).

DE VIS, C. W., 1884. Myology of *Chlamydosaurus Kingii*. *Proc. Linn. Soc. New South Wales*. **8**(3): 300-20.

DINARDO, J., 1985. A note on longevity in *Egernia cunninghami* with a new captive longevity record. *Herpetofauna* **15**(1): 14-15.

DIXON, J. R., 1964. The systematics and distribution of lizards of the genus *Phyllodactylus* in North and Central America. *Res. Cent. New Mexico State Univ. Sci. Bull.* 64-1; 139 pp.

DIXON, J. R. AND KLUGE, A. G., 1964. A new gekkonid lizard genus from Australia. *Copeia* **1964**(1): 174-80.

DONE, B. S. AND HEATWOLE, H., 1977a. Social behavior of some Australian skinks. *Copeia* **1977**(3): 419-30.

DONE, B. S. AND HEATWOLE, H., 1977b. Effects of hormones on the aggressive behaviour and social organization of the scincid lizard, *Sphenomorphus kosciuskoi*. *Z. Tierpsychol.* **44**(1): 1-12.

DONNELLAN, S. C., 1985. The evolution of sex chromosomes in scincid lizards. Unpublished Ph.D. Thesis, Macquarie University; 142 pp.

DOUGLAS, A. M. AND RIDE, W. D. L., 1962. The results of an expedition to Bernier and Dorre Islands, Shark Bay, Western Australia. Reptiles. *Western Australia Fisheries Dept. Fauna Bull.* **2**: 113-19.

DRYDEN, G. L. AND TAYLOR, E. H., 1969. Reptiles from the Mariana and Caroline Islands. *Univ. Kans. Sci. Bull.* **48**(8): 269-79.

DUFAURE, J. P. AND HUBERT, J., 1961. Table de développement du lézard vivipare: *Lacerta (Zootoca) vivipara* Jaquin. *Archs. Anat. Microsc. Morph. Exp.* **50:** 309-327.

DUNSON, W. A., 1974. Salt gland secretion in a mangrove monitor lizard. *Comp. Biochem. Physiol.* **47A:** 1245-55.

DUNSON, W. A., 1982. Low water vapour conductance of hard-shelled eggs of the gecko lizards *Hemidactylus* and *Lepidodactylus. J. Exp. Zool.* **219:** 377-79.

EDWARDS, A. R., 1978. Studies of thermoregulation in the sleepy lizard, *Trachydosaurus rugosus* (Gray). Honours Thesis, School of Biological Sciences, Flinders University; 53 pp.

EHMANN, H., 1976a. The reptiles of the Mt. Lofty Ranges, South Australia. Part I. *Herpetofauna* **8**(1): 2-5.

EHMANN, H., 1076b. The reptiles of the Mount Lofty Ranges South Australia, Part 2. *Herpetofauna* **8**(2): 5-13.

EHMANN, H., 1976c. Cover photograph — stellate knob-tailed gecko *(Nephrurus stellatus). Herpetofauna* **8**(2): 22.

EHMANN, H., 1979. The rediscovery of the bronzeback legless lizard. *Wildlife in Australia.* **16**(1): 13-14.

EHMANN, H., 1980. Diurnal perching by the southern spiny-tailed gecko *Diplodactylus intermedius. Herpetofauna* **12**(1): 37.

EHMANN, H., 1981. The natural history and conservation of the bronzeback *(Ophidiocephalus taeniatus* Lucas and Frost) (Lacertilia, Pygopodidae). Pp. 7-13 *in* Proceedings of the Melbourne Herpetological Symposium ed by C. B. Banks and A. A. Martin. Zoology Board Victoria, Melbourne; 199 pp.

EHMANN, H., 1982. The natural history and conservation status of the Adelaide pigmy bluetongue lizard *Tiliqua adelaidensis. Herpetofauna* **14**(1): 61-76.

EHMANN, H. AND METCALFE, D., 1978. The rediscovery of *Ophidiocephalus taeniatus* Lucas and Frost (Pygopodidae, Lacertilia) the bronzeback. *Herpetofauna* **9**(2): 8-10.

EIDENMÜLLER, B. AND HORN, H.-G., 1985. Eingene Nachzuchten und der gegenwärtige Stand der Nachzucht von *Varanus (Odatria) storri* Mertens, 1966 (Sauria: Varanidae). *Salamandra* **21**(1): 55-61.

ESTES, R., 1983. Sauria terrestria, Amphisbaenia. Handbuch der Paläoherpetologie **10**A: 1-249.

ESTES, R., 1984. Fish, amphibians and reptiles from the Etadunna Formation, Miocene of South Australia. *Aust. Zool.* **21**(4): 335-43.

ESTES, R. AND WILLIAMS, E. E., 1984. Ontogenetic variation in the molariform teeth of lizards. *J. Vert. Paleon.* **4**(1): 96-107.

ESTES, R., FRAZZETTA, T. H. AND WILLIAMS, E. E., 1970. Studies on the fossil snake *Dinilysia patagonica* Woodward: Part 1. Cranial morphology. *Bull. Mus. Comp. Zool.* **140**(2): 25-74.

FEDER, M. E. AND FEDER, J. H., 1981. Diel variation of oxygen consumption in three species of Philippine gekkonid lizards. *Copeia* **1981**(1): 204-09.

FEJERVARY, G. J. de, 1918. Contributions to a monograph on fossil Varanidae and on Megalanidae. *Mus. Nat'l. Hungarici Ann.* **16:** 341-467. Not seen.

FERGUSSON, B. AND ALGAR, D., 1986. Home range and activity patterns of pregnant female skinks, *Tiliqua rugosa. Aust. Wildl. Res.* **13**(2): 287-94.

FERGUSSON, B., BRADSHAW, S. D. AND CANNON, J. R., 1985. Hormonal control of femoral gland secretions in the lizard, *Amphibolorus ornatus. General Comp. Endocr.* **57**(3): 371-76.

FIELD, R., 1980. The pink-tongued skink *(Tiliqua gerrardii)* in captivity. *Herpetofauna* **11**(2): 6-10.

FIRTH, B. T., 1979. Panting thresholds of lizards: role of the eyes in panting in a gecko, *Oedura tryoni. Comp. Biochem. Physiol.* **64A**(1): 121-23.

FIRTH, B. T. AND HEATWOLE, H., 1976. Panting thresholds of lizards: the role of the pineal complex in panting responses in an agamid, *Amphibolurus muricatus. Gen. Comp. Endocrin.* **29:** 388-401.

FITZGERALD, M., 1983a. Some observations on the reproductive biology of the common scaly-foot *Pygopus lepidopodus. Herpetofauna* **14**(2): 79-80.

FITZGERALD, M., 1983b. A note on water collection by the bearded dragon *Amphibolurus vitticeps. Herpetofauna* **14**(2): 93.

FITZGERALD, M., 1985. A New South Wales record for the freckled tree monitor *Varanus tristis orientalis. Herpetofauna* **15**(1): 23.

FLEAY, D., 1931. Blue-tongued lizards. *Vict. Nat.* **48**(1): 9-10.

FLEAY, D., 1937. Black snakes in combat. *Proc. Roy. Zool. Soc. New South Wales.* **1936-37:** 40-42.

FLEAY, D., 1950. Goannas: giant lizards of the Australian bush. *Animal Kingdom* **53**(3): 92-96.

FLEAY, D., 1951. Savage battles between snakes. *Walkabout* **17**(5): 10-14.

FLEAY, D., 1953. Tree goanna's interesting nest site. *Walkabout* **19**(7): 25.

FLOWER, S. S., 1925. Contributions to our knowledge of the duration of life in vertebrate animals. — III. Reptiles. *Proc. Zool. Soc. Lond.* **1925**(3): 911-81.

FLOWER, S. S., 1937. Further notes on the duration of life in animals. — III. Reptiles. *Proc. Zool. Soc. Lond.* **107**(A): 1-39.

FLYNN, T. T., 1923. On the occurrence of a true allantoplacenta of the conjoint type in an Australian lizard. *Rec. Aust. Mus.* **14**(1): 72-77.

FORD, J., 1963a. The reptilian fauna of the islands between Dongara and Lancelin, Western Australia. *West. Aust. Nat.* **8**(6): 135-42.

FORD, J., 1963b. The distribution and variation of the skinks *Egernia pulchra* and *E. bos* in Western Australia. *West. Aust. Nat.* **9**(2): 25-29.

FORD, J., 1965a. The reptilian fauna of the islands between Dongara and Lancelin, Western Australia: additional notes. *West. Aust. Nat.* **9**(7): 174-75.

FORD, J., 1965b. The skink *Egernia pulchra* in the Stirling Range. *West. Aust. Nat.* **9**(7): 175-76.

FORD, J., 1968. Distribution and variation of the skink *Ctenotus labillardieri* (Gray) of southwestern Australia. *J. Roy. Soc. West. Aust.* **51**(3): 68-75.

FOWLER, J., 1974. The hybridisation of an alpine water skink and a common water skink. *South Australian Herpetologist* **2**(1): 13.

FOWLER, J., 1978. Breeding. *South Aust. Herp. Group Newsletter* 1978(April): 4.

FOWLER, J., 1985. The hybridisation of an alpine water skink and a common water skink. *South Aust. Herp. Group Newsletter* 1985(February): 4.

FOX, W., 1948. Effect of temperature on development of scutellation in the garter snake, *Thamnophis elegans atratus. Copeia* **1948**(4): 252-62.

FOX, W., GORDON, C. AND FOX, M. H., 1961. Morphological effects of low temperatures during the embryonic development of the garter snake, *Thamnophis elegans. Zoologica* (N.Y.) **46**(2): 57-71.

FRANCIS, M., 1981. Observations on heat regulating behaviour in captive specimens of *Diporiphora winneckei* (Lucas and Frost). *Herpetofauna* **12**(2): 35-36.

FRANKENBERG, E. AND WERNER, Y. L., 1984. The defensive vocal "distress" repertoire of gekkonid lizards: intra- and inter-specific variation. *Amphibia-Reptilia* **5**(2): 109-24.

FRASER, S. P., 1980. Insights into the physiological and behavioural thermoregulatory abilities of small lizards. Unpublished Honours Thesis, Zoology Department, Sydney University; 58 pp.

FRASER, S. AND GRIGG, G. C., 1984. Control of thermal conductance is insignificant to thermoregulation in small reptiles. *Physiol Zoöl.* **57**(4): 392-400.

FRAUCA, H., 1966. Harry Frauca's book of reptiles. Jacaranda Press, Brisbane; 100 pp.

FROUDIST, J., 1970. Thermoregulation in geckoes, with respect to its possible influence on nocturnal habit. Unpublished B.Sc. Honours Thesis, University Western Australia; 42 pp.

FUGLER, C. M., 1966. *Lepidodactylus lugubris* Duméril and Bibron in western South America. *J. Ohio Herp. Soc.* **5**(4): 162.

FUHN, I. E., 1967. *Pseudemoia*, eine neue monotypische Gattung aus Südostaustralien *(Ablepharus/Emo/aspenceri)* Lucas and Frost, 1894. *Zool. Anz.* **179**(3/4): 243-47.

FUKADA, H., 1965. Breeding habits of some Japanese reptiles (critical review). *Bull. Kyoto Gakugei Univ.* Ser. B, No. 27: 65-82.

FYFE, G., 1979a. Notes on the black-headed monitor *(Varanus tristis)*. *Vict. Herp. Soc. Newsletter* **13**: 16.

FYFE, G., 1979b. Spotlighting for reptiles. *Vict. Herp. Soc. Newsletter* **14**: 19.

FYFE, G., 1980. Notes on the black-headed monitor *(Varanus tristis)*. *Slerpetofauna* **1**(1): 15.

FYFE, G., 1981a. Range extension for *Hemidactylus frenatus*, the Asian house gecko. *Herpetofauna* **13**(1): 33.

FYFE, G., 1981b. Nocturnal sightings of two dragon species. *Herpetofauna* **13**(1): 34.

FYFE, G., 1985. Some notes on sympatry between *Tiliqua occipitalis* and *Tiliqua multifasciata* in the Ayers Rock region and their association with aboriginal people of the area. *Herpetofauna* **15**(1): 18-19.

GADOW, H., 1901. Amphibia and reptiles. Cambridge Natural History series. Macmillan and Co., Ltd, London; 668 pp.

GALLIFORD, M., 1978. A brief study of *Nephrurus stellatus* in captivity. *South Aust. Herp. Group Newsletter* 1978(mid-year?): 1-2.

GALLIFORD, M., 1981. Notes of the starred knob-tailed gecko, *Nephrurus stellatus*, caught spotlighting. *Herpetofauna* **12**(2): 33-34.

GANS, C., 1961. The feeding mechanism of snakes and its possible evolution. *Amer. Zool.* **1**(2): 217-27.

GANS, C., 1969. Comments on inertial feeding. *Copeia* **1969**(4): 855-57.

GANS, C., MERLIN, R. AND BLUMER, W. F. C., 1982. The water-collecting mechanism of *Moloch horridus* re-examined. *Amphibia-Reptilia* **3**(1): 57-64.

GARLAND, T., jun., 1985. Ontogenetic and individual variation in size, shape and speed in the Australian agamid lizard *Amphibolurus nuchalis*. *J. Zool.* (Lond.) *(A)* **207**(3): 425-39.

GEORGES, A., 1979. Head-body temperature differences in the Australian blue-tongued lizard, *Tiliqua scincoides* during radiant heating. *J. Thermal Biology* **4**(3): 213-17.

GERMAN, P., 1986. Communal egg laying in the weasel skink *(Lampropholis mustelina)*. *Thylacinus* **11**(3): 21.

GIBBONS, J. R. H., 1977. Comparative ecology and behaviour of lizards of the *Amphibolurus decresii* species complex. Unpublished Ph.D. Dissertation, University of Adelaide. Not seen.

GIBBONS, J. R. H., 1979. The hind leg pushup display of the *Amphibolurus decresii* species complex (Lacertilia: Agamidae). *Copeia* **1979**(1): 29-40.

GIBBONS, J. R. H. AND LILLYWHITE, H. B., 1981. Ecological segregation, colour matching, and speciation in lizards of the *Amphibolurus decresii* species complex (Lacertilia: Agamidae). *Ecology* **62**(6): 1573-84.

GIBBONS, J. R. H. AND ZUG, G. R., 1987. *Gehyra, Hemidactylus* and *Nactus* (Pacific Geckos). Eggs and Hatchlings. *Herpetol. Rev.* **18**(2): 35-36.

GIDDINGS, S., 1980. Breeding water dragons *Physignathus lesueurii lesueurii* and *Physignathus lesueurii howittii* in captivity. *South Aust. Herp. Group Newsletter* 1980(Feb.): 2-3.

GIDDINGS, S., 1983a. Easter trip to Noonning homestead. *South Aust. Herp. Group Newsletter* 1983(Sept.): 2-3.

GIDDINGS, S., 1983b. An observed oviposition of an eastern water dragon *(Physignathus lesueurii)*. *South Aust. Herp. Group Newsletter* 1983(Sept.): 8-9.

GIDDINGS, S., 1984. An observation of parturition of a blotched-bluetongue. *(Tiliqua nigrolutea)*. *South Aust. Herp. Group Newsletter* 1984(May): 4.

GILL, B., 1986. Collins handguide to the frogs and reptiles of New Zealand. Collins, Auckland; 112 pp.

GILLAM, M. W., CAWOOD, I. S. AND HONNER, G. J., 1978. New reptile records from central Australia, Northern Territory. *Herpetofauna* **9**(2): 18-25.

GLAUERT, L., 1959. Herpetological miscellanea. IX. — *Ablepharus wotjulum*, a new skink from west Kimberley. *West. Aust. Nat.* **6**(8): 192-93.

GLAZEBROOK, R., 1977. Old man goanna. *North Qld Nat.* **44**(170): 4-6.

GOIN, C. J., GOIN, O. B. AND ZUG, G. R., 1978. Introduction to herpetology. Third Edition. W. H. Freeman and Co., San Francisco; 378 pp.

GORMAN, G. C. AND GRESS, F., 1970. Sex chromosomes of a pygopodid lizard, *Lialis burtonis*. *Experientia* **26**: 206-07.

GOW, G., 1979. Notes on the biology of *Nephrurus asper* Günther 1876. *North. Terr. Nat.* **1**(2): 19-20.

GOW, G. F., 1982. Notes on the reproductive biology of the pygmy mulga goanna *Varanus gilleni* Lucas & Frost 1895. *North. Terr. Nat.* No. 5: 4-5.

GREEN, B., 1972a. Water and electrolyte balance in the sand goanna *Varanus gouldii*. Unpublished Ph.D. Thesis, University of Adelaide. Not seen.

GREEN, B., 1972b. Water losses of the sand goanna *(Varanus gouldii)* in its natural environments. *Ecology* **53**: 452-57.

GREEN, B. AND KING, D., 1978. Home range and activity patterns of the sand goanna, *Varanus gouldii* (Reptilia: Varanidae). *Aust. J. Wildl. Res.* **5**: 417-24.

GREEN, D., 1973a. The reptiles of the outer north-western suburbs of Sydney. *Herpetofauna* **6**(2): 2-5.

GREEN, D., 1973b. Observations on the southern leaf-tailed gecko *Phyllurus platurus* (Shaw). *Herpetofauna* **6**(2): 21-24.

GREEN, R. H., 1965. Two skink lizards newly recorded from Tasmania. *Rec. Queen Vict. Mus. Launceston* No. 9: 1-4.

GREEN, R. H., 1984. The fauna of Ordnance Point, north-western Tasmania. *Rec. Queen Vict. Mus. Launceston* No. 84: 32-52.

GREEN, R. H. AND MCGARVIE, A. M., 1971. The birds of King Island with reference to other western Bass Strait islands and annotated lists of the vertebrate fauna. *Rec. Queen Vict. Mus.* No. 40: 1-42.

GREENE, H. W., 1983. Dietary correlates of the origin and radiation of snakes. *Amer. Zool.* **23**(2): 431-41.

GREENE, H. W., 1986. Diet and arboreality in the emerald monitor, *Varanus prasinus*, with comments on the study of adaptation. *Fieldiana (Zool.)* N.S. No. 31; 12 pp.

GREER, A. E., 1967. A new generic arrangement for some Australian scincid lizards. *Breviora* No. 267; 19 pp.

GREER, A. E. 1970. A subfamilial classification of scincid lizards. *Bull. Mus. Comp. Zool.* **139**(3): 151-84.

GREER, A. E., 1977. On the adaptive significance of the loss of an oviduct in reptiles. *Proc. Linn. Soc. New South Wales* **101**(4): 242-49.

GREER, A. E., 1979a. *Eremiascincus*, a new generic name for some Australian sand swimming skinks (Lacertilia: Scincidae). *Rec. Aust. Mus.* **32**(7): 321-38.

GREER, A. E., 1979b. A phylogenetic subdivision of Australian skinks. *Rec. Aust. Mus.* **32**(8): 339-71.

GREER, A. E., 1979c. A new *Sphenomorphus* (Lacertilia: Scincidae) from the rainforests of northeastern Queensland. *Rec. Aust. Mus.* **32**(9): 373-88.

GREER, A. E., 1979d. A new species of *Lerista* (Lacertilia: Scincidae) from northern Queensland, with remarks on the origin of the genus. *Rec. Aust. Mus.* **32**(10): 383-88.

GREER, A. E., 1980a. Critical thermal maximum temperatures in Australian scincid lizards: their ecological and evolutionary significance. *Aust. J. Zool.* **28**: 91-102.

GREER, A. E., 1980b. A new species of *Morethia* (Lacertilia: Scincidae) from northern Australia, with comments on the biology and relationships of the genus. *Rec. Aust. Mus.* **33**(2): 89-122.

GREER, A. E., 1982. A new species of *Leiolopisma* (Lacertilia: Scincidae) from Western Australia, with notes on the biology and relationships of other Australian species. *Rec. Aust. Mus.* **34**(12): 549-73.

GREER, A. E., 1983a. A new species of *Lerista* from Groote Eylandt and the Sir Edward Pellew Group in northern Australia. *J. Herp.* **17**(1): 48-53.

GREER, A. E., 1983b. The Australian scincid lizard genus *Calyptotis* De Vis: resurrection of the name, description of four new species, and discussion of relationships. *Rec. Aust. Mus.* **35**(1): 29-59.

GREER, A. E., 1983c. On the adaptive significance of the reptilian spectacle: the evidence from scincid, teiid, and lacertid lizards. Pp. 213-21 *in* Advances in Herpetology and Evolutionary Biology ed by G. J. Rhodin and K. Miyata. Mus. Comp. Zool., Cambridge, Massachusetts; 725 pp.

GREER, A. E., 1985a. The relationships of the lizard genera *Anelytropsis* and *Dibamus. J. Herp.* **19**(1): 116-56.

GREER, A. E., 1985b. Facial tongue-wiping in xantusiid lizards: its systematic implications. *J. Herp.* **19**(1): 174-75.

GREER, A. E., 1985c. A new species of *Sphenomorphus* from northeastern Queensland. *J. Herp.* **19**(4): 469-73.

GREER, A. E., 1986a. Lygosomine (Scincidae) monophyly: a third, corroborating character and a reply to critics. *J. Herp.* **20**(1): 123-26.

GREER, A. E., 1986b. On the absence of visceral fat bodies within a major lineage of scincid lizards. *J. Herp.* **20**(2): 265-67.

GREER, A. E., 1986c. Diagnosis of the *Lerista bipes* species-group (Lacertilia: Scincidae), with a description of a new species and an updated diagnosis of the genus. *Rec. West. Aust. Mus.* **13**(1): 121-27.

GREER, A. E. In press. Limb reduction in the scincid lizard genus *Lerista*. Intrapopulation and interspecific variation in the number of phalanges and presacral vertebrae. *J. Herp.* **21**(4).

GREER, A. E. AND COGGER, H. G., 1985. Systematics of the reduce-limbed and limbless skinks currently assigned to the genus *Anomalopus* (Lacertilia: Scincidae). *Rec. Aust. Mus.* **37**(1): 11-54.

GREER, A. E. AND KLUGE, A. G., 1980. A new species of *Lampropholis* (Lacertilia: Scincidae) from the rainforests of northeastern Queensland. *Occ. Pap. Mus. Zool. Univ. Michigan* No. 691; 12 pp.

GREER, A. E. AND MYS, B. In press. Resurrection of *Lipinia rouxi* (Hediger 1934) (Reptilia: Lacertilia: Scincidae), another skink to have lost the left oviduct. *Amphibia-Reptilia* **8**(4).

GREER, A. E. AND PARKER, F., 1974. The *fasciatus* species group of *Sphenomorphus* (Lacertilia: Scincidae): notes on eight previously described species and descriptions of three new species. *Papua New Guinea Sci. Soc. Proc.* **25**: 31-61.

GREER, A. E., MCDONALD, K. R. AND LAWRIE, B. C., 1983. Three new species of *Lerista* (Scincidae) from northern Queensland with a diagnosis of the *wilkinsi* species group. *J. Herp.* **17**(3): 247-55.

GRIGG, G. C. AND HARLOW, P., 1981. A fetal-maternal shift of blood oxygen affinity in an Australian viviparous lizard, *Sphenomorphus quoyii* (Reptilia, Scincidae). *J. Comp. Physiol.* **142**: 495-99.

GRIGG, G. C., DRANE, C. R. AND COURTICE, G. P., 1979. Time constants of heating and cooling in the eastern water dragon, *Physignathus lesueurii* and some generalizations about heating and cooling in reptiles. *J. Therm. Biol.* **4**: 95-103.

GRIGG, G., SHINE, R. AND EHMANN, H. (eds), 1985. The biology of Australasian frogs and reptiles. Surrey Beatty & Sons Pty Ltd, Chipping Norton, New South Wales; 527 pp.

GROOM, S., 1973a. Notes on the keeping and distribution of the jacky lizard (*Amphibolurus muricatus*). *Herpetofauna* **5**(4): 3-5.

GROOM, S., 1973b. Further notes on the jacky lizard *Amphibolurus muricatus* in captivity. *Herpetofauna* **6**(1): 6.

HAACKE, W., 1883. Zur Naturgeschichte der Stummelschwanzeidechsen. *Zool. Gart.* 1883(Aug.): pp.? Not seen.

HAACKE, W., 1885. Über eine neue Art uterinaler Brutpflege bei Reptilien. *Zool. Anz.* **8**: 435-39.

HAACKE, W. D., 1977. If floated its way across the world — and any scientist with sense leaves it well alone! *Afr. Wildl.* **31**(1): 30-31.

HAACKE, W. D., 1982. Australian reptiles through South African eyes. *Transvaal Mus. Bull.* No. 18: 28-30.

HALL, T. S., 1905. A lizard mimicking a poisonous snake. *Vict. Nat.* **22**(4): 74.

HALLIDAY, T. R. AND ADLER, K. (eds), 1986. The encyclopaedia of reptiles and amphibians. George Allen & Unwin, London; 143+xvi pp.

HAMEL, H. T., CALDWELL, F. T. AND ABRAMS, R. M., 1967. Regulation of body temperature in the blue-tongued lizard. *Science* **156**: 1260-62.

HARCOURT, N., 1986. A review of the frilled lizard *(Chlamydosaurus kingii)* in captivity. *Thylacinus* **11**(3): 16-18.

HARDY, C. J. AND HARDY, C. M., 1977. Tail regeneration and other observations in a species of agamid lizard. *Aust. Zool.* **19**(2): 141-48.

HARDY, G. S., 1977. The New Zealand Scincidae (Reptilia: Lacertilia); a taxonomic and zoogeographic study. *New Zealand J. Zool.* **4**: 221-325.

HARLOW, P. AND VAN DER STRAATEN, M., 1976. Reptiles of the Oxford Falls area. *Herpetofauna* **8**(1): 6-7.

HARRIS, B. F. AND JOHNSTON, P. G., 1977. Electrophoretic evidence for the specific status of the lizards *Lampropholis guichenoti* (Duméril and Bibron 1839) and *L. delicata* (De Vis 1887). *Aust. Zool.* **19**(2): 149-54.

HARRIS, M., 1985. Goanna oil survives the test of time well. The Week-end Australian, 12-13 January, 1985.

HARRISON, L. AND WEEKES, H. C., 1925. On the occurrence of placentation in the scincid lizard *Lygosoma entrecasteauxi. Proc. Linn. Soc. New South Wales* **50**(4): 470-86.

HARVEY, C., 1983. A new species of *Nephrurus* (Reptilia: Gekkonidae) from South Australia. *Trans Roy. Soc. South Aust.* **107**(3/4): 231-35.

HARVEY, C. AND MOVER, A., 1978. Darwin trip report — January 1978. *South Aust. Herp. Group Newsletter* 1978 (mid-year?): 2-4.

HASEGAWA, M., 1984. Biennial reproduction in the lizard *Eumeces okadae* on Miyake-Jima, Japan. *Herpetologica* **40**(2): 194-99.

HAY, M., 1972. The breeding of *Physignathus lesueurii* L. in captivity. *Herpetofauna* **5**(1): 2-3.

HEATWOLE, H. 1970. Thermal ecology of the desert dragon *Amphibolurus inermis. Ecol. Monogr.* **40**: 425-57.

HEATWOLE, H., 1976. Reptile ecology. University Queensland Press, St. Lucia; 178 pp.

HEATWOLE, H. AND FIRTH, B. T., 1982. Voluntary maximum temperature of the Jackie lizard, *Amphibolurus muricatus. Copeia* **1982**(4): 824-29.

HEATWOLE, H., FIRTH, B. T. AND STODDART, H., 1975. Influence of season, photoperiod and thermal acclimation on the panting threshold of *Amphibolurus muricatus. J. Exp. Zool.* **191**(2): 183-92.

HEATWOLE, H., FIRTH, B. T. AND WEBB, G. J. W., 1973. Panting thresholds of lizards — I. Some methodological and internal influences on the panting threshold of an agamid, *Amphibolurus muricatus. Comp. Biochem. Physiol.* **46A**: 799-826.

HECHT, M. K., 1975. The morphology and relationships of the largest known terrestrial lizard, *Megalania prisca* Owen, from the Pleistocene of Australia. *Proc. Roy. Soc. Vict.* **87**(1 & 2): 239-50.

HEDIGER, H., 1934. Beitrag zur Herpetologie und Zoogeographie Neu Britanniens und einiger umliegender Gebiete. *Zool. Jahrb. (Syst.)* **66**(1/2): 1-152.

HERMES, N., 1981. Mertens water monitor feeding on trapped fish. *Herpetofauna* **13**(1): 34.

HEWER, A. M., 1948. Tasmanian lizards. *Tas. Nat.* **1**(3): 8-11.

HEWER, A. AND MOLLISON, B. C., 1974. Reptiles and amphibians of Tasmania. *Tasmanian Yearbook,* No. 8 (1974): 51-60.

HICKMAN, J. L., 1960. Observations on the skink lizard *Egernia whitii* (Lacépède). *Pap. Proc. Roy. Soc. Tasmania* **94**:111-18.

HILL, J., 1923. An exciting contest between two lizards. *Vict. Nat.* **40**(5): 96-97.

HILLER, U., 1976. Comparative studies on the functional morphology of two gekkonid lizards. *J. Bombay Nat. Hist. Soc.* **73**(2): 278-82.

HITZ, R., 1983. Pflege und Nachzucht von *Trachydosaurus rugosus* Gray, 1827 im Terrarium (Sauria: Scincidae). *Salamandra* **19**(4): 198-210.

HOFFSTETTER, R. AND GASC, J. P., 1969. Vertebrae and ribs of modern reptiles. Pp. 201-310 *in* Biology of the Reptilia, Vol. 1, Morph. A ed by C. Gans, A. d'A. Bellairs, T. S. Parsons. Academic Press, London; 373 pp.

HOLDER, L. A., 1960. The comparative morphology of the axial skeleton in the Australian Gekkonidae. *J. Linn. Soc. (Zool.)* **44**: 300-35.

HOLMES, R. AND LIGHT, A., 1983. A serendipitous age estimation of a lizard, *Tiliqua rugosa* (Lacertilia: Scincidae). *West. Aust. Nat.* **15**(7): 159-60.

HOLMES, R. S., KING, M. AND KING, D., 1975. Phenetic relationships among varanid lizards based upon comparative electrophoretic data and karyotypic analyses. *Biochem. Syst. Ecol.* **3**: 257-62.

HONEGGER, R. E., 1964. Herpetologisches aus dem Züricher Zoo I. Beiträge zur Haltung und Zucht verschiedener Reptilien. *Die Aquar. Terr. Zeit.* **17**(11): 339-42.

HONEGGER, R. E., 1966. Beobachtungen an der Herpetofauna der Seychellen. *Salamandra* **2**(1/2): 21-36.

HORN, H.-G., 1978. Nachzucht von *Varanus gilleni. Salamandra* **14**(1): 29-32.

HORN, H.-G., 1980. Bisher unbekannte Details zur Kenntnis von *Varanus varius* auf Grund von feldherpetologischen und terraristischen Beobachtungen (Reptilia: Sauria: Varanidae). *Salamandra* **16**(1): 1-18.

HORN, H.-G., 1981. *Varanus spenceri,* nicht *Varanus giganteus:* eine Richtigstellung (Reptilia: Sauria: Varanidae). *Salamandra* **17**(1/2): 78-81.

HORN, H.-G., 1985. Beiträge zum Verhalten von Waranen: Die Ritualkämpfe von *Varanus komodoensis* Ouwens, 1912 und *V. semiremex* Peters, 1869 sowie die Imponierphasen der Ritualkämpfe von *V. timorensis timorensis* (Gray 1831) und *V. t. similis* Mertens, 1958 (Sauria: Varanidae). *Salamandra* **21**(2/3): 169-79.

HORN, H.-G. AND SCHÜRER, U., 1978. Bemerkungen zu *Varanus (Odatria) glebopalma* Mitchell 1955 (Reptilia: Sauria: Varanidae). *Salamandra* **14**(3): 105-16.

HORNER, P. G., 1984. Notes on the scincid lizard *Cryptoblepharus litoralis* (Mertens 1958) in the Northern Territory. *North. Terr. Nat.* **7**: 4-7.

HORNER, P. AND KING, M., 1985. A new species of *Ctenotus* (Scincidae, Reptilia) from the Northern Territory. *The Beagle, Occ. Pap. North. Terr. Mus. Arts Sci.* **2**(1): 143-48.

HORTON, D., 1968. Evolution of the genus *Egernia* (Sauria: Scincidae). Unpublished M.Sc. Thesis; University of New England; 166 pp.

HORTON, D. R., 1972. Evolution of the genus *Egernia* (Lacertilia: Scincidae). *J. Herp.* **6**(2): 101-09.

HOSER, R. T., 1983. Notes on egg laying in the scalyfoot *(Pygopus lepidopodus)* and other reptiles. *Herptile* **8**(4): 134-36.

HOSER, R. T., 1984. Preferred activity temperatures of nocturnal reptiles in the Sydney area. *Herptile* **9**(1): 12-13.

HOSMER, W., 1956. *Rhynchoedura ornata* (Gekkonidae) and *Ablepharus kinghorni* (Scincidae) in S.W. Queensland. *North Qld Nat.* **114**: 17-18.

HOUSTON, T. F., 1974. Revision of the *Amphibolurus decresii* complex (Lacertilia: Agamidae) of South Australia. *Trans. Roy. Soc. South Aust.* **98**(2): 49-60.

HOUSTON, T. F., 1977. A new species of *Diporiphora* from South Australia and geographic variation in *D. winneckei* Lucas and Frost (Lacertilia: Agamidae). *Trans. Roy. Soc. South Aust.* **101**(8): 199-205.

HOUSTON, T. F., 1978. Dragon lizards and goannas of South Australia. South Aus. Mus., Adelaide; 84 pp.

HOUSTON, T. F. AND TYLER, M. J., 1979. Reptiles and amphibians. Pp. 115-22 *in* Natural history of Kangaroo Island ed by M. J. Tyler, C. R. Twidale and J. K. Ling. Royal Society South Australia, Adelaide; 184 pp.

HOW, R. A. AND KITCHENER, D. J., 1983. The biology of the gecko *Oedura reticulata* Bustard, in a small habitat isolate in the Western Australian wheatbelt. *Aust. Wildlife Res.* **10**(3): 543-56.

HOW, R. A., DELL, J. AND WELLINGTON, B. D., 1986. Comparative biology of eight species of *Diplodactylus* gecko in Western Australia. *Herpetologica* **42**(4): 471-82.

HUDSON, P., 1977. An account of egg laying by the thorny devil *Moloch horridus* (Gray). *Herpetofauna* **9**(1): 23-24.

HUDSON, P., 1979. Notes on the behavioural antics of the painted dragon *Amphibolurus pictus* Peters. *Herpetofauna* **11**(1): 26.

HUDSON, P., 1981. Observations on egg laying by the marbled gecko, *Phyllodactylus marmoratus* (Fitzinger). *Herpetofauna* **13**(1): 32-33.

HUDSON, P., MIRTSCHIN, P. AND GARRETT, C., 1981. Notes on Flinders Island (S.A.). Its reptiles and birds. *South Aust. Nat.* **56**(2): 21-31.

HUEY, R. B. AND PIANKA, E. R., 1981. Ecological consequences of foraging mode. *Ecology* **62**(4): 991-99.

HUMPHRIES, R. B., 1972. An investigation of the agamid genus *Amphibolurus* — comparative karyology and relationships. Unpublished B.Sc. Honours Thesis, University of Western Australia. Not seen.

HUMPHREYS, W. F., 1976. Spider induced egg mortality in a skink population. *Copeia* **1976**(2): 404.

HUNSAKER, D. II AND BREESE, P., 1967. Herpetofauna of the Hawaiian Islands. *Pacific Science* **21**(3): 423-28.

HUSBAND, G., 1979a. Range extension for *Chelosania brunnea*. *Herpetofauna* **10**(2): 29-30.

HUSBAND, G. A., 1979b. Notes on a nest and hatchlings of *Varanus acanthurus*. *Herpetofauna* **11**(1): 29-30.

HUSBAND, G., 1980a. Unusual burrowing behaviour in *Pygopus lepidopodus*. *Herpetofauna* **12**(1): 36.

HUSBAND, G. A., 1980b. A note on egglaying by *Hemidactylus frenatus* (house gecko) in Darwin. *Herpetofauna* **12**(1): 36.

HUTCHINSON, M. N., 1981. The systematic relationships of the genera *Egernia* and *Tiliqua* (Lacertilia: Scincidae). A review and immunological reassessment. Pp. 176-93 *in* Proceedings of the Melbourne Herpetological Symposium ed by C. B. Banks and A. A. Martin. Zoological Board of Victoria, Melbourne; 199 pp.

HUTCHINSON, M. N., 1983. The generic relationships of the Australian lizards of the family Scincidae. A review and immunological reassessment. Unpublished Ph.D. Thesis, La Trobe University, Bundoora, Victoria; 294 pp.

HUXLEY, T. H., 1887. Preliminary note on the fossil remains of a chelonian reptile, *Ceratochelys sthenurus*, from Lord Howe's Island. Australia. *Proc. Roy. Soc.* (Lond.) **42**: 232-38.

INGRAM, G. J., 1977. Three species of small lizards — two of them new Genus *Menetia* (Lacertilia, Scincidae) in Queensland. *Vict. Nat.* **94**(5): 184-87.

INGRAM, G. P., 1978. A new species of gecko, genus *Crytodactylus*, from Cape York Peninsula, Queensland, Australia. *Vict. Nat.* **95**: 142-46.

INGRAM, G. J., 1979a. The occurrence of lizards of the genus *Emoia* (Lacertilia, Scincidae) in Australia. *Mem. Qld Mus.* **19**(3): 431-37.

INGRAM, G. J., 1979b. Two new species of skinks, genus *Ctenotus* (Reptilia Lacertilia, Scincidae), from Cape York Peninsula, Queensland, Australia. *J. Herp.* **13**(3): 279-82.

INGRAM, G. AND EHMANN, H., 1981. A new species of scincid lizard of the genus *Leiolopisma* (Scincidae: Lygosominae) from southeastern Queensland and northeastern New South Wales. *Mem. Qld Mus.* **20**(2): 307-10.

INGRAM, G. AND RAWLINSON, P., 1981. Five new species of skinks (genus *Lampropholis*) from Queensland and New South Wales. *Mem. Qld Mus.* **20**(2): 311-17.

IRVINE, W., 1957. Notes on the genus *Varanus* (monitor lizards). *North Qld Nat.* No. 119: 1-2.

IRWIN, B., 1986. Captive breeding of two species of monitor. *Thylacinus* **11**(2): 4-5.

JAMES, C., 1983. Reproduction in lizards from the wet-dry tropics of Australia. Unpublished B.Sc. Honours Thesis, University of Sydney; 87 pp.

JAMES, C. D., MORTON, S. R., BRAITHWAITE, R. W. AND WOMBEY, J. C., 1984. Dietary pathways through lizards of the Alligator Rivers Region, Northern Territory. Office of the Supervising Scientist, Bondi Junction, New South Wales, Technical Memorandum 6; 11 pp.

JAMES, C. AND SHINE, R., 1985. The seasonal timing of reproduction: a tropical-temperate comparison in Australian lizards. *Oecologia* **67**: 464-74.

JANESCH, W., 1906. Über *Archaeophis proavus* Mass., eine Schlange aus dem Eocän des Monte Bolca. *Beitr. Paläont. Geol. Österreich -Ungarns.* **19**(1): 1-33.

JENKINS, R. AND BARTELL, R., 1980. A field guide to reptiles of the Australian High Country. Inkata Press, Melbourne; 278 pp.

JOHN-ADLER, H. B., GARLAND, T. JR. AND BENNETT, A. F., 1986. Locomotory capacities, oxygen consumption, and the cost of locomotion of the shingle-back lizard *(Trachydosaurus rugosus).* *Physiol. Zoöl.* **59**(5): 523-31.

JOHNSON, C. R., 1972. Head-body temperature differences in *Varanus gouldii* (Sauria: Varanidae). *Comp. Biochem. Physiol.* **43**A: 1025-29.

JOHNSON, C. R., 1975. Defensive display behaviour in some Australian and Papuan-New Guinean pygopodid lizards, boid, colubrid and elapid snakes. *Zool. J. Linn. Soc.* **56**: 265-82.

JOHNSON, C. R., 1976. Some behavioural observations on wild and captive sand monitors, *Varanus gouldii* (Sauria: Varanidae). *Zool. J. Linn. Soc.* **59**: 377-80.

JOHNSON, C. R., 1977. Thermoregulation in four Australian lizards of the genus *Egernia* (Sauria: Scincidae). *Zool. J. Linn. Soc.* **60**: 381-90.

JOHNSTON, G. R., 1979. The eggs, incubation and young of the bearded dragon *Amphibolurus vitticeps* Ahl. *Herpetofauna* **11**(1): 5-8.

JOHNSTON, G. R., 1981. A note on *Moloch horridus* Gray, 1841. *Herpetofauna* **13**(1): 29.

JOHNSTON, G. R., 1982. The herpetofauna of the Middleback Range area, South Australia 1. An annoted checklist. *Herpetofauna* **14**(1): 52-60.

JOHNSTON, G. R. AND ELLINS, P., 1979. The reptiles of the Sir Joseph Banks Islands, South Australia. *Herpetofauna* **10**(2): 9-12.

JOHNSTONE, R. E., 1983. Herpetofauna of the Hamersley Range National Park. Pp. 7-11 *in* A faunal survey of the Hamersley Range National Park Western Australia 1980 ed by B. G. Muir. National Parks Authority of Western Australia, Bulletin No. 1; 36pp.

JOSS, J. M. P., 1985. Ovarian steroid production in oviparous lizards of the genus *Lampropholis* (Scincidae). Pp. 319-26 *in* The biology of Australasian frogs and reptiles ed by G. Grigg, R. Shine and H. Ehmann. Surrey Beatty and Sons Pty Ltd, Chipping Norton, NSW; 527pp.

JOSS, J. M. P. AND MINARD, J. A., 1985. On the reproductive cycles of *Lampropholis guichenoti* and *L. delicata* (Squamata: Scincidae) in the Sydney region. *Aust. J. Zool.* **33**: 699-704.

KÄSTLE, W., 1969. Ein Schlangen-Flossenfuss *(Delma f. fraseri)* in Gefangenschaft. *Aquar. Terr. Zeit.* **22**(3): 85-86.

KEAST, A., 1959. The reptiles of Australia. Pp. 115-35 *in* Biogeography and ecology in Australia ed by A. Keast, R. L. Crocker and C. S. Christian. W. Junk, The Hague; 640pp.

KENNERSON, K. J., 1979. Remarks on the longevity of *Varanus varius. Herpetofauna* **10**(2): 32.

KENNERSON, K., 1980. Carrion diet in the lace monitor *Varanus varius. Herpetofauna* **12**(1): 36.

KENNERSON, K. J. AND COCHRANE, G. J., 1981. Avid appetite for dandelion blossoms *Taraxarun officinale* by a western bearded dragon *Amphibolurus vitticeps* Ahl. *Herpetofauna* **12**(2): 34-35.

KENT, D. S., 1987. Notes on the biology and osteology of *Amphibolurus diemensis* (Gray 1841), the mountain dragon. *Vic. Nat.* **104**(4): 101-04.

KERFOOT, W. C., 1970. The effect of functional changes upon the variability of lizard and snake body scale numbers. *Copeia* **1970**(2): 252-60.

KERSHAW, J. A., 1927. Victorian reptiles. *Vict. Nat.* **43**(12): 335-44.

KING, D., 1964. The osteology of the water skink, *Lygosoma (Sphenomorphus) quoyii. Aust. J. Zool.* **12**(2): 201-16.

KING, D., 1980. The thermal biology of free-living sand goannas *(Varanus gouldii)* in southern Australia. *Copeia* **1980**(4): 755-67.

KING, D. AND GREEN, B., 1979. Notes on diet and reproduction of the sand goanna, *Varanus gouldii rosenbergi. Copeia* **1979**(1): 64-70.

KING, D. AND RHODES, L., 1982. Sex ratio and breeding season of *Varanus acanthurus. Copeia* **1984**(4): 784-87.

KING, M., 1973a. Chromosomes of two Australian lizards of the families Scincidae and Gekkonidae. *Cytologia* **38**: 205-10.

KING, M., 1973b. Karyotypic studies of some Australian Scincidae (Reptilia). *Aust. J. Zool.* **21**: 21-32.

KING, M., 1977a. Reproduction in the Australian gecko *Phyllodactylus marmoratus* (Gray). *Herpetologica* **33**(1): 7-13.

KING, M., 1977b. Chromosome and morphometric variation in the gekko *Diplodactylus vittatus* (Gray). *Aust. J. Zool.* **25**: 43-57.

KING, M., 1977c. The evolution of sex chromosomes in lizards. Pp. 55-60 *in* Reproduction and evolution. Proceedings of the fourth symposium on comparative biology of reproduction ed by J. H. Calaby and C. G. Tyndale-Biscoe. Australian Academy of Sciences, Canberra.

KING, M., 1978. A new chromosome form of *Hemidactylus frenatus* (Duméril and Bibron). *Herpetologica* **34**(2): 216-18.

KING, M., 1979. Karyotypic evolution in *Gehyra* (Gekkonidae: Reptilia). I. The *Gehyra variegata-punctata* complex. *Aust. J. Zool.* **27**: 373-93.

KING, M., 1981a. Chromosomal change and speciation in lizards. Pp. 262-86 *in* Evolution and speciation ed by W. Atchley and D. Woodruff. Cambridge University Press; 436pp.

KING, M., 1981b. Notes on the distribution of *Gehyra nana* Storr and *Gehyra punctata* (Fry) in Australia. *Aust. J. Herp.* **1**(2): 55-56.

KING, M., 1982a. Karyotypic variation in *Gehyra* (Gekkonidae: Reptilia). II. A new species from the Alligator Rivers region in northern Australia. *Aust. J. Zool.* **30**: 93-101.

KING, M., 1982b. A new species of *Gehyra* (Reptilia: Gekkonidae) from central Australia. *Trans. Roy. Soc. South Aust.* **106**(4): 155-58.

KING, M., 1983a. Karyotypic evolution in *Gehyra* (Gekkonidae: Reptilia) III. The *Gehyra australis* complex. *Aust. J. Zool.* **31**: 723-41.

KING, M., 1983b. The *Gehyra australis* species complex (Sauria: Gekkonidae). *Amphibia-Reptilia* **4**: 147-69.

KING, M., 1984a. Karyotypic evolution in *Gehyra* (Gekkonidae: Reptilia) IV. Chromosome change and speciation. *Genetica* **64**: 101-14.

KING, M., 1984b. A new species of *Gehyra* (Reptilia: Gekkonidae) from northern Western Australia. *Trans. Roy. Soc. South Aust.* **108**(2): 113-17.

KING, M., 1984c. Three new species of *Oedura* (Reptilia: Gekkonidae) from the Mitchell Plateau of North Western Australia. *Amphibia-Reptilia* **5**(3-4): 329-37.

KING, M., 1985. Chromosome markers and their use in phylogeny and systematics. Pp. 165-75 *in* The biology of Australasian frogs and reptiles ed by G. Grigg, R. Shine and H. Ehmann. Surrey Beatty & Sons Pty Ltd, Chipping Norton, NSW; 527 pp.

KING, M. AND GOW, G., 1983. A new species of *Oedura* (Gekkonidae: Reptilia) from the Alligator Rivers region of northern Australia. *Copeia* **1983**(2): 445-49.

KING, M. AND HAYMAN, D., 1978. Seasonal variation of chiasma frequency in *Phyllodactylus marmoratus* (Gray) (Gekkonidae — Reptilia). *Chromosoma* (Berlin) **69**: 131-54.

KING, M. AND KING, D., 1975. Chromosomal evolution in the lizard genus *Varanus* (Reptilia). *Aust. J. Biol. Sci.* **28**: 89-108.

KING, M. AND KING, D., 1977. An additional chromosome race of *Phyllodactyllus marmoratus* (Gray) (Reptilia: Gekkonidae) and its phylogenetic implications. *Aust. J. Zool.* **25**: 667-72.

KING, M. AND ROFE, R., 1976. Karyotypic variation in the Australian gecko *Phyllodactylus marmoratus* (Gray) (Gekkonidae: Reptilia). *Chromosoma* (Berlin) **54**: 75-87.

KING, M., BRAITHWAITE, R. W. AND WOMBEY, J. C., 1982. A new species of *Diplodactylus* (Reptilia: Gekkonidae) from the Alligator Rivers Region, Northern Territory. *Trans. Roy. Soc. South Aust.* **106**(1): 15-18.

KING, M., MENGDEN, G. A. AND KING, D., 1982. A pericentric-inversion polymorphism and a ZZ/ZW sex-chromosome system in *Varanus acanthurus* Boulenger analyzed by G- and C- banding and Ag staining. *Genetica* **58**: 39-45.

KINGHORN, J. R., 1923. A new *Varanus* from Coquet Island, Queensland. *Rec. Aust. Mus.* **14**(2): 135-37.

KINGHORN, J. R., 1924. A short review of the lizards belonging to the genus *Lialis* Gray. *Rec. Aust. Mus.* **14**(3): 184-88.

KINGHORN, J. R., 1931. Herpetological notes. No. 2. *Rec. Aust. Mus.* **18**(3): 85-91.

KLAGES, H. G., 1982. Pflege und Nachzucht der australischen Bodenagame *Amphibolurus nuchalis* (Reptilia: Sauria: Agamidae). *Salamandra* **18**(1/2): 65-70.

KLUGE, A. G., 1962a. A new species of gekkonid lizard, genus *Diplodactylus*, from the Carnarvon region, Western Australia. *West. Aust. Nat.* **8**(3): 73-75.

KLUGE, A. G., 1962b. A new species of gekkonid lizard, genus *Diplodactylus* (Gray), from the southern interior of Western Australia. *West. Aust. Nat.* **8**(4): 97-101.

KLUGE, A. G., 1967a. Higher taxonomic categories of gekkonid lizards and their evolution. *Bull. Amer. Mus. Nat. Hist.* **135**(1): 1-60.

KLUGE, A. G., 1967b. Systematics, phylogeny, and zoo-geography of the lizard genus *Diplodactylus* Gray (Gekkonidae). *Aust. J. Zool.* **15**: 1007-108.

KLUGE, A. G., 1968. Phylogenetic relationships of the gekkonid lizard genera *Lepidodactylus* Fitzinger, *Hemiphyllodactylus* Bleeker, and *Pseudogekko* Taylor. *Philippine J. Sci.* **95**(3): 331-52.

KLUGE, A. G., 1974. A taxonomic revision of the lizard family Pygopodidae. *Misc. Publs Mus. Zool. Univ. Mich.* No. 147; 221 pp.

KLUGE, A. G., 1976. Phylogenetic relationships in the lizard family Pygopodidae: an evaluation of theory, methods and data. *Misc. Publs Mus. Zool. Univ. Mich.* No. 152; 72 pp.

KLUGE, A. G., 1982. The status of the parthenogenetic gekkonid lizard, *Gehyra variegata ogasawarasimae* Okada. *J. Herpetol.* **16**(1): 86-87.

KLUGE, A. G., 1983. Cladistic relationships among gekkonid lizards. *Copeia* **1983**(2): 465-75.

KLUGE, A. G. AND ECKARDT, M. J., 1969. *Hemidactylus garnotii* Duméril and Bibron, a triploid all-female species of gekkonid lizard. *Copeia* **1969**(4): 651-64.

KOCH, L. E., 1970. Predation of the scorpion, *Urodacus hoplurus*, by the lizard, *Varanus gouldi*. *West. Aust. Nat.* **11**(5): 120-21.

KOPSTEIN, F., 1926. Reptilien von den Molukken und den benachbartetn Inseln. *Zool. Meded.* **9**: 71-112.

KREFFT, G., 1866. On the vertebrated animals of the lower Murray and Darling, their habits, economy and geographical distribution. *Trans. Philos. Soc. New South Wales* 1862-1865: 1-33.

LAMBERT, M. R. K., 1985. A few of the herpetofauna in the Commonwealth (Oriental and Australian zones). *Brit. Herp. Soc. Bull.* **14**: 15-19.

LANGEBARTEL, D. A., 1968. The hyoid and its associated muscles in snakes. *Illinois Biological Monographs* No. 38; 156 pp.

LA RIVERS, I., 1948. Some Hawaiian ecological notes. *The Wasman Collector* **7**(3): 85-110. Not seen.

LAWTON, D., 1982. Incubation of Jacky Dragon (*Amphibolorus muricatus*) eggs with notes on growth of the young. *Thylacinus* **7**(1): 17-21.

LEE, A. K. AND BADHAM, J. A., 1963. Body temperature, activity, and behaviour of the agamid lizard, *Amphibolurus barbatus*. *Copeia* **1963**(2): 387-94.

LE SOUËF, D., 1918. The blue-tongued lizard. *Vict. Nat.* **35**(1): 15.

LE SOUEF, A. S. AND MCFADYEN, E., 1937. Notes on the *Moloch horridus*. *Proc. Roy. Zool. Soc. New South Wales* 1936-37: 29-31.

LICHT, P., DAWSON, W. R., SHOEMAKER, V. H. AND MAIN, A. R., 1966a. Observations on the thermal relations of Western Australian lizards. *Copeia* **1966**(1): 97-110.

LICHT, P., DAWSON, W. R. AND SHOEMAKER, V. H., 1966b. Heat resistance of some Australian lizards. *Copeia* **1966**(2): 162-69.

LIMPUS, C. J., 1982. The reptiles of Lizard Island. *Herpeto-fauna* **13**(2): 1-6.

LINTON, E. H., 1929. *Pygopus*, the mud-dweller. *Vict. Nat.* **45**(10): 248-52.

LITTLEJOHN, M. J. AND RAWLINSON, P. A., 1971. Amphibians and reptiles of Victoria. *Victorian Year Book* No. 85: 1-36.

LONGLEY, G., 1938. Notes on a pink-tongued skink (*Hemisphaeriodon gerrardii*). *Proc. Roy. Zool. Soc. New South Wales* 1937-38: 19-21.

LONGLEY, G., 1939. The blue tongued lizard (*Tiliqua scincoides*). *Proc. Roy. Zool. Soc. New South Wales* 1938-39: 39-42.

LONGLEY, G., 1940. Notes on certain lizards. *Proc. Roy. Zool. Soc. New South Wales* 1939-40: 34-39.

LONGLEY, G., 1941. Notes on some Australian lizards. *Proc. Roy. Zool. Soc. New South Wales* 1940-41: 30-35.

LONGLEY, G., 1943. Notes on *Goniocephalus cristipes*. *Proc. Roy. Zool. Soc. New South Wales*. 1942-43: 30-31.

LONGLEY, G., 1944a. A note on the shingle back lizard (*Trachysaurus rugosus*). *Proc. Roy. Zool. Soc. New South Wales* 1943-44: 20.

LONGLEY, G., 1944b. Notes on a hybrid blue tongue lizard. *Proc. Roy. Zool. Soc. New South Wales* 1943-44: 23-24.

LONGLEY, G., 1945a. Notes on the lace monitor (*Varanus varius*). *Proc. Roy. Zool. Soc. New South Wales* 1944-45: 20-21.

LONGLEY, G., 1945b. Notes on Burton's legless lizard (*Lialis burtonis*). *Proc. Roy. Zool. Soc. New South Wales* 1944-45: 33-34.

LONGLEY, G., 1946a. The giant skink (*Egernia major*) in the vivarium. *Proc. Roy. Zool. Soc. New South Wales* 1945-46: 33-34.

LONGLEY, G., 1946b. Observations on a young frilled lizard, *Chlamydosaurus kingii*. *Proc. Roy. Zool. Soc. New South Wales* 1945-46: 35-36.

LONGLEY, G., 1947a. Gould's monitor *(Varanus gouldii)* in the vivarium. *Proc. Roy. Zool. Soc. New South Wales* 1946-47: 27-28.

LONGLEY, G., 1947b. Notes on the hatching of eggs of the water dragon *(Physignathus lesueurii)*. *Proc. Roy. Zool. Soc. New South Wales* 1946-47: 29.

LONGLEY, G., 1947c. Breeding Cunningham's skink *(Egernia cunninghamii)* in the vivarium. *Proc. Roy. Zool. Soc. New South Wales. 1946-47:* 30.

LONGMAN, H. A. 1915. Reptiles from Queensland and the Northern Territory. *Mem. Qld Mus.* **3:** 30-34.

LONGMAN, H. A., 1918. Notes on some Queensland and Papuan reptiles. *Mem. Qld Mus.* **6:** 37-44.

LÖNNBERG, E. AND ANDERSSON, L. G., 1913. Results of Dr Mjöbergs Swedish scientific expeditions to Australia 1910-13. III. Reptiles. *Kungl. Svenska Vetenskapsakad. Handl.* **52**(3): 1-17.

LOVERIDGE, A., 1934. Australian reptiles in the Museum of Comparative Zoology, Cambridge, Massachusetts. *Bull. Mus. Comp. Zool.* **77**(6): 243-383.

LOVERIDGE, A., 1938. On some reptiles and amphibians from the central region of Australia. *Trans. Roy. Soc. South Aust.* **62**(2): 183-91.

LOVERIDGE, A., 1939. A new skink *(Leiolopisma hawaiiensis)* from Honolulu. *Proc. Biol. Soc. Wash.* **52:** 1-2.

LOVERIDGE, A., 1949. On some reptiles and amphibians from the Northern Territory. *Trans. Roy. Soc. South Aust.* **72**(2): 208-15.

LOW, T., 1978. The reptiles of Magnetic Island, Nth Queensland. *Herpetofauna* **9**(2): 10-14.

LOW, T., 1980. A new species of gecko, genus *Gehyra* (Reptilia: Gekkonidae) from Queensland. *Vict. Nat.* **96:** 190-96.

LUCAS, A. H. S. AND FROST, C., 1894. The lizards indigenous to Victoria. *Proc. Roy. Soc. Vict.* **6**(NS): 24-92.

LUCAS, A. H. S. AND FROST, C., 1896. Reptilia. Pp. 112-51 *in* Report on the work of the Horn scientific expedition to Central Australia. Part II — Zoology ed by B. Spencer. Melville, Mullen and Slade, Melbourne; 431 pp.

LUDOWICI, P. A., 1973. Notes on an undescribed species of legless lizards of the genus *Delma*. *Roy. Zool. Soc. Bull. Herp.* **1**(2): 9.

LUDOWICI, P. A., 1975. Notes on an undescribed species of legless lizard of the genus *Delma*. *Herpetofauna* **7**(2): 20-21.

LUNNEY, D. AND BARKER, J., 1986. Survey of reptiles and amphibians of the coastal forests near Bega, N.S.W. *Aust. Zool.* **22**(3): 1-9.

MACKAY, R. D., 1959. Reptiles of Lion Island, New South Wales. *Aust. Zool.* **12**(4): 308-09.

MACKENZIE, W. C. AND OWEN, W. J., 1923. Studies on the comparative anatomy of the alimentary canal of Australian reptiles. *Proc. Roy. Soc. Vict.* **36**(1): 41-49.

MACLEAN, S., 1980. Ultrastructure of epidermal sensory receptors in *Amphibolurus barbatus* (Lacertilia: Agamidae). *Cell Tissue Research* **210**(3): 435-45.

MADIGAN, C. T., 1930. Lake Eyre, South Australia. *Geogr. J.* **76**(3): 215-40.

MADIGAN, C. T., 1936. Central Australia. Oxford University Press, London; 267 pp.

MAHENDRA, B. C., 1938. Some remarks on the phylogeny of the Ophidia. *Anat. Anz.* **86:** 347-56.

MANSERGH, I., 1982. Notes on the range extension of the alpine water skink *(Sphenomorphus kosciuskoi)* in Victoria. *Vict. Nat.* **99**(3): 123-24.

MARCELLINI, D. L., 1971. Activity patterns of the gecko *Hemidactylus frenatus*. *Copeia* **1971**(4): 631-35.

MARCELLINI, D. L., 1972. The function of a vocal display of the lizard *Hemidactylus frenatus* (Sauria: Gekkonidae). *Anim. Behav.* **25**(2): 414-17.

MARCELLINI, D. L., 1974. Acoustic behaviour of the gekkonid lizard, *Hemidactylus frenatus*. *Herpetologica* **30**(1): 44-52.

MARCELLINI, D. L., 1976. Some aspects of the thermal ecology of the gecko *Hemidactylus frenatus*. *Herpetologica* **32**(3): 341-45.

MARCELLINI, D. L., 1977. The function of a vocal display of the lizard *Hemidactylus frenatus* (Sauria: Gekkonidae). *Anim. Behav.* **25**(2): 414-17.

MARCELLINI, D. L. 1978. The acoustic behaviour of lizards. Pp. 287-300 *in* Behaviour and neurology of lizards ed by N. Greenberg and P. D. MacLean. U.S. Department of Health, Education and Welfare, Publication No. (Adm) 77-491; 352 pp.

MARKWELL, K., 1985. The artificial incubation of lace monitor *(Varanus varius)* eggs. *Herpetofauna* **15**(1): 16-17.

MARTIN, K., 1972. Captivity observations of some Australian legless lizards. *Herpetofauna* **5**(3): 5-6.

MARTIN, K., 1973. An interesting rain forest inhabitant. *Herpetofauna* **6**(2): 2.

MARTIN, K., 1975. Reptiles of the Alice Springs area. *Herpetofauna* **7**(2): 6-7.

MARYAN, B., 1984a. Observation of active *Delma fraseri* during night time hours. *South Aust. Herp. Group Newsletter* 1984(Feb.): 2.

MARYAN, B., 1984b. Unusual defensive behaviour in captive *Delma australis* and in the wild (Kluge). *South Aust. Herp. Group Newsletter* 1984(May): 7.

MARYAN, B., 1984c. *Lialis burtonis* Gray (diurnal and nocturnal). *South Aust. Herp. Group Newsletter* 1984(May): 8.

MARYAN, B., 1985. Unusual defensive behaviour by *Delma australis* in captivity and in the wild. *Herpetofauna* **15**(2): 51.

MATTHEY, R., 1949. Les chromosomes des vertébrés. *Libr. Univ.* F. Rouge, Lausanne, Paris.

MATZ, G., 1968. Les scinques australiens. *Aquarama* **2**(1): 27-29.

MAU, K.-G., 1978. Nachweis natürlicher Parthenogenese bei *Lepidodactylus lugubris* durch Gefangenschaftsnachzucht (Reptilia: Sauria: Gekkonidae). *Salamandra* **14**(2): 90-97.

MAYHEW, W. W., 1963. Observations on captive *Amphibolurus pictus* an Australian agamid lizard. *Herpetologica* **19**(2): 81-88.

MCCOY, C. J. AND BUSACK, S. D., 1970. The lizards *Hemidactylus frenatus* and *Leiolopisma metallica* on the island of Hawaii. *Herpetologica* **26**(3): 303.

MCCOY, M., 1980. Reptiles of the Solomon Islands. Wau Ecology Institute, Handbook No. 7; 80 pp.

MCDONALD, K. R., 1977. Observations on the skink *Anomalopus reticulatus* (Günther) (Lacertilia: Scincidae) *Vict. Nat.* **94**(3): 99-103.

McDowell, S. B., 1972. The evolution of the tongue of snakes, and its bearing on snake origins. Pp. 191-273 *in* Evolutionary biology, vol. 6 ed by T. Dobzhansky, M. K. Hecht and W. C. Steere. Appleton-Century-Crofts, New York; 445 pp.

McDowell, S. B. and Bogert, C. M., 1954. The systematic position of *Lanthanotus* and the affinities of the anguinomorphan lizards. *Bull. Amer. Mus. Nat. Hist.* **105**(1): 1-142.

McGregor, R. C., 1904. Notes on Hawaiian reptiles from the island of Maui. *Proc. U.S. Nat'l. Mus.* **28**: 115-18.

McPhee, D. R., 1979. The observer's book of snakes and lizards of Australia. Methuen, Sydney; 157 pp.

Mebs, D., 1973. Drohreaktionen beim Blattschwanzgecko, *Phyllurus platurus*. *Salamandra* **9**(2): 71-74.

Mebs, D., 1974. Haltungserfahrungen mit *Tiliqua casuarinae* (Sauria, Scincidae). *Salamandra* **10**(3/4): 104-06.

Meredith, J., 1954. Reptiles of the Central Tablelands, N.S.W. *Reptilia* **1**(2): 6-12.

Meredith, J. and Cann, C., 1952. Reptiles of the Central Tablelands. *Wildlife* **15**(3): 223-29.

Mertens, R., 1926. Über einige Eidechsen in Gefangeschaft. *Bl. Aquar.-Terrar.-Kunde.* **37**: 94-104. Not seen.

Mertens, R., 1931. *Ablepharus boutonii* (Desjardin) und seine geographische Variation. *Zool. Jb. Syst.* **61**: 63-210.

Mertens, R., 1933. Weitere Mitteilungen über die Rassen von *Ablepharus boutonii* (Desjardin), I. *Zool. Anz.* **105** (3/4): 92-96.

Mertens, R., 1934. Weitere Mitteilungen über die Rassen von *Ablepharus boutonii* (Desjardin), II. *Zool. Anz.* **108** (1/2): 40-43.

Mertens, R., 1942a. Die Familie der Warane (Varanidae). Erster Teil: Allgemeines. *Abh. Senckenberg. Naturf. Ges.* **462**: 1-116.

Mertens, R., 1942b. Die Familie der Warane (Varanidae). Zweiter Teil: der Schädel. *Abh. Senckenberg. Naturf. Ges.* **465**: 117-234.

Mertens, R., 1942c. Die Familie der Warane (Varanidae). Dritter Teil: Taxonomie. *Abh. Senckenberg. Naturf. Ges.* **466**: 235-391.

Mertens, R., 1957a. Ein neuer melanistischer Waran aus dem südlichen Australien. *Zool. Anz.* **159**(1/2): 17-20.

Mertens, R., 1957b. Two new goannas from Australia. *West. Aust. Nat.* **5**(7): 183-85.

Mertens, R., 1958a. Neue Eidechsen aus Australien. *Senck. Biol.* **39**(1/2): 51-56.

Mertens, R., 1958b. Bemerkungen über die Warane Australiens. *Senck. Biol.* **39**(5/6): 229-64.

Mertens, R., 1963. Ein wenig bekannter Zwergwaran, *Varanus acanthurus primordius*. *Die Aquar. Terrar. Zeitsch.* **16**(12): 377-79.

Mertens, R., 1964. Weitere Mitteilungen über die Rassen von *Ablepharus boutonii* (Desjardin), III. *Zool. Anz.* **173**(2): 99-110.

Mertens, R., 1965. Zur Kenntnis der australischen Eidechsen-familie der Pygopodidae. *Verhandl. Zool.-Bot. Gesellsch. Wien* **105-06**: 56-66.

Mertens, R., 1966. Beobachtungen an Flossenfussechsen (Pygopodidae). *Die Aguar. und Terrar. Zeitsch.* **19**: 54-58.

Mertens, T. J., 1986. Reptiles and amphibians (reprinted from Bowman Park Management Study). *South Aust. Herp. Group Newsletter* No. 55: 1-8.

Miles, T., 1973. Measurements and notes on adult and juvenile pink tongue skinks *(Tiliqua gerrardii)*. *Herpetofauna* **6**(1): 16-17.

Milewski, A. V., 1981. A comparison of reptile communities in relation to soil fertility in the mediterranean and adjacent arid parts of Australia and southern Africa. *J. Biogeog.* **8**(6): 493-503.

Millar, B., 1978. Cox's Scrub Trip Report November 27th, 1977. *South Aust. Herpetologist* 1978(April): 5.

Millen, G., 1976. Cover photograph — mating pair of *Emoia cyanogaster*. *Herpetofauna* **8**(1): 1.

Miller, B., 1978. Birth of desert skinks. *South Aust. Herpetologist* 1978(April): 5.

Miller, B., 1980. The occurrence of the marbled velvet gecko *Oedura marmorata* (Gray) in South Australia. *Herpetofauna* **12**(1): 13-15.

Miller, M. R., 1966. The cochlear duct of lizards. *Proc. Calif. Acad. Sci.* **33**(11): 255-359.

Milton, D., 1980a. Some aspects of the population dynamics of *Lampropholis guichenoti* in Toohey Forest near Brisbane. *Herpetofauna* **11**(2): 19-23.

Milton, D., 1980b. An example of community egglaying in *Oedura tryoni* (De Vis). *Herpetofauna* **11**(2): 28-29.

Milton, D. A., 1987. Reproduction of two closely related skinks, *Egernia modesta* and *E. whitii* (Lacertilia: Scincidae) in south-east Queensland. *Aust. J. Zool.* **35**: 35-41.

Milton, D. A. and Hughes, J. M., 1986. Habitat selection by two closely related skinks, *Egernia modesta* Storr and *Egernia whitii* Lacepede (Lacertilia: Scincidae). *Aust. Wildl. Res.* **13**(2): 295-300.

Milton, D. A., Hughes, J. M. and Mather, P. B., 1983. Electrophoretic evidence for the specific distinctness of *Egernia modesta* and *E. whitii* (Lacertilia: Scincidae). *Herpetologica* **39**(2): 100-05.

Mincham, H., 1966. The white lizard of Lake Eyre. *South Aust. Nat.* **41**(2): 41.

Mincham, H., 1970. Reptiles of Australia and New Zealand. Rigby, Adelaide; 63 pp.

Minton, S. A., 1966. A contribution to the herpetology of West Pakistan. *Bull. Amer. Mus. Nat. Hist.* **134**: 27-184.

Minton, S. A. and Minton, M. R., 1981. Toxicity of some Australian snake venoms for potential prey species of reptiles and amphibians. *Toxicon* **19**(6): 749-55.

Mitchell, F. J., 1948. A revision of the lacertilian genus *Tympanocryptis*. *Rec. South Aust. Mus.* **9**(1): 57-86.

Mitchell, F. J., 1953. A brief revision of the four-fingered members of the genus *Leiolopisma* (Lacertilia). *Rec. South Aust. Mus.* **11**(1): 75-90.

Mitchell, F. J., 1955. Preliminary account of the Reptilia and Amphibia collected by the National Geographical Society — Commonwealth Government — Smithsonian Institution expedition to Arnhem Land (April to November 1948). *Rec. South Aust. Mus.* **11**(4): 373-408.

Mitchell, F. J., 1958. Adaptive convergence in Australian reptiles. *Aust. Mus. Mag.* **12**(10): 314-17.

Mitchell, F. J., 1959. Communal egg-laying in the lizard *Leiolopisma guichenoti* (Duméril and Bibron). *Trans. Roy. Soc. South Aust.* **82**: 121-22.

MITCHELL, F. J., 1965a. The affinities of *Tympanocryptis maculosa* Mitchell (Lacertilia — Agamidae). *Rec. South Aust. Mus.* **15**(1): 179-91.

MITCHELL, F. J., 1965b. Australian geckos assigned to the genus *Gehyra* Gray (Reptilia, Gekkonidae). *Senck. Biol.* **46**(4): 287-319.

MITCHELL, F. J., 1973. Studies on the ecology of the agamid lizard *Amphibolurus maculosus* (Mitchell). *Trans. Roy. Soc. South Aust.* **97**(1): 47-76.

MOODY, S. M., 1980. Phylogenetic and historical biogeographical relationships of the genera in the family Agamidae (Reptilia: Lacertilia). Unpublished Ph.D. Thesis, University of Michigan, Ann Arbor; 373 pp.

MORITZ, C., 1983. Parthenogenesis in the endemic Australian lizard *Heteronotia binoei* (Gekkonidae). *Science* **220**: 735-37.

MORITZ, C., 1984a. The origin and evolution of parthenogenesis in *Heteronotia binoei* (Gekkonidae). I. Chromosome banding studies. *Chromosoma* **89**: 151-62.

MORITZ, C., 1984b. The evolution of a highly variable sex chromosome in *Gehyra purpurascens* (Gekkonidae). *Chromosoma* **90**: 111-19.

MORITZ, C., 1986. The population biology of *Gehyra* (Gekkonidae): chromosome changes and speciation. *Syst. Zool.* **35**(1): 46-67.

MORITZ, C. AND KING, D., 1985. Cytogenetic perspectives on parthenogenesis in the Gekkonidae. Pp. 327-37 *in* Biology of Australasian frogs and reptiles ed by G. Grigg, R. Shine and H. Ehmann. Surrey Beatty and Sons Pty Ltd, Chipping Norton, NSW; 527 pp.

MORLEY, T. P., 1985. Some observations on reproduction in the three lined skink *Leiolopisma trilineata*. *Vict. Nat.* **102**(2): 63-64.

MORLEY, T. P. AND MORLEY, P. T., 1985. An inventory of the reptiles of Danggali Conservation Park. *Herpetofauna* **15**(2): 32-36.

MORRIS, P. B., TRITTON, W. AND TRITTON, R., 1963. *Hemisphaeriodon gerrardii*. *Herpetofauna* **1**(3): 17-18.

MORRISON, P. C., 1948. Vegetarian lizards. *Wild Life* **10**(3): 132.

MORRISON, C., 1950. Fairy floss gecko. *Wild Life* **12**(11): 517 (inadvertently carried over to **12**(12): 561).

MORRISON, C., 1951. That squirting gecko. Remarks by L. Glauert on stinging effect in eye of *Strophurus* tail exudate. *Wild Life* **13**(1): 62.

MUDRACK, W., 1969. Paarung und Eiablage bei *Varanus storri* Mertens, 1966. *Aquaterra* **6**: 25-28. Not seen.

MUDRACK, W., 1974. Der Rosazungenskink-eine terraristische Kostbarkeit. *Aquar. Mag.* **8**(10): 407-11.

MÜNSCH, W., 1981. Durch Nachzucht erhalten: der Stachelskink. *Aquar. Mag.* **15**(6): 375-77.

MURPHY, J., 1972. Notes on Indo-australian varanids in captivity. *Int. Zoo Yearb.* **12**: 199-202.

MURPHY, J. B. AND LAMOREAUX, W. E., 1978. Threatening behaviour in Mertens' water monitor *Varanus mertensi* (Sauria: Varanidae). *Herpetologica* **34**(2): 202-05.

MURPHY, J. B. AND MITCHELL, L. A., 1974. Ritualized combat behaviour of the pygmy mulga monitor lizard, *Varanus gilleni* (Sauria: Varanidae). *Herptologica* **30**(1): 90-97.

MURPHY, J. B., BARKER, D. G. AND TRYON, B. W., 1978. Miscellaneous notes on the reproductive biology of reptiles. 2. Eleven species of the family Boidae, genera *Candoia, Corallus, Epicrates* and *Python. J. Herp.* **12**(3): 385-90.

MURRAY, P. J., 1980. The small vertebrate community at Badgingarra, Western Australia. Unpublished B.Sc. Thesis, School of Environmental and Life Sciences, Murdoch University; 103 pp.

MUTTON, L. A., 1970. The response of geckos to sunlight. Unpublished Honours Thesis, University of Western Australia, Perth. Not seen.

NANKIVELL, R., 1976. Breeding of the larger spiny-tailed skink. *West. Aust. Nat.* **13**(6): 146-47.

NAYLOR, L. M., 1980. The maintenance of a group of prickly forest skinks (*Tropidophorus queenslandiae* De Vis), in captivity. *Thylacinus* **5**(4): 5-6.

NEILL, W. T., 1957. Notes on the pygopodid lizards, *Lialis burtoni* and *L. jicari*. *Copeia* **1957**(3): 230-32.

NIEKISCH, M., 1975. Pflege und Nachtzucht von *Egernia cunninghami*. *Salamandra* **11**(3/4): 130-35.

NIEKISCH, M., 1980. Terraristische Beobachtungen zur Biologie von *Egernia cunninghami*. *Salamandra* **16**(3): 162-76.

NOPSCA, F., 1908. Zur Kenntnis der fossilen Eidechsen. *Beitr. Paläont. Geol. Österreich-Ungarns* **21**(1-2): 33-62.

NOPSCA, F., 1923. *Eidolosaurus* und *Pachyophis*. Zwei neue Neucom-reptilien. *Palaeontographica* **65**: 99-154.

NORTHCUTT, R. G., 1978. Forebrain and midbrain organization in lizards and its phylogenetic significance. Pp. 11-64 *in* Behaviour and neurology of lizards ed by N. Greenberg and P. D. Maclean. Department of Health Education Welfare, Washington DC, Publ. No. (ADM) 77-491; 352 pp.

OLIVER, J. A. AND SHAW, C. E., 1953. The amphibians and reptiles of the Hawaiian Islands. *Zoologica* **38**(2): 65-95.

ORMSBY, A. I., 1961. A curious breeding record of a legless lizard *(Pygopus lepidopodus)*. *Proc. Roy. Zool. Soc. New South Wales* 1958-1959: 59.

OWEN, R., 1858. Description of some remains of a gigantic land-lizard (*Megalania prisca*, Owen) from Australia. *Philos. Trans. Roy. Soc. London.* **149**: 43-48.

OWEN, R., 1880. Description of some remains of the gigantic land-lizard (*Megalania prisca*, Owen) from Australia. Part II. *Philos. Trans. Roy. Soc. London* **171**(3): 1037-50.

OWEN, R., 1881. Description of some remains of the gigantic land-lizard (*Megalania prisca*, Owen), from Australia. Part III. *Philos. Trans. Roy. Soc. London* **172**(2): 547-56.

OWEN, R., 1887a. Description of fossil remains, including foot-bones, of *Megalania prisca*. — Part IV. *Philos. Trans. Roy. Soc. London* **177**(1): 327-30.

OWEN, R., 1887b. Description of fossil remains of two species of a megalanian genus *(Meiolania)* from "Lord Howe's Island". *Philos. Trans. Roy. Soc. London* **177**(2): 471-80.

OWEN, R., 1888. On parts of the skeleton of *Meiolania platyceps* (Ow.). *Philos. Trans. Roy. Soc. Lond.* **179**(B): 181-91.

PARKER, H. W., 1956. The lizard genus *Aprasia;* its taxonomy and temperature correlated variation. *Bull. Brit. Mus. (Nat. Hist.)* **3**(9): 365-85.

PARMENTER, C. J. AND HEATWOLE, H., 1975. Panting thresholds of lizards. IV. The effect of dehydration on the panting threshold of *Amphibolurus barbatus* and *Amphibolurus muricatus*. *J. Exp. Zool.* **191**: 327-32.

PATCHELL, F. C. AND SHINE, R., 1986a. Food habits and reproductive biology of the Australian legless lizards (Pygopodidae). *Copeia* **1986**(1): 30-39.

PATCHELL, F. C. AND SHINE, R., 1986b. Feeding mechanisms in pygopodid lizards: how can *Lialis* swallow such large prey? *J. Herp.* **20**(1): 59-64.

PATCHELL, F. C. AND SHINE, R., 1986c. Hinged teeth for hard-bodied prey: a case of convergent evolution between snakes and legless lizards. *J. Zool.* (Lond.) (A)**208**: 269-75.

PATON, J., 1965. *Moloch horridus. South Aust. Nat.* **40**(2): 25-26.

PEARSON, O. AND REEVES, G., 1978. Reptiles of Hinchinbrook Island. *Qld Nat.* **22**(1-4): 69-71.

PENGILLEY, R., 1972. Systematic relationships and ecology of some lygosomine lizards from southeastern Australia. Unpublished Ph.D. Thesis, Australian National University; 313 pp.

PENGILLEY, R., 1981. Notes on the biology of *Varanus spenceri* and *V. gouldii*, Barkly Tablelands, Northern Territory. *Aust. J. Herp.* **1**(1): 23-26.

PENGILLEY, R., 1982. Note on the reproductive biology of the ring-tailed dragon (*Chelosania brunnea*). *North. Terr. Nat.* **5**: 6.

PERNETTA, J. C. AND BLACK, D., 1983. Species of gecko (*Lepidodactylus*) in the Port Moresby area, with the description of a new species. *J. Herp.* **17**(2): 121-28.

PETERS, U., 1968. *Moloch horridus, Varanus spenceri, V. mitchelli, Egernia bungana* und *Heteronotia binoei* im Taronga-Zoo, Sydney. *Die Aquar. Terr. Zeit.* **21**(8): 252-54.

PETERS, U., 1969a. Beobachtungen am Mangrove-Waran, *Varanus (Odatria) semiremex* Peters, 1869. *Aqua. Terra.* **6**: 61-63.

PETERS, U., 1969b. Observations on *Varanus mertensi* and *Varanus mitchelli* in captivity. *Bull. Zoo Management* **2**(2): 20-22.

PETERS, U., 1969c. Zum ersten Mal in Gefangenschaft: Eiablage und Schlupf von *Varanus spenceri. Aquar. Terrar.* **16**(9): 306-07.

PETERS, U., 1969d. Fang und Haltung von *Varanus storri, V. timorensis similis* und *V. semiremex. Aquar. Terrar.* **16**(10): 338-40.

PETERS, U., 1969e. Zum ersten Male nachgezüchtet: Spencers Waran. *Aguar. Mag.* **1969**(10): 412-13.

PETERS, U., 1970a. Raubechsen. Drei Baumwarane aus Australien. *Aquar. Mag.* **1970**(1): 22-25.

PETERS, U., 1970b. Der Dickschwanz-Gecko *Gymnodactylus milii* (Bory). *Aqua. Terra.* **6**: 190.

PETERS, U., 1970c. Geckos aus Neusüdwales. *Aquar. Terrar.* **17**(11): 370-72.

PETERS, U., 1971a. The first hatching of *Varanus spenceri* in captivity. *Bull. Zoo Management* **3**(2): 17-18.

PETERS, U., 1971b. *Varanus mertensi* in Gefangenschaft. *Aquar. Terrar.* **18**(6): 192-93. Not seen.

PETERS, U., 1971c. Das Echsenporträt: der Wüstenskink. *Aquar. Mag.* **12**: 525.

PETERS, U., 1971. Bermerkungen über Mitchell's Wasserwaran *Varanus (Varanus) mitchelli* Mertens, 1958. *Aqua. Terra.* **8**: 75-77.

PETERS, U., 1973a. A contribution to the ecology of *Varanus (Odatria) storri. Koolewong* **2**(4): 12-13.

PETERS, U., 1973b. Ein Beitrag zur Ökologie Von *Varanus (Odatria) storri* Mertens 1966. *Aquarium* **53**: 462-63.

PETERS, U., 1974. The dragons in Taronga's reptile department. *Koolewong* **3**(3): 5-6.

PETERS, U., 1975. The dragons in Taronga's reptile department. *Koolewong* **4**(1): 4-6.

PETERS, U. W., 1976a. A hybrid birth at Taronga's reptile department. *South Aust. Nat.* **50**(3): 53-54.

PETERS, U. W., 1976b. Seltsame Reptilien Australiens. *Das Aquarium* **90**: 557-62.

PETERS, U., 1983. Australian dragons. *Zoonooz* **56**(6): 7-9.

PETZOLD, H.-G., 1965. Über die Widerstandsfähigkeit von Geckonen-Eiern und einige andere Beobachtungen an *Hemidactylus frenatus* Dum. and Bibr. 1836. *Zool. Gart.* **31**(5): 262-65.

PHILIPP, G. A., 1979. Sperm storage in *Moloch horridus. West. Aust. Nat.* **14**(6): 161.

PHILIPP, G. A., 1980. An observation of predatory behaviour by a pygopodid lizard on a scorpion. *West. Aust. Nat.* **14**(8): 240.

PHILLIPS, J. A., 1986. Ontogeny of metabolic processes in blue-tongued skinks, *Tiliqua scincoides. Herpetologica* **42**(4): 405-12.

PIANKA, E. R., 1968. Notes on the biology of *Varanus eremius. West. Aust. Nat.* **11**(2): 39-44.

PIANKA, E. R., 1969a. Habitat specificity, speciation, and species density in Australian desert lizards. *Ecology* **50**(3): 498-502.

PIANKA, E. R., 1969b. Sympatry of desert lizards (*Ctenotus*) in Western Australia. *Ecology* **50**(6): 1012-30.

PIANKA, E. R., 1969c. Notes on the biology of *Varanus caudolineatus* and *Varanus gilleni. West. Aust. Nat.* **11**(4): 76-82.

PIANKA, E. R., 1970a. Notes on *Varanus brevicauda. West. Aust. Nat.* **11**(5): 113-16.

PIANKA, E. R., 1970b. Notes on the biology of *Varanus gouldii flavirufus. West. Aust. Nat.* **11**(6): 141-44.

PIANKA, E. R., 1971a. Comparative ecology of two lizards. *Copeia* **1971**(1): 129-38.

PIANKA, E. R., 1971b. Notes on the biology of *Varanus tristis. West. Aust. Nat.* **11**(8): 180-83.

PIANKA, E. R., 1971c. Ecology of the agamid lizard *Amphibolurus isolepis* in Western Australia. *Copeia* **1971**(3): 527-36.

PIANKA, E. R., 1971d. Notes on the biology of *Amphibolurus cristatus* and *Amphibolurus scutulatus. West. Aust. Nat.* **12**(2): 36-41.

PIANKA, E. R., 1972. Zoogeography and speciation of Australian desert lizards: an ecological perspective. *Copeia* **1972**(1): 127-45.

PIANKA, E. R., 1973. The structure of lizard communities. *Ann. Rev. Ecol. Syst.* **4**: 53-74.

PIANKA, E. R., 1981. Diversity and adaptive radiations of Australian desert lizards. Pp. 1377-92 *in* Ecological biogeography of Australia ed by A. Keast. W. Junk, The Hague; 2142 pp.

PIANKA, E. R., 1982. Observations on the ecology of *Varanus* in the Great Victoria Desert. *West. Aust. Nat.* **15**(2): 1-8.

PIANKA, E. R., 1986. Ecology and natural history of desert lizards. Analyses of the ecological niche and community structure. Princeton Univ. Press, Princeton, N. J.; 208 pp.

PIANKA, E. R. AND GILES, W. F., 1982. Notes on the biology of two species of nocturnal skinks, *Egernia inornata* and *Egernia striata*, in the Great Victoria Desert. *West. Aust. Nat.* **15**(2): 44-49.

PIANKA, E. R., HUEY, R. B. AND LAWLER, L. R., 1979. Niche segregation in desert lizards. Pp. 67-115 *in* Analysis of ecological systems ed by D. J. Horn, G. R. Stairs and R. D. Mitchell. Ohio State University Press, Columbus, Ohio; 312 pp.

PIANKA, E. R. AND PIANKA, H. D., 1970. The ecology of *Moloch horridus* (Lacertilia: Agamidae) in Western Australia. *Copeia* **1970**(1): 90-103.

PIANKA, E. R. AND PIANKA, H. D., 1976. Comparative ecology of twelve species of nocturnal lizards (Gekkonidae) in the Western Australian desert. *Copeia* **1976**(1): 125-42

PICKWORTH, B., 1981. Observations of behaviour patterns displayed by a pair of bearded dragons, *(Amphibolurus barbatus)* (Cuvier). *Herpetofauna* **12**(2): 13-15

POLLECK, R., 1980. Der Rostkopfwaran *(Varanus semiremex). Aqu.-Mag.* (Stuttgart) **14**: 39. Not seen.

POPPER, J. S., 1980. The foraging behaviour of *Sphenomorphus quoyii* (Scincidae). Unpublished B.Sc. Hons. Thesis, Flinders University. Not seen.

PORTER, K. R., 1972. Herpetology. W. B. Saunders Co., Philadelphia; 524 pp.

PORTER, W. P., 1967. Solar radiation through the living body walls of vertebrates with emphasis on desert reptiles. *Ecological Monographs* **37**(4): 273-96.

POUGH, F. H., 1969. The morphology of undersand respiration in reptiles. *Herpetologica* **25**(3): 216-23.

POUGH, F. H., 1973. Lizard energetics and diet. *Ecology* **54**(4): 837-44.

POUGH, F. H., 1983. Feeding mechanisms, body size, and the ecology and evolution of snakes. Introduction to the symposium. *Amer. Zool.* **23**(2): 339-42.

POWELL, H., HEATWOLE, H. AND HEATWOLE, M., 1977. Winter aggregations of *Leiolopisma guichenoti. British J. Herp.* **5**(11): 789-91.

RAGE, J. C., 1977. La position phylétique de *Dinilysia patagonica*, serpent du Crétacé supérieur. *C. R. Acad. Sci. Paris* **284**: 1765-68.

RAGE, J. C., 1982. La phylogénie des Lépidosauriens (Reptilia): une approche cladistique. *C. R. Acad. Sci. Paris* **294**: 563-66.

RANKIN, P., 1972. Untitled notes on *Egernia cunninghami. Herpetofuauna* **4**(4): 5.

RANKIN, P. R., 1973a. Lizard mimicking a snake — juvenile *Tiliqua casaurinae* (Duméril and Bibron). *Herpetofauna* **5**(4): 13-14.

RANKIN, P. R., 1973b. The barred sided skink *Sphenomorphus tenuis tenuis* (Gray) in the Sydney region. *Herpetofauna* **6**(1): 8-14.

RANKIN, P. R., 1973c. Untitled observations on *Cryptoblepharus virgatus* (as *C. boutonii). Herpetofauna* **6**(1): 25.

RANKIN, P. R., 1976. A note on a possible diversionary defence mechanism in the worm lizard, *Aprasia inaurita* Kluge. *Herpetofauna* **8**(2): 18-19.

RANKIN, P. R., 1977. Burrow plugging in the netted dragon *Amphibolurus nuchalis* with reports on the occurrence in three other Australian agamids. *Herpetofauna* **9**(1): 18-22.

RANKIN, P. R., 1978a. Notes on the biology of the skink *Sphenomorphus pardalis* (Macleay) including a captive breeding record. *Herpetofauna* **10**(1): 4-7.

RANKIN, P. R., 1978b. A new species of lizard (Lacertilia: Scincidae) from the Northern Territory, closely allied to *Ctenotus decaneurus* Storr. *Rec. Aust. Mus.* **31**(10): 395-409.

RANKIN, P. R., 1979. A taxonomic revision of the genus *Menetia* (Lacertilia, Scincidae) in the Northern Territory. *Rec. Aust. Mus.* **32**(14): 491-99.

RANKIN, P. R. AND GILLAM, M. W., 1979. A new lizard in the genus *Ctenotus* (Lacertilia: Scincidae) from the Northern Territory with notes on its biology. *Rec. Aust. Mus.* **32**(15): 501-11.

RAWLINSON, P. A., 1969. The reptiles of East Gippsland. *Proc. Roy. Soc. Vict.* **82**(1): 113-28.

RAWLINSON, P. A., 1974a. Biogeography and ecology of the reptiles of Tasmania and the Bass Strait Area. Pp. 291-338 *in* Biogeography and ecology in Tasmania ed by W. D. Williams. W. Junk, The Hague; 498 pp.

RAWLINSON, P. A., 1974b. Revision of the endemic southeastern Australian lizard genus *Pseudemoia* (Scincidae: Lygosominae). *Mem. Nat'l Mus. Vict.* **35**: 87-96.

RAWLINSON, P. A., 1974c. Natural history of Curtis Island, Bass Strait. 4. The reptiles of Curtis and Rodondo Islands. *Pap. Proc. Roy. Soc. Tas.* **107**: 153-70.

RAWLINSON, P. A., 1975. Two new lizard species from the genus *Leiolopisma* (Scincidae: Lygosominae) in southeastern Australia and Tasmania. *Mem. Nat'l Mus. Vict.* **36**: 1-15.

RAWLINSON, P. A., 1976. The endemic Australian lizard genus *Morethia* (Scincidae: Lygosominae) in southern Australia. *Mem. Nat'l Mus. Vict.* **37**: 27-41.

REBOUÇAS-SPIEKER, R. AND VANZOLINI, P. E., 1978. Parturition in *Mabuya macrorhyncha* Hoge, 1946 (Sauria, Scincidae), with a note on the distribution of maternal behaviour in lizards. *Papéis Avulsos Zool.* **32**(8): 95-99.

RESE, R., 1984. Der Zwergwaran *Varanus storri* Mertens 1966. *Sauria* (Berlin) **1984**(1): 33-34. Not seen.

RICE, G. E. AND BRADSHAW, S. D., 1980. Changes in dermal reflectance and vascularity and their effects on thermoregulation in *Amphibolurus nuchalis* (Reptilia: Agamidae). *J. Comp. Physiol.* **135**(B): 139-46.

RICHARDSON, K. C. AND HINCHLIFFE, P. M., 1983. Caudal glands and their secretions in the western spiny-tailed gecko, *Diplodactylus spinigerus. Copeia* **1983**(1): 161-69.

RIEPPEL, O., 1971. Der physiologische Farbwechsel bei *Gehyra variegata* (Duméril und Bibron) 1836. *Aquar. Terrar.* **8**: 78-80.

RIEPPEL, O., 1978. The evolution of the naso-frontal joint in snakes and its bearing on snake origins. *Z. Zool. Syst. Evolut.-forsch.* **16**(1): 14-27.

RIEPPEL, O., 1980. The sound-transmitting apparatus in primitive snakes and its phylogenetic significance. *Zoomorphology* **96**: 45-62.

RIEPPEL, O., 1981. The hyobranchial skeleton in some little known lizards and snakes. *J. Herp.* **15**(4): 433-40.

RIEPPEL, O., 1984a. The cranial morphology of the fossorial lizard genus *Dibamus* with a consideration of its phylogenetic relationships. *J. Zool.* **204**: 289-327.

RIEPPEL, O., 1984b. Miniaturization of the lizard skull: its functional and evolutionary implications. *Symp. Zool. Soc. Lond.* No. **52**: 503-20.

ROBERTS, B., 1985. Trip report. Uro Bluff, Kingoonya, Cariewerloo. Oct. 1984. *South Aust. Herp. Group Newsletter* 1985(July): 6-9.

ROBERTSON, P., 1976. Aspects of the ecology and reproduction of two species of thigmothermic skinks, *Anotis maccoyi* and *Hemiergis decresiensis* in southeastern Australia. Unpublished B.Sc. Honours Thesis, University of Melbourne. Not seen.

ROBERTSON, P., 1980. Captivity mating of pink tongue skinks *(Tiliqua gerrartlii)*. *Vict. Herp. Soc. Newsletter* No. 18: 11-12.

ROBERTSON, P., 1981. Comparative reproductive ecology of two southeastern Australian skinks. Pp. 25-37 *in* Proceedings of the Melbourne Herpetological Symposium ed by C. B. Banks and A. A. Martin. Zoological Board of Victoria, Melbourne; 199 pp.

ROBINSON, A. C., 1980. Notes on the mammals and reptiles of Pearson, Dorothee and Greenly Islands, South Australia. *Trans. Roy. Soc. South Aust.* **104**(5 & 6): 93-99.

ROBINSON, A. C., MIRTSCHIN, P. J., COPLEY, P. B., CANTY, P. D. AND JENKINS, R. B., 1985. The Reevesby Island goanna — a problem in conservation management. *South Aust. Nat.* **59**(4): 56-62.

ROBINSON, M., 1974. The mating behaviour of Burton's legless lizard *(Lialis burtonis)*. *Roy. Zool. Soc. Bull. Herp.* **1**(3): 14-16.

ROBSON, P., 1968. A rare agamid. *Wildlife in Aust.* **5**(3): 84.

ROESCH, K. E., 1956. Ein Beitrag zur Frage der Fortpflanzung und Aufzucht der Stütz-Eidechse *(Tiliqua rugosa)*. *Die Aquar. Terr. Zeit.* **9**: 270-73. Not seen.

ROMER, A. S., 1956. Osteology of the reptiles. University of Chicago Press, Chicago; 772 pp.

ROSE, A. B., 1974. Gut contents of some amphibians & reptiles. *Herpetofauna* **7**(1): 4-8.

ROSE, S., 1985. Captive breeding of Cunningham's skink. *The Vipera* **1**(8): 14-22.

ROSENBERG, H. I. AND RUSSELL, A. P., 1980. Structural and functional aspects of tail squirting: a unique defense mechanism of *Diplodactylus* (Reptilia: Gekkonidae). *Canadian J. Zool.* **58**(5): 865-81.

ROSENBERG, H. I., RUSSELL, A. P. AND KAPOOR, M., 1984. Preliminary characterization of the defensive secretion of *Diplodactylus* (Reptilia: Gekkonidae). *Copeia* **1984**(4): 1025-28.

RÖSLER, H., 1980. Ein Australier, der Samtgecko *Oedura monilis* (De Vis, 1888). *Elaphe* **1980**(4): 53-56.

RÖSLER, H., 1981. De australische bladstaartgekko, *Phyllurus cornutus*. *Lacerta* **39**(9): 128-29.

ROUNSEVELL, D. E., 1978. Communal egg-laying in the three-lined skink *Leilopisma trilineata*. *Tasmanian Naturalist* No. 52: 1-2.

ROUNSEVELL, D., BROTHERS, N. AND HOLDSWORTH, M., 1985. The status and ecology of the Pedra Branca skink *Pseudemoia palfreymani*. Pp. 477-80 *in* The biology of Australasian frogs and reptiles ed by G. Grigg, R. Shine, H. Ehmann. Surrey Beatty & Sons Pty Ltd, Chipping Norton, NSW; 527 pp.

RÜEGG, R., 1974. Nachzucht beim Timor-Baumwaran, *Varanus timorensis similis* Mertens 1958. *Aquarium mit Aqua Terra* **8**(62): 360-63.

RUSSELL, A. P., 1976. Some comments concerning interrelationships amongst gekkonine geckos. Pp. 217-44 *in* Morphology and biology of reptiles ed by A. d'A. Bellairs and C. B. Cox. *Linnean Society Symposium Series* No. 3., 290 pp.

RUSSELL, A. P., 1977. The phalangeal formula of *Hemidactylus* Oken, 1817 (Reptilia, Gekkonidae): a correction and a functional explanation. *Zbl. Vet. Med. C. Anat. Hist. Embryol.* **6**: 332-38.

RUSSELL, A. P., 1979. Parallelism and integrated design in the foot structure of gekkonine and diplodactyline geckos. *Copeia* **1979**(1): 1-21.

RUSSELL, A. P., 1980. *Underwoodisaurus* Wermuth 1965, a junior synonym of *Phyllurus* Schinz 1822. *J. Herp.* **14**(4): 415-16.

RUSSELL, A. P. AND ROSENBERG, H. I., 1981. Subgeneric classification in the gekkonid genus *Diplodactylus*. *Herpetologica* **37**(2): 86-92.

RUSSELL, A. P. AND ROSENBERG, H. I., 1981. Self-grooming in *Diplodactylus spinigerus* (Reptilia: Gekkonidae), with a brief review of such behaviour in reptiles. *Canadian J. Zool.* **59**(3): 564-66.

SABATH, M. D., 1981. Gekkonid lizards of Guam, Mariana Islands: reproduction and habitat preference. *J. Herp.* **15**(1): 71-75.

SADLIER, R., 1981. A report on the reptiles encountered in the Jabiru Project area. Report to Supervising Scientist, Alligators Rivers Region Research Institute. Not seen.

SADLIER, R. A., 1984. A new Australian scincid lizard, *Menetia concinna*, from the Alligator Rivers Region, Northern Territory. *Rec. Aust. Mus.* **36**(1-2): 45-49.

SADLIER, R. A., 1985. A new Australian scincid lizard, *Ctenotus coggeri*, from the Alligator Rivers Region, Northern Territory. *Rec. Aust. Mus.* **36**(3-4): 153-56.

SADLIER, R., WOMBEY, J. C. AND BRAITHWAITE, R. W., 1985. *Ctenotus kurnbudj* and *Ctenotus gagudju*, two new lizards (Scincidae) from the Alligator Rivers region of the Northern Territory. *The Beagle, Occ. Pap. North. Terr. Mus. Arts Sci.* **2**(1): 95-103.

SAHI, D., 1980. Some observations on the common house gecko *(Hemidactylus frenatus)* of southern India. *J. Bombay Nat. Hist. Soc.* **76**(3): 521-23.

SAINT-GIRONS, H., LEMIRE, M. AND BRADSHAW, S. D., 1977. Structure de la glande nasale externe de *Tiliqua rugosa* (Reptilia, Scincidae) et rapports avec sa fonction. *Zoomorphologie* **88**: 277-88.

SAINT-GIRONS, H., RICE, G. E. AND BRADSHAW, S. D., 1981. Histologie comparée et ultrastructure de la glande nasale externe de quelques Varanidae (Reptilia: Lacertilia). *Ann. Sci. Natur. Zool.* 13(**3**): 15-21.

SATRAWAHA, R. AND BULL, C. M., 1981. The area occupied by an omnivorous lizard, *Trachydosaurus rugosus*. *Aust. Wildl. Res.* **8**(2): 435-42.

SAVILLE-KENT, W., 1895. Observations on the frilled lizard, *Chlamydosaurus kingi*. *Proc. Zool. Soc.* 1895: 712-19.

SAVILLE-KENT, W., 1897. The naturalist in Australia. Chapman and Hall, Ltd, London; 302 pp.

SAVITZKY, A. H., 1980. The role of venom delivery strategies in snake evolution. *Evolution* **34**(6): 1194-204.

SAVITZKY, A. H., 1981. Hinged teeth in snakes: an adaptation for swallowing hard-bodied prey. *Science* **212**: 346-49.

SCANLON, J. D., 1986. Herpetological records from northern N.S.W. *Herpetofauna* **16**(2): 38-39.

SCHAFER, S., 1979. Beards and blue-tongues. *Zoonooz* **52** (4): 14.

SCHMIDA, G. E., 1971. Der getupfelte Baumwaran Australiens, *Varanus timorensis similis*. *Die Aquar. Terr. Zeit.* **24**(5): 168-70.

SCHMIDA, G. E., 1973a. Geckos aus Nordwest-australien. *Die Aquar. Terr. Zeit.* **26**(8): 280-81.

SCHMIDA, G. E., 1973b. Geckos aus Nordwest-australien II. *Die Aquar. Terr. Zeit.* **26**(9): 316-17.

SCHMIDA, G. E., 1974. Der Kurzschwanzwaran *(Varanus brevicauda)*. *Die Aquar. Terr. Zeit.* **27**(11): 390-94.

SCHMIDA, G. E., 1975. Däumlinge aus der Familie der Riesen. Kurzschwanz-, Gillen- und Schwanzstrichwaran. *Aquar. Mag.* **9**(1): 8-11.

SCHMIDA, G. , 1985. The cold-blooded Australians. Doubleday, Sydney; 280pp.

SCHNEE, P., 1901. Biologische Notizen über *Lygosoma cyanurum* Less. sowie *Lepidodactylus lugubris* D. & B. *Zeit. Naturw.* **74**: 273-83

SCHNEE, P., 1902. Die Kriechtiere der Marshall-Inseln. *Zool. Gart.* **43**: 354-62.

SCHULZ, M., 1985. The occurrence of the mourning skink, *Egernia coventryi* Storr, in saltmarsh in Westernport Bay, Victoria. *Vict. Nat.* **102**(5): 148-52.

SCHÜRER, U. AND HORN, H.-G., 1976. Freiland — und Gefangenschaftsbeobachtungen am australischen Wasserwaran, *Varanus mertensi*. *Salamandra* **12**(4): 176-88.

SCHUSTER, M. N., 1981a. Relict lizards and rainforest refugia in eastern Australia: an eco-historical interpretation. *Qld Geog. J.* (3)**6**: 49-56.

SCHUSTER, M. N., 1981b. Origins and historical biogeography of eastern Australian rainforest Scincidae. Pp. 17-21 *in* Proceedings of the Melbourne Herpetological Symposium ed by C. B. Banks and A. A. Martin. Zoological Board of Victoria, Melbourne; 199 pp.

SCHWANER, T. D., 1980. Reproductive biology of lizards on the American Samoan Islands. *Occ. Pap. Mus. Nat. Hist. Univ. Kansas* No. 86; 53 pp.

SCHWANER, T. D., MILLER, B. AND TYLER, M. J., 1985. Reptiles and amphibians. Pp. 159-68 *in* Natural history of Eyre Peninsula ed by C. R. Twidale, M. J. Tyler and M. Davies. Royal Society of South Australia, Adelaide; 229 pp.

SCOTT, T. C., 1962. Association of young and adult water dragons *(Physignathus longirostris)*. *West. Aust. Nat.* **8**(3): 79.

SENN, D. G. AND NORTHCUTT, R. G., 1973. The forebrain and midbrain of some squamates and their bearing on the origin of snakes. *J. Morph.* **140**(2): 135-52.

SERVENTY, V. N., 1951. Natural history notes from the south coast. *West. Aust. Nat.* **3**(2): 34-36.

SHEA, G., 1980. Notes on ecdysis in *Tiliqua rugosa* (Gray). *Herpetofauna* **12**(1): 32-33.

SHEA, G., 1981. Notes on the reproductive biology of the eastern blue-tongue skink, *Tiliqua scincoides* (Shaw). *Herpetofauna* **12**(2): 16-23.

SHEA, G. M., 1982. Insular range extensions for the New Guinea bluetongue, *Tiliqua gigas* (Boddaert) (Lacertilia: Scincidae). *Herpetofauna* **13**(2): 7-11.

SHEA, G. M., 1984. Egg deposition site in the gecko *Diplodactylus williamsi*. *Vict. Nat.* **101**(5): 198-99.

SHEA, G., 1986. Island herpetofaunas in New South Wales: a review. *Herpetofauna* **16**(2): 30-38.

SHEA, G. M., 1987a. *Delma nasuta* (Lacertilia: Pygopodidae), an addition to the herpetofauna of New South Wales and Victoria, with a note on rapid colour change in this species. *Vict. Nat.* **104**(1): 5-8.

SHEA, G. M., 1987b. Oviparity in *Leiolopisma jigurru* and a brief review of reproductive mode in *Leiolopisma*. *Herp. Rev.* **18**(2): 29-32.

SHEA, G. In press. Notes on the biology of *Paradelma orientalis*. *Herpetofauna*.

SHEA, G. M. AND MILLER, B., 1986. The occurrence of *Hemiergis initialis* (Werner, 1910) (Lacertilia: Scincidae) in South Australia. *Trans. Roy. Soc. South Aust.* **110**(1/2): 89-90.

SHEA, G. M. AND PETERSON, M., 1981. Observations on sympatry of the social lizards *Tiliqua multifasciata* Sternfeld and *T. occipitalis* (Peters). *Aust. J. Herp.* **1**(1): 27-28.

SHEA, G. M. AND PETERSON, M., 1985. The Blue Mountains water skink, *Sphenomorphus leuraensis* (Lacertilia: Scincidae): a redescription, with notes on its natural history. *Proc. Linn. Soc. New South Wales* **108**(2): 141-48.

SHEA, G. M. AND REDDACLIFF, G. L., 1986. Ossifications in the hemipenes of varanids. *J. Herp.* **20**(4): 566-68.

SHEA, G., MILLGATE, M. AND PECK, S. In press. A range extension for the rare skink *Anomalopus mackayi*. *Herpetofauna*.

SHINE, R., 1971. The ecological energetics of the scincid lizard *Egernia cunninghami* (Gray, 1832). Unpublished Honours Thesis, Australia National University, Canberra; 92 pp.

SHINE, R., 1980. "Costs" of reproduction in reptiles. *Oecologia* **46**: 92-100.

SHINE, R., 1983. Reptilian viviparity in cold climates: testing the assumptions of an evolutionary hypothesis. *Oecologia* **57**: 397-405.

SHINE, R., 1985. The evolution of viviparity in reptiles: an ecological analysis. Pp. 605-94 *in* Biology of the Reptilia, vol. 15 ed by C. Gans and F. Billett. John Wiley & Sons, Inc., New York; 731 pp.

SHINE, R., 1986a. Evolutionary advantages of limblessness: evidence from the pygopodid lizards. *Copeia* **1986**(2): 525-29.

SHINE, R., 1986b. Food habits, habitats and reproductive biology of four sympatric species of varanid lizards in tropical Australia. *Herpetologica* **42**(3): 346-60.

SHUGG, H., 1983. Bobtails eating *Patersonia* flowers. *West. Aust. Nat.* **15**(6): 148.

SHUTE, C. C. D. AND BELLAIRS, A. d'A., 1953. The cochlear apparatus of Gekkonidae and Pygopodidae and its bearing on the affinities of these groups of lizards. *Proc. Zool. Soc. London* **123**(3): 695-709.

SIEBENROCK, F., 1895. Das Skelet der Agamidae. *Sitzungsber. Kaiser. Akad. Wissensch.* **104**(9): 1089-196.

SIMBOTWE, M. P., 1985. Sexual dimorphism and reproduction of *Lampropholis guichenoti* (Lacertilia: Scincidae). Pp. 11-16 *in* The biology of Australasian frogs and reptiles ed by G. Grigg, R. Shine and H. Ehmann. Surrey Beatty and Sons Pty Ltd, Chipping Norton, NSW; 527 pp.

SIMPSON, K. N. G., 1973. Amphibians, reptiles and mammals of the Murray River region between Mildura and Renmark, Australia. *Mem. Nat'l Mus. Vict.* **34**: 275-79.

SLATER, P. AND LINDGREN, E., 1955. A visit to Queen Victoria Spring, January 1955. *West. Aust. Nat.* **5**(1): 10-18.

SMALES, I., 1981. The herpetofauna of Yellingbo State Faunal Reserve. *Vict. Nat.* **98**(6): 234-46.

SMITH, H. M. AND GRANT, C., 1961. The mourning gecko in the Americas. *Herpetologica* **17**(1): 68.

SMITH, J., 1974. Hatching bearded dragon eggs. *South Australian Herpetologist* **2**(1): 12.

SMITH, J., 1979. Notes on incubation and hatching of eggs of the eastern water dragon. *Herpetofauna* **10**(2): 12-14.

SMITH, J., 1984. Hatching bearded dragon eggs. *South Aust. Herp. Group Newsletter* 1984(November): 13.

SMITH, J. AND SCHWANER, T. D., 1981. Notes on reproduction by captive *Amphibolurus nullarbor* (Sauria: Agamidae). *Trans. Roy Soc. South Aust.* **105**: 215-16.

SMITH, L. A., 1976. The reptiles of Barrow Island. *West. Aust. Nat.* **13**(6): 125-36.

SMITH, L. A. AND CHAPMAN, A., 1976. Reptiles and frogs of Tarin Rock and North Tarin Rock Reserves. *Rec. West. Aust. Mus.* Suppl. No. 2: 85-86.

SMITH, L. A. AND JOHNSTONE, R. E., 1981. Amphibians and reptiles of Mitchell Plateau and adjacent coast and lowlands, Kimberley, Western Australia. Pp. 215-27 *in* Biological survey of Mitchell Plateau and Admiralty Gulf, Kimberley, Western Australia. Western Australian Museum, Perth; pp.?

SMITH, M. A., 1935. The fauna of British India. Reptilia and Amphibia. Vol. II. — Sauria. Taylor and Francis Ltd, London; 440 pp.

SMITH, M. J., 1976. Small fossil vertebrates from Victoria Cave, Naracoorte, South Australia. IV. Reptiles. *Trans. Roy. Soc. South Aust.* **100**(1): 39-51.

SMITH, M. J., 1982. Reptiles from late Pleistocene deposits on Kangaroo Island, South Australia. *Trans. Roy. Soc. South Aust.* **106**(2): 61-66.

SMYTH, M., 1968. The distribution and life history of the skink *Hemiergis peronii* (Fitzinger). *Trans. Roy. Soc. South Aust.* **92**: 51-58.

SMYTH, M., 1971. Pearson Island Expedition 1969. — 5 Reptiles. *Trans. Roy. Soc. South Aust.* **95**(3): 147-48.

SMYTH, M., 1972. The genus *Morethia* (Lacertilia, Scincidae) in South Australia. *Rec. South Aust. Mus.* **16**(12): 1-14.

SMYTH, M., 1974. Changes in the fat stores of the skinks *Morethia boulengeri* and *Hemiergis peronii* (Lacertilia). *Aust. J. Zool.* **22**(2): 135-45.

SMYTH, M. AND SMITH, M. J., 1968. Obligatory sperm storage in the skink *Hemiergis peronii*. *Science* **161**: 575-76.

SMYTH, M. AND SMITH, M. J., 1974. Aspects of the natural history of three Australian skinks, *Morethia boulengeri*, *Menetia greyii* and *Lerista bougainvillii*. *J. Herp.* **8**(4): 329-35.

SNYDER, J. O., 1919. Notes on Hawaiian lizards. *Proc. U.S. Nat'l Mus.* **54**: 19-25.

SONNEMANN, N., 1974. Notes on *Delma fraseri* in the northeast of Victoria. *Herpetofauna* **7**(1): 15.

SPELLERBERG, I. F., 1971. Thermoregulation and temperature tolerances in the *Sphenomorphus quoyi* species complex (Lacertilia: Scincidae) of southeast Australia. Unpublished Ph.D. Thesis, La Trobe University, Bundoora, Victoria. Not seen.

SPELLERBERG, I. F., 1972a. Temperature tolerances of southeast Australian reptiles examined in relation to reptile thermoregulatory behaviour and distribution. *Oecologia* **9**: 23-46.

SPELLERBERG, I. F., 1972b. Thermal ecology of allopatric lizards (*Sphenomorphus*) in southeast Australia. I. The environment and lizard critical temperatures. *Oecologia* **9**: 371-83.

SPELLERBERG, I. F., 1972c. Thermal ecology of allopatric lizards (*Sphenomorphus*) in southeast Australia. II. Physiological aspects of thermoregulation. *Oecologia* **9**: 385-98.

SPELLERBERG, I. F., 1972d. Thermal ecology of allopatric lizards (*Sphenomorphus*) in southeast Australia. III. Behavioural aspects of thermoregulation. *Oecologia* **11**: 1-16.

SPORN, C. C., 1955. The breeding of the mountain devil in captivity. *West. Aust. Nat.* **5**(1): 1-5.

SPORN, C. C., 1958. Further observations on the mountain devil in captivity. *West. Aust. Nat.* **6**(6): 136-37.

SPORN, C. C., 1965. Additional observations on the life history of the mountain devil, *Moloch horridus*, in captivity. *West. Aust. Nat.* **9**(7): 157-59.

SPRACKLAND, R. G., 1980. Some notes on Storr's dwarf spiny-tailed monitor. *Kansas Herp. Soc. Newsletter* **40**: 7-9.

STAMMER, D., 1976. Reptiles. Pp. 131-59 *in* H. Horton, Around Mount Isa. A guide to the flora and fauna. University Queensland Press, St. Lucia; 181 pp.

STAMMER, D., 1981. Some notes on the cane toad (*Bufo marinus*). *Aust. J. Herp.* **1**(2): 61.

STAMMER, D., 1983. Letter to editor. *Varanus tristis orientalis.* *Herpetofauna* **14**(2): 91.

STEBBINS, R. C. AND BARWICK, R. E., 1968. Radiotelemetric study of thermoregulation in a lace monitor. *Copeia* **1968**(3): 541-47.

STEINER, H. AND ANDERS, G., 1946. Zur Frage der Entstehung von Rudimenten. Die Reduktion der Gliedmassen von *Chalcides tridactylus* Laur. *Rev. Suisse Zool.* **53**: 537-46.

STEPHENSON, G., 1977. Notes on *Tiliqua gerrardii* in captivity. *Herpetofauna* **9**(1): 4-5.

STEPHENSON, N. G., 1960. The comparative osteology of Australian geckos and its bearing on their morphological status. *J. Linn. Soc. Lond. (Zool.)* **44**(297): 278-99.

STEPHENSON, N. G., 1962. The comparative morphology of the head skeleton, girdles and hind limbs in the Pygopodidae. *J. Linn. Soc. (Lond.) (Zool.)* **44**(300): 627-44.

STETTLER, P. H., 1966. Einige Gefagenschaftbeobactungen an *Varanus semiremex* Peters, 1869. *Aquaterra* **3**: 95-98.

STIRLING, E. C., 1912. Observations on the habits of the large central Australian monitor (*Varanus giganteus*), with a note on the "fat bodies" of this species. *Trans. Roy. Soc. South Aust.* **36**: 26-33.

STIRLING, E. C. AND ZIETZ, A., 1893. Vertebrata. Reptilia. *Trans. Roy. Soc. South Aust.* **16**(2): 159-76.

STIRNBERG, E. AND HORN, H.-G., 1981. Eine unerwartete Nachzucht im Terrarium: *Varanus (Odatria) storri* (Reptilia: Sauria: Varanidae). *Salamandra* **17**(1/2): 55-62.

STOKLEY, P. S., 1947. The post-cranial skeleton of *Aprasia repens*. *Copeia* **1947**(1): 22-28.

STORR, G. M., 1960. *Egernia bos* a new skink from the south coast of Western Australia. *West. Aust. Nat.* **7**(4): 99-103.

STORR, G. M., 1961. *Ablepharus boutonii clarus*, a new skink from the Esperance District, Western Australia. *West. Aust. Nat.* **7**(7): 176-78.

STORR, G. M., 1963. The gekkonid genus *Nephrurus* in Western Australia, including a new species and three new subspecies. *J. Roy. Soc. West. Aust.* **46**(3): 85-90.

STORR, G. M., 1964a. *Ctenotus*, a new generic name for a group of Australian skinks. *West. Aust. Nat.* **9**(4): 84-85.

STORR, G. M., 1964b. The agamid lizards of the genus *Tympanocryptis* in Western Australia. *J. Roy. Soc. West. Aust.* **47**(2): 43-50.

STORR, G. M., 1964c. Some aspects of the geography of Australian reptiles. *Senck. Biol.* **45**(3/5): 577-89.

STORR, G. M., 1965. The *Amphibolurus maculatus* species-group (Lacertilia, Agamidae) in Western Australia. *J. Roy. Soc. West. Aust.* **48**(2): 45-54.

STORR, G. M., 1966a. The *Amphibolurus reticulatus* species-group (Lacertilia, Agamidae) in Western Australia. *J. Roy. Soc. West. Aust.* **49**(1): 17-25.

STORR, G. M., 1966b. Rediscovery and taxonomic status of the Australian lizard *Varanus primordius. Copeia* **1966**(3): 583-84.

STORR, G. M., 1967a. The genus *Sphenomorphus* (Lacertilia, Scincidae) in Western Australia and the Northern Territory. *J. Roy. Soc. West. Aust.* **50**(1): 10-20.

STORR, G. M., 1967b. Geographic races of the agamid lizard *Amphibolurus caudicinctus. J. Proc. Roy. West. Aust.* **50**(2): 49-56.

STORR, G. M., 1968. Revision of the *Egernia whitei* species-group (Lacertilia: Scincidae). *J. Roy. Soc. West. Aust.* **51**(2): 51-62.

STORR, G. M., 1969. The genus *Ctenotus* (Lacertilia, Scincidae) in the Eastern Division of Western Australia. *J. Roy. Soc. West. Aust.* **51**(4): 97-109.

STORR, G. M., 1970. The genus *Ctenotus* (Lacertilia, Scincidae) in the Northern Territory. *J. Roy. Soc. West. Aust.* **52**(4): 97-108.

STORR, G. M., 1971a. The genus *Ctenotus* (Lacertilia, Scincidae) in South Australia. *Rec. South Aust. Mus.* **16**(6): 1-15.

STORR, G. M., 1971b. The genus *Lerista* (Lacertilia, Scincidae) in Western Australia. *J. Roy. Soc. West. Aust.* **54**(3): 59-75.

STORR, G. M., 1972a. Revisionary notes on the *Sphenomorphus isolepis* complex (Lacertilia, Scincidae). *Zool. Meded.* **47**: 1-5.

STORR, G. M., 1972b. The genus *Morethia* (Lacertilia, Scincidae) in Western Australia. *J. Roy. Soc. West. Aust.* **55**(3): 73-79.

STORR, G. M., 1973. The genus *Ctenotus* (Lacertilia, Scincidae) in the South-west and Eucla Divisions of Western Australia. *J. Roy. Soc. West. Aust.* **56**(3): 86-93.

STORR, G. M., 1974a. Revision of the *Sphenomorphus richardsonii* species-group (Lacertilia, Scincidae). *Rec. West. Aust. Mus.* **3**(1): 66-70.

STORR, G. M., 1974b. The genus *Notoscincus* (Lacertilia, Scincidae) in Western Australia and Northern Territory. *Rec. West. Aust. Mus.* **3**(2): 111-14.

STORR, G. M., 1974c. Agamid lizards of the genera *Caimanops*, *Physignathus* and *Diporiphora* in Western Australia and Northern Territory. *Rec. West. Aust. Mus.* **3**(2): 121-46.

STORR, G. M., 1974d. The genus *Carlia* (Lacertilia, Scincidae) in Western Australia and Northern Territory. *Rec. West. Aust. Mus.* **3**(2): 151-65.

STORR, G. M., 1975a. The genus *Ctenotus* (Lacertilia, Scincidae) in the Kimberley and North-west Divisions of Western Australia. *Rec. West. Aust. Mus.* **3**(3): 209-43.

STORR, G. M., 1975b. The genus *Hemiergis* (Lacertilia, Scincidae) in Western Australia. *Rec. West. Aust. Mus.* **3**(4): 251-60.

STORR, G. M., 1975c. The genus *Proablepharus* (Scincidae, Lacertilia) in Western Australia. *Rec. West. Aust. Mus.* **3**(4): 335-38.

STORR, G. M., 1976a. The genus *Cryptoblepharus* (Lacertilia, Scincidae) in Western Australia. *Rec. West. Aust. Mus.* **4**(1): 53-63.

STORR, G. M., 1976b. The genus *Omolepida* (Lacertilia, Scincidae) in Western Australia. *Rec. West. Aust. Mus.* **4**(2): 163-70.

STORR, G. M., 1976c. The genus *Menetia* (Lacertilia, Scincidae) in Western Australia. *Rec. West. Aust. Mus.* **4**(2): 189-200.

STORR, G. M., 1976d. Revisionary notes on the *Lerista* (Lacertilia, Scincidae) of Western Australia. *Rec. West. Aust. Mus.* **4**(3): 241-56.

STORR, G. M., 1977. The *Amphibolurus adelaidensis* species group (Lacertilia, Agamidae) in Western Australia. *Rec. West. Aust. Mus.* **5**(1): 73-81.

STORR, G. M., 1978a. The genus *Egernia* (Lacertilia, Scincidae) in Western Australia. *Rec. West. Aust. Mus.* **6**(2): 147-87.

STORR, G. M., 1978b. Taxonomic notes on the reptiles of the Shark Bay region, Western Australia. *Rec. West. Aust. Mus.* **6**(3): 303-18.

STORR, G. M., 1978c. Notes on the *Ctenotus* (Lacertilia, Scincidae) of Queensland. *Rec. West. Aust. Mus.* **6**(3): 319-32.

STORR, G. M., 1978d. *Ctenotus rubicundus,* a new scincid lizard from Western Australia. *Rec. West. Aust. Mus.* **6**(3): 332-35.

STORR, G. M., 1978e. Seven new gekkonid lizards from Western Australia. *Rec. West. Aust. Mus.* **6**(3): 337-52.

STORR, G. M., 1979a. Two new *Diporiphora* (Lacertilia, Agamidae) from Western Australia. *Rec. West. Aust. Mus.* **7**(2): 255-63.

STORR, G. M., 1979b. The *Diplodactylus vittatus* complex (Lacertilia, Gekkonidae) in Western Australia. *Rec. West. Aust. Mus.* **7**(4): 391-402.

STORR, G. M., 1979c. *Ctenotus greeri,* a new scincid lizard from Western Australia. *Rec. West. Aust. Mus.* **8**(1): 143-46.

STORR, G. M., 1980a. The monitor lizards (genus *Varanus* Merrem, 1820) of Western Australia. *Rec. West. Aust. Mus.* **8**(2): 237-93.

STORR, G. M., 1980b. A new *Lerista* and two new *Ctenotus* (Lacertilia: Scincidae) from Western Australia. *Rec. West. Aust. Mus.* **8**(3): 441-47.

STORR, G. M., 1981a. Three new agamid lizards from Western Australia. *Rec. West. Aust. Mus.* **8**(4): 599-607.

STORR, G. M., 1981b. Ten new *Ctenotus* (Lacertilia: Scincidae) from Australia. *Rec. West. Aust. Mus.* **9**(2): 125-46.

STORR, G. M., 1982a. Four new *Lerista* (Lacertilia, Scincidae) from Western and South Australia. *Rec. West. Aust. Mus.* **10**(1): 1-9.

STORR, G. M., 1982b. Two new *Gehyra* (Lacertilia: Gekkonidae) from Australia. *Rec. West. Aust. Mus.* **10**(1): 53-59.

STORR, G. M., 1982c. Taxonomic notes on the genus *Tympanocryptis* Peters (Lacertilia: Agamidae). *Rec. West. Aust. Mus.* **10**(1): 61-66.

STORR, G., 1982d. Revision of the bearded dragons (Lacertilia: Agamidae) of Western Australia with notes on the dismemberment of the genus *Amphibolurus*. *Rec. West. Aust. Mus.* **10**(2): 199-214.

STORR, G. M., 1983. Two new lizards from Western Australia (genera *Diplodactylus* and *Lerista*). *Rec. West. Aust. Mus.* **11**(1): 59-62.

STORR, G. M., 1984a. Revision of the *Lerista nichollsi* complex (Lacertilia: Scincidae). *Rec. West. Aust. Mus.* **11**(2): 109-18.

STORR, G. M., 1984b. A new *Ctenotus* (Lacertilia: Scincidae) from Western Australia. *Rec. West. Aust. Mus.* **11**(2): 191-93.

STORR, G. M., 1984c. Note on *Tympanocryptis lineata macra* (Lacertilia: Agamidae). *Rec. West. Aust. Mus.* **11**(3): 317.

STORR, G. M. 1984d. A new *Lerista* (Lacertilia, Scincidae) from Western Australia. *Rec. West. Aust. Mus.* **11**(3): 287-90.

STORR, G. M., 1985a. Two new skinks (Lacertilia: Scincidae) from Western Australia. *Rec. West. Aust. Mus.* **12**(2): 193-96.

STORR, G. M., 1985b. Revision of *Lerista frosti* and allied species (Lacertilia: Scincidae). *Rec. West. Aust. Mus.* **12**(3): 307-16.

STORR, G. M., 1986a. Two new members of the *Lerista nichollsi* complex (Lacertilia: Scincidae). *Rec. West. Aust. Mus.* **13**(1): 47-52.

STORR, G. M., 1986b. A new species of *Lerista* (Lacertilia: Scincidae) with two subspecies from central Australia. *Rec. West. Aust. Mus.* **13**(1): 145-49.

STORR, G. M., 1987. The genus *Phyllodactylus* (Lacertilia: Gekkonidae) in Western Australia. *Rec. West. Aust. Mus.* **13**(2): 275-84.

STORR, G. M. AND HANLON, T. M. S., 1980. Herpetofauna of the Exmouth Region, Western Australia. *Rec. West. Aust. Mus.* **8**(3): 423-39.

STORR, G. M. AND HAROLD, G., 1980. Herpetofauna of the Zuytdorp coast and hinterland, Western Australia. *Rec. West. Aust. Mus.* **8**(3): 359-75.

STORR, G. M. AND SMITH, L. A., 1981. Amphibians and reptiles. Pp. 54-59 *in* Wildlife of the Edgar Ranges area, south-west Kimberley, Western Australia, ed by N. L. McKenzie. *Wildlife Resources Bulletin Western Australia* No. 10; 71 pp.

STORR, G. M., SMITH, L. A. AND JOHNSTONE, R. E., 1981. Lizards of Western Australia. I. Skinks. University Western Australia Press, Nedlands, Western Australia; 200 pp.

STORR, G. M., SMITH, L. A. AND JOHNSTONE, R. E., 1983. Lizards of Western Australia. II. Dragons and monitors. Western Australia Museum, Perth; 114 pp.

SWAN, G., 1972. Observations on birth of bluetongues. *Herpetofauna* **5**(2): 9.

SWANSON, S., 1976. Lizards of Australia. Angus and Robertson, Sydney; 80 pp.

SWANSON, S., 1979. Some rock-dwelling reptiles of the Arnhem Land escarpment. *North. Terr. Nat.* **1**: 14-18.

TASOULIS, T., 1985. Observations on the lace monitor *Varanus varius. Herpetofauna* **15**(1): 25.

TAYLOR, J. A., 1985a. Burrow construction and utilization by the lizard *Ctenotus taeniolatus. Herpetofauna* **15**(2): 44-47.

TAYLOR, J. A., 1985b. Reproductive biology of the Australian lizard *Ctenotus taeniolatus. Herpetologica* **41**(4): 408-18.

TAYLOR, J. A., 1986. Food and foraging behaviour of the lizard, *Ctenotus taeniolatus. Aust. J. Ecol.* **11**: 49-54.

THOMPSON, J., 1977a. Embryo-maternal relationships in a viviparous skink, *Sphenomorphus quoyii* (Lacertilia: Scincidae). Pp. 279-80 *in* Reproduction and evolution, ed by J. M. Calaby and C. H. Tyndale-Biscoe. Australian Academy of Sciences, Canberra. Not seen.

THOMPSON, J., 1977b. Embryo-maternal relationships in a viviparous skink, *Sphenomorphus quoyii* (Duméril and Bibron). Unpublished Ph.D. Thesis, University of Sydney; 158 pp.

THOMSON, D. F. AND HOSMER, W., 1963. A preliminary account of the herpetology of the Great Sandy Desert of central western Australia. Reptiles and amphibians of the Bindibu Expedition. *Proc. Roy. Soc. Vict.* **77**(1): 217-37.

THROCKMORTON, G. S. AND CLARKE, L. K., 1981. Intracranial joint movements in the agamid lizard *Amphibolurus barbatus. J. Exp. Zool.* **216**: 25-35.

THROCKMORTON, G. S., DE BAVAY, J., CHAFFEY, W. MERROTSY, B., NOSKE, S. AND NOSKE, R., 1985. The mechanism of frill erection in the bearded dragon *Amphibolurus barbatus* with comments on the jacky lizard *A. muricatus* (Agamidae). *J. Morph.* **183**: 285-92.

THURSTON, H. R., 1973. Two lizards. *North. Qld Nat.* **41**: 3.

TILLEY, S. J., 1984. Skeletal variation in the Australian *Sphenomorphus* group (Lacertilia: Scincidae). Unpubl. Ph.D. Thesis, Department of Zoology, La Trobe University, Bundoora, Victoria; 364 pp.

TIMMS, B. V., 1977. Notes on the she-oak skink. *Hunter Natural History* **9**(2): 99-102.

TINKLE, D. W., WILBUR, H. M. AND TILLEY, S. G., 1970. Evolutionary strategies in lizard reproduction. *Evolution* **24**(1): 55-74.

TRUE, M. AND REIDY, P., 1981. Some herpetofauna of the Perth, West Australia, region. *Newsletter Vict. Herp. Soc.* **21**: 2-8.

TSCHAMBERS, B., 1949. Birth of the Australian blue-tongued lizard, *Tiliqua scincoides* (Shaw). *Herpetologica* **5**(6): 141-42.

TUBB, J. A., 1938. The Sir Joseph Banks Islands. II. Reptilia, Part 1: General. *Proc. Roy. Soc. Vict.* **50**(2): 383-93.

TYLER, M. J., 1960. Observations on diet and size variation of *Amphibolurus adelaidensis* (Gray) (Reptilia: Agamidae) on the Nullarbor Plain. *Trans. Roy. Soc. South Aust.* **83**: 111-17.

TYLER, M. J., 1961. On the diet and feeding habits of *Hemidactylus frenatus* (Duméril and Bibron) (Reptilia: Gekkonidae) at Rangoon, Burma. *Trans. Roy. Soc. South Aust.* **84**: 45-49.

TYLER, M. J., GROSS, G. F., RIX, C. E. AND INNS, R. W., 1976. Terrestrial fauna and aquatic vertebrates. Pp. 121-29 *in* Natural history of the Adelaide region, ed by C. R. Twidale, M. J. Tyler and B. P. Webb. Royal Society of South Australia, Adelaide; 189 pp.

UNDERWOOD, G., 1957. On lizards of the family Pygopodidae. A contribution to the morphology and phylogeny of the Squamata. *J. Morph.* **100**(2): 207-68.

UNDERWOOD, G., 1970. The eye. Pp. 1-97 in Biology of the Reptilia, Vol. 2, Morph. B., ed by C. Gans and T. S. Parsons. Academic Press, New York; 374 pp.

UNDERWOOD, G., 1976a. A systematic analysis of boid snakes. Pp. 151-75 in Morphology and biology of reptiles, ed by A.d'A. Bellairs and C. B. Cox. *Linnean Society Symposium Series* No. 3; 290 pp.

UNDERWOOD, G., 1976b. Simplication and degeneration in the course of evolution of squamate reptiles. *Colloques internat. C.N.R.S.* No. 266: 341-52.

VERON, J. E. N., 1969a. The reproductive cycle of the water skink, *Sphenomorphus quoyii. J. Herp.* **3**(1-2): 55-63.

VERON, J. E. N., 1969b. An analysis of stomach contents of the water skink, *Sphenomorphus quoyi. J. Herp.* **3**(3-4): 187-89.

VERON, J. AND HEATWOLE, H., 1970. Temperature relations of the water skink, *Sphenomorphus quoyi. J. Herp.* **4**(3-4): 141-53.

VESTJENS, W. J. M., 1977. Reptilian predation on birds and eggs at Lake Cowal, N.S.W. *Emu* **77**(1): 36-37.

VINCENT, J., 1981. Co-operative feeding behaviour between a yellow-faced honeyeater and a lace goanna. *The Clementis* (Bairnsdale Field Nat. Club) **20**: 10-11.

VITT, L. J. AND CONGDON, J. D., 1978. Body shape, reproductive effort, and relative clutch mass in lizards: resolution of a paradox. *Amer. Nat.* **112**: 595-608

VITT, L. J., 1986. Reproductive tactics of sympatric gekkonid lizards with a comment on the evolutionary and ecological consequences of invariant clutch size. *Copeia* **1986**(3): 773-86.

WAITE, E. R., 1929. The reptiles and amphibians of South Australia. Government Printer, Adelaide; 270 pp.

WAKEFIELD, N. A., 1956. Blue-tongued lizards and instinct. *Vict. Nat.* **72**(9): 143-44.

WALLS, G. L., 1940. Ophthalmological implications for the early history of the snakes. *Copeia* **1940**(1): 1-8.

WALLS, G. L., 1942. The vertebrate eye and its adaptive radiation. Cranbrook Institute of Science, Bloomfield Hills, Michigan; 785 pp.

WAPSTRA, M. AND WAPSTRA, E., 1986. Breeding record of the delicate skink *Leiolopisma delicata* in southern Tasmania. *Tasmanian Nat.* **87**: 5-6.

WARBURG, M. R., 1965a. Studies on the environmental physiology of some Australian lizards from arid and semi-arid habitats. *Aust. J. Zool.* **13**: 563-75.

WARBURG, M. R., 1965b. The influence of ambient temperature and humidity on the body temperature and water loss from two Australian lizards, *Tiliqua rugosa* (Gray) (Scincidae) and *Amphibolurus barbatus* Cuvier (Agamidae). *Aust. J. Zool.* **13**: 331-50.

WARBURG, M. R., 1966. On the water economy of several Australian geckos, agamids, and skinks. *Copeia* **1966**(2): 230-35.

WEAVERS, B. W., 1983. Thermal ecology of *Varanus varius* (Shaw), the lace monitor. Unpublished Ph.D. Thesis, Australian National University, Canberra; pp.?

WEBB, G., 1973. Notes on *Lialis burtonis* (Burton's legless lizard) in captivity. *Roy. Zool. Soc. Bull. Herp.* **1**(1): 4-5.

WEBB, G. A., 1983. Diet in a herpetofaunal community on the Hawkesbury Sandstone Formation in the Sydney area. *Herpetofauna* **14**(2): 87-91.

WEBB, G. A., 1984. The distribution, ecology and status of the southern angle-headed dragon *(Gonocephalus spinipes)* in New South Wales. Forestry Commission of New South Wales, Sydney, N.S.W. Report No. 981; 13 pp.

WEBB, G. A. AND SIMPSON, J. A., 1986. Some unusual food items for the southern blotched blue-tongue lizard *Tiliqua nigrolutea* (Quoy and Gaimard) at Bombala, New South Wales. *Herpetofauna* **16**(2): 44-49.

WEBB, G. J. W., JOHNSON, C. R. AND FIRTH, B. T., 1972. Head-body temperature differences in lizards. *Physiol. Zoöl.* **45**(2): 130-42.

WEBBER, P., 1978. To spy on a desert skink. *Aust. Nat. Hist.* **19**(8): 270-75.

WEBBER, P., 1979. Burrow density, position and relationship of burrows to vegetation coverage shown Rosen's desert skink *Egernia inornata* (Lacertilia: Scincidae). *Herpetofauna* **10**(2): 16-20.

WEBER, E. AND WERNER, Y. L., 1977. Vocalizations of two snake-lizards (Reptilia: Sauria: Pygopodidae). *Herpetologica* **33**(3): 353-63.

WEEKES, H. C., 1927a. A note on reproductive phenomena in some lizards. *Proc. Linn. Soc. New South Wales* **52**:(2): 25-32.

WEEKES, H. C., 1927b. Placentation and other phenomena in the scincid lizard *Lygosoma (Hinulia) quoyi. Proc. Linn. Soc. New South Wales* **52**(4): 499-554.

WEEKES, H. C., 1929. On placentation in reptiles. No. i. 1. *Denisonia superba* and *D. suta*; 2. *Lygosoma (Liolepisma) weeksae. Proc. Linn. Soc. New South Wales* **54**(2): 34-60.

WEEKES, H. C., 1930. On placentation in reptiles. II. *Proc. Linn. Soc. New South Wales* **55**(5): 550-76.

WEEKES, H. C., 1934. The corpus luteum in certain oviparous and viviparous reptiles. *Proc. Linn. Soc. New South Wales* **59**(5-6): 380-91.

WEEKES, H. C., 1935. A review of placentation among reptiles with particular regard to the function and evolution of the placenta. *Proc. Zool. Soc. Lond.* **1935**: 625-45.

WEIGEL, J., 1984. Devils, dragons and bloodsuckers. *Wildlife Aust.* **21**(3): 13-17.

WEIGEL, J., 1985. A preliminary description of a new dwarf rock goanna *Varanus minor* sp. nov. *Newsletter of the Reptile Keepers' Association* **7**: 5-8.

WELLS, R., 1971. Hibernation — bearded dragons. *Herpetofauna* **3**(1): 4-6 pp.

WELLS, R., 1972. Notes on *Goniocephalus boydii* (Macleay). *Herpetofauna* **5**(2): 24.

WELLS, R., 1975. Notes on an unidentified skink of the genus *Carlia* from Black Mountain, N.E. Queensland. *Herpetofauna* **7**(2): 11.

WELLS, R., 1979a. New reptile records for the Northern Territory. *North. Terr. Nat.* **1**(2): 3-4.

WELLS, R., 1979b. A large aggregation of skink eggs. *Herpetofauna* **11**(1): 19.

WELLS, R., 1981. Utilization of the same site for communal egg-laying by *Lampropholis delicata* and *L. guichenoti. Aust. J. Herp.* **1**(1): 35-36.

WELLS, R. AND HUSBAND, G., 1979. Comments on the reproduction of *Pygopus lepidopodus* (Lacepede). *Herpetofauna* **11**(1): 22-25.

WELLS, R. W. AND WELLINGTON, C. R., 1984. A synopsis of the class Reptilia in Australia. *Aust. J. Herp.* **1**(3-4): 73-129.

WELLS, R. W. AND WELLINGTON, C. R., 1985. A classification of the Amphibia and Reptilia of Australia. *Aust. J. Herp.* Suppl. Ser. No. 1: 1-61.

WENDT, S. (ed), 1987. Australia's wonderful wildlife. *Australian Women's Weekly,* Australian Consolidated Press, Sydney; 129pp.

WERMUTH, H., 1965. Liste der rezenten Amphibien und Reptilien. Gekkonidae, Pygopodidae, Xantusiidae. *Das Tierreich* **80:** 1-246.

WERMUTH, H., 1967. Liste der rezenten Amphibien und Reptilien. Agamidae. *Das Tierreich* **86:** 1-127.

WERNER, Y. L., 1956. Chromosome numbers of some male geckos (Reptilia: Gekkonoidea). *Bull. Res. Coun. Israel* **5B**(3-4): 319.

WERNER, Y. L., 1969. Eye size in geckos of various ecological types (Reptilia: Gekkonidae and Sphaerodactylidae). *Israel J. Zool.* **18**(2-3): 291-316.

WERNER, Y. L., 1973. Auditory sensitivity and vocalization in lizards (Reptilia: Gekkonoidea, Iguanidae, Pygopodidae and Scincidae). *Israel J. Zool.* **22**(2-4): 204-05.

WERNER, Y. L., 1980. Apparent homosexual behaviour in an all-female population of a lizard, *Lepidodactylus lugubris* and its probable interpretation. *Z. Tierpsychol.* **54:** 144-50.

WERNER, Y. L. AND WHITAKER, A. H., 1978. Observations and comments on the body temperatures of some New Zealand reptiles. *N.Z. Sci.* **5:** 375-93.

WEST, J. A., 1979. The occurrence of some exotic reptiles and amphibians in New Zealand. *Herpetofauna* **10**(2): 4-9.

WEVER, E. G., 1974. The ear of *Lialis burtonis* (Sauria: Pygopodidae), its structure and function. *Copeia* **1974**(2): 297-305.

WEVER, E. G., 1978. The reptile ear. Its structure and function. Princeton University Press, Princeton, NJ; 1024 pp.

WHITAKER, A. H., 1970. A note on the lizards of the Tokelau Islands, Polynesia. *Herpetologica* **26**(3): 355-58.

WHITE, J., 1976. Reptiles of the Corunna Hills. *Herpetofauna* **8**(1): 21-23.

WHITE, J., 1979. The road to Mokari. *Herpetofauna* **11**(1): 13-16.

WHITE, S. R., 1948. Observations on the mountain devil *(Moloch horridus). West. Aust. Nat.* **1**(4): 78-81.

WHITE, S. R., 1949. Some notes on the netted dragon lizard. *West. Aust. Nat.* **1**(8): 157-61.

WILEY, E. O., 1981. Phylogenetics. The theory and practice of phylogenetic systematics. John Wiley and Sons, New York; 439 pp.

WILHOFT, D. C., 1960. Observations on adults and juveniles of *Hemisphaeriodon gerrardii* in captivity. *North Qld Nat.* **28:** 3-4.

WILHOFT, D. C., 1961. Temperature responses in two tropical Australian skinks. *Herpetologica* **17**(2): 109-13.

WILHOFT, D. C., 1963a. Reproduction in the tropical Australian skink, *Leiolopisma rhomboidalis. Amer. Midl. Nat.* **70**(2): 442-61.

WILHOFT, D. C., 1963b. Gonadal histology and seasonal changes in the tropical Australian lizard, *Leiolopisma rhomboidalis. J. Morph.* **113**(2): 185-204.

WILHOFT, D. C., 1964. Seasonal changes in the thyroid and interrenal glands of the tropical Australian lizard, *Leiolopisma rhomboidalis. Gen. Comp. Endocrin* **4:** 42-53.

WILHOFT, D. C. AND REITER, E. O., 1965. Sexual cycle of the lizard, *Leiolopisma fuscum,* a tropical Australian skink. *J. Morph.* **116**(3): 379-88.

WILLIAMS, C. K., 1965. Competition between species of lizards in the field, viewed through the aspect of thermoregulation. Unpublished B.Sc. Honours Thesis, University of Western Australia, Nedlands; 28 pp.

WILLIAMS, E. E. AND PETERSON, J. A., 1982. Convergent and alternative designs in the digital adhesive pads of scincid lizards. *Science* **215:** 1509-11.

WILLIAMS, L. E., 1964. Incubation of lizards' eggs. *South Aust. Nat.* **38**(4): 65.

WILSON, K. J., 1974. The relationship of oxygen supply for activity to body temperature in four species of lizards. *Copeia* **1974**(4): 920-34.

WILSON, K. J. AND LEE, A. K., 1974. Energy expenditure of a large herbivorous lizard. *Copeia* **1974**(2): 338-48.

WITTEN, G. J., 1972. A study of *Amphibolurus nobbi* Witten 1972. Unpublished M.Sc. Thesis, University of New England, Armidale, N.S.W.; 43 pp. Not seen.

WITTEN, G. J., 1974. Population movements of the agamid lizard *Amphibolurus nobbi. Aust. Zool.* **18**(2): 129-32.

WITTEN, G. J., 1978. A triploid male individual *Amphibolurus nobbi nobbi* Witten (Lacertilia: Agamidae). *Aust. Zool.* **19**(3): 305-08.

WITTEN, G. J., 1982a. Comparative morphology and karyology of the Australian members of the family Agamidae and their phylogenetic implications. Unpublished Ph.D. Thesis, Dept. of Anatomy, Sydney University; 272 pp.

WITTEN, G. J., 1982b. Phyletic groups within the family Agamidae (Reptilia: Lacertilia) in Australia. Pp. 225-28 *in* Evolution of the flora and fauna of arid Australia, ed by W. R. Barker and P. J. M. Greenslade. Peacock Publications, Frewville, South Australia.

WITTEN, G. J., 1983. Some karotypes of Australian agamids (Reptilia: Lacertilia). *Aust. J. Zool.* **31:** 533-40.

WITTEN, G. J., 1984. Relationships of *Tympanocryptis aurita* Storr, 1981. *Rec. West. Aust. Mus.* **11**(4): 399-401.

WITTEN, G. J., 1985. Relative growth in Australian agamid lizards: adaptation and evolution. *Aust. J. Zool.* **33**(3): 349-62.

WITTEN, G. J. AND COVENTRY, A. J., 1984. A new lizard of the genus *Amphibolurus* (Agamidae) from southern Australia. *Proc. Roy. Soc. Vict.* **96**(3): 155-59.

WITTEN, G. J. AND HEATWOLE, H., 1978. Preferred temperature of the agamid lizard *Amphibolurus nobbi nobbi. Copeia* **1978**(2): 362-64.

WOODWARD, H., 1874. New facts bearing on the inquiry concerning forms intermediate between birds and reptiles. *Quart. J. Geol. Soc.* **30:** 8-15.

WORRELL, E., 1956. A new water-monitor from northern Australia. *Aust. Zool.* **12**(3): 201-02.

WORRELL, E., 1958. Song of the snake. Angus and Robertson, Sydney; 210 pp.

WORRELL, E., 1963. Reptiles of Australia. Angus and Robertson, Sydney; 207 pp.

WRIGHT, P., 1982. Observations of predator/prey relationships between praying mantids and geckos. *North. Terr. Nat.* **5:** 10-11.

YASHIRO, H., 1931. Observation on *Hemidactylus frenatus. Amoeba* **3**(3): 51-53. In Japanese, not seen.

ZIETZ, F. R., 1914. Lacertilia. *Trans. Roy. Soc. South Aust.* **38:** 441-44.

ZIETZ, F. R., 1915. Lacertilia. *Trans. Roy. Soc. South Aust.* **39:** 766-69.

ZIMMERMANN, H., 1985. Geburt des Stachelschwanz —
Skinks. *Aqua. Mag.* **19**(9): 372-74.

ZUG, G. F., 1985. *Cyrtodactylus pelagicus;* correct usage. *Herp. Review* **16**(3): 67.

ZWINENBERG, A. J., 1974. Boyd's forest dragon, *Goniocephalus boydii* (Macleay). *Bull. Maryland Herpet. Soc.* **10**(4): 99-101.

ZWINENBERG, A. J., 1977. Die Australische Wasseragame (*Physignathus lesueurii*). *Aquaria* **24**: 93-97. Not seen.

Author's Errata for "The Biology and Evolution of Australian Lizards"

Pp. vii and x, captions for Figs 1, 14, 27, 31, 70, 76 and 81. See corresponding captions in text for complete and correct versions.

P. xv, Fig. 1 caption, line 1. Insert "and temperature" after "rainfall".

P. xvii, paragraph 1, line 6. Read "leg" for "tail".

P. 1, paragraph 4, line 2. Read "475" for "476".

P. 3, next to last line. Read "related" for "realted".

P. 4, paragraph 6. Insert a paragraph break at the end of the first sentence.

P. 5, paragraph 4, line 7. Read "to" for "on".

P. 5, paragraph 5, last line. Delete the comma after "basking".

P. 6, paragraph 2, line 20. Insert a semi-colon before "other".

P. 6, paragraph 2, line 22. Replace the comma before "*Tiliqua*" with a semi-colon.

P. 6, paragraph 2, line 26. Replace the comma before "*Ctenotus*" with a semi-colon.

P. 6, paragraph 4, line 6. Read "Storr 1964c" for "Storr 1964".

P. 15, last word. Read "*Ctenophorus*" for "*Ctenophorous*".

P. 16, paragraph 3, last line. Read "Anonymous" for "Anonymons".

P. 30, paragraph 3. Substitute "the latinized Greek word" for "Latin".

P. 32, paragraph 7, line 5. Read "northwestern" for "northeastern".

P. 34, Fig. 18 caption. Delete "lack of an external ear opening,".

P. 36, Fig. 21 caption, line 3. Read "2.3.4.<u>4</u>.3" for "2.3.4.4.<u>3</u>".

P. 40. Read "*fionni*" for "*fioni*" and read "*yinnietherra*" for "*yinnietharra*".

P. 55, paragraph 5, line 3. Delete second "and".

P. 55, paragraph 5, line 5. Replace the semi-colon with ", and".

P. 62, paragraph 2, lines 10, 11. Read "*Lepidodactylus*" for "*Lepido*dactylus".

P. 62, Fig. 29. Read "ceratobranchial" for "certobranchial".

P. 65, paragraph 6, line 6. Insert "a" before "very".

P. 66, paragraph 4, line 4. Read "afternoon" for "afternnon".

P. 73, paragraph 4, line 10. Read "tailless" for "tailess" (twice)

P. 77, paragraph 1, line 2. Read "centred" for "centered".

P. 77, paragraph 9, line 8. Read "*lindneri*" for "*linderi*".

P. 78, paragraph 3, line 1. Insert "mouth" before "lining".

P. 78, paragraph 9, line 13. Read "*pulcher*" for "*pucher*".

P. 79, paragraph 6, line 5. Close the parentheses after "*Rhynchoedura*".

P. 81, paragraph 3, line 10. Romanize the first "and" and delete the second "and".

P. 88. Read "*Rhynchoedura*" for "*Rhnychoedura*".

P. 90, references for *Oedura monilis*. Read "Rösler 1980 (C)" for "Rössler 1980".

P. 90, reference for *Oedura reticulata*. Read "W" for "w".

P. 92, Table 8 heading. Insert "(mm)" after "width".

Pp. 92, 93. Read "Rösler" for "Rössler".

P. 101, paragraph 3, line 16. Read "*Aclys*" for "*Acyls*".

P. 109, first line. Read "*Ophidiocephalus*" for "*Ophidocephalus*".

P. 111, paragraph 2, line 9. Read "clinally" for "clinically".

P. 113, Table 4. Read "*plebeia*" for "*plebia*".

P. 115, reference for *Pletholax gracilis*. Read "Bamford" for "Bramford".

P. 118, Fig. 55 caption. Read "*Hemiergis*" for "*Hermiergis*".

P. 118, paragraph 3, line 11. Read "*tympanum*" for "*tympanus*".

P. 123, Fig. 123 caption. Add "(southern population)" at end of caption.

P. 125, paragraph 4, line 13. Delete "*C. triumviratus*".

P. 131, line 4. Read "Hickman" for "Hickmann".

P. 137, paragraph 4, lines 2 and 4. Read "inguinal" for "inquinal".

P. 138, first line. Read "*albofasciolatus*" for "*albofasiolatus*".

P. 140, paragraph 3, line 12. Read "lightning" for "lightening".

P. 142, paragraph 1, line 3. Read "perches" for "perchs".

P. 145, paragraph 3, line 13. Delete "nearly".

P. 153, paragraph 5, line 16. Replace the first comma with a semi-colon.

P. 152, paragraph 6, line 5. Read "Ph.D." for "PH.D".

P. 155, paragraph 5, line 5. Read "*lesueurii*" for "*lesuerii*".

P. 155, paragraph 7, line 7. Read "*arnhemensis*" for "*arnhemiensis*".

P. 156, paragraph 6, line 10. Delete "described".

P. 157, paragraph 3, line 2. Read "an" for "a".

P. 158, paragraph 2, line 6. Read "inguinal" for "inquinal".

P. 162, paragraph 5, last line. Replace the semi-colon with a comma.

P. 175. Read "*Claireascincus*" for "*Carinascincus*".

P. 180, Table 9. Place the Schafer 1979 reference under *Tiliqua scincoides*.

P. 196, paragraph 2, line 12. Begin parentheses before "Schürer".

P. 197, paragraph 2, line 11. Read "Pianka" for "ianka".

P. 199, paragraph 2, line 55. Read "quarters" for "quaters".

P. 204, paragraph 7, next to last line. Read "water" for "oxygen".

P. 216, paragraph 6, line 15. Read "lizards" for "lizard".

P. 220, Table 1 heading. For "all other lizards and snakes" read "all other squamates".

P. 221, Atenor, A. 1974. Delete "*of*".

P. 221, Arena, P. C. 1986. Read "Murdoch" for "Murdock".

P. 221, Baker, J. K. 1979. Italicize "Pacific Science".

P. 221, Bamford, M. J. 1980. Read "Murdoch" for "Murdock".

P. 223, Brotzler, A. 1965. Italicize "Freunde Kölner Zoo".

P. 225, Cogger, H. C. 1975. Read "*Pseudothecadactylus*" for "*Pseudothecadatylus*".

P. 235, Maclean, S. 1980. Italicize "Cell Tissue Research".

P. 236, Mertens, R. 1966. Read "*Aquar.*" for "*Aguar.*".

P. 240, Rose, S. 1985. Italicize "The Vipera".

P. 241, Rosenberg, H. L., Russell, A. P. and Kapoor, M. 1984. Italicize "Canadian".

Index

Bold page entries pertain to illustrations.